LIGHT PROPAGATION IN GAIN MEDIA

Over the past two decades, optical amplifiers have become of key importance in modern communications. In addition to this, the technology has applications in cutting-edge research such as biophotonics and lab-on-a-chip devices. This book provides a comprehensive treatment of the fundamental concepts, theory, and analytical techniques behind modern optical amplifier technology.

The book covers all major optical amplification schemes in conventional materials, including the Raman and parametric gain processes. The final chapter is devoted to optical gain in metamaterials, a topic that has been attracting considerable attention in recent years. The authors emphasize analytical insights to give a deeper, more intuitive understanding of various amplification schemes. The book assumes background knowledge of electrical engineering or applied physics, including exposure to electrodynamics and wave motion, and is ideal for graduate students and researchers in physics, optics, bio-optics, and communications.

MALIN PREMARATNE is Research Director and Associate Professor in the Advanced Computing and Simulation Laboratory, Department of Electrical and Computer Systems Engineering, Monash University, Clayton, Australia. He guides the research program in theory, modeling, and simulation of light propagation in guided and scattering media.

GOVIND P. AGRAWAL is Professor of Optics and Physics in the Institute of Optics, University of Rochester, USA. His current research interests include optical communications, nonlinear optics, and laser physics.

LIGHT PROPAGATION IN GAIN MEDIA

Optical Amplifiers

MALIN PREMARATNE

Monash University
Australia

GOVIND P. AGRAWAL

The Institute of Optics
University of Rochester
USA

CAMBRIDGE
UNIVERSITY PRESS

CAMBRIDGE UNIVERSITY PRESS
Cambridge, New York, Melbourne, Madrid, Cape Town, Singapore,
São Paulo, Delhi, Dubai, Tokyo, Mexico City

Cambridge University Press
The Edinburgh Building, Cambridge CB2 8RU, UK

Published in the United States of America by Cambridge University Press, New York

www.cambridge.org
Information on this title: www.cambridge.org/9780521493482

© M. Premaratne and G. P. Agrawal 2011

First published 2011

Printed in the United Kingdom at the University Press, Cambridge

A catalogue record for this publication is available from the British Library

ISBN 978-0-521-49348-2 Hardback

For Anne, Sipra, Caroline, and Claire
–Govind P. Agrawal

For Erosha, Gehan, and Sayumi
–Malin Premaratne

Contents

Preface

Everything should be made as simple as possible, but not simpler.
–Albert Einstein

An optical fiber amplifier is a key component for enabling efficient transmission of wavelength-division multiplexed (WDM) signals over long distances. Even though many alternative technologies were available, erbium-doped fiber amplifiers won the race during the early 1990s and became a standard component for long-haul optical telecommunications systems. However, owing to the recent success in producing low-cost, high-power, semiconductor lasers operating near 1450 nm, the Raman amplifier technology has also gained prominence in the deployment of modern lightwave systems. Moreover, because of the push for integrated optoelectronic circuits, semiconductor optical amplifiers, rare-earth-doped planar waveguide amplifiers, and silicon optical amplifiers are also gaining much interest these days.

Interestingly, even though completely unrelated to the conventional optical communications technology, optical amplifiers are also finding applications in biomedical technology either as power boosters or signal-processing elements. Light is increasingly used as a tool for stretching, rotating, moving, or imaging cells in biological media. The so-called lab-on-chip devices are likely to integrate elements that are both acoustically and optically active, or use optical excitation for sensing and calibrating tasks. Most importantly, these new chips will have optical elements that can be broadly used for processing different forms of signals.

There are many excellent books that cover selective aspects of active optical devices including optical amplifiers. This book is not intended to replace these books but to complement them. The book grew out of the realization that there is a need for coherent presentation of the theory behind various optical amplifiers from a common conceptual standpoint, while highlighting the capabilities of established methods and the limitations of current approaches.

Most of the existing literature on amplifiers covers the advances in amplifier technology related to their fabrication and operation. While it is important to be aware of these developments, this awareness is being achieved at the cost of skipping the fundamentals and the material related to the important physical concepts and underlying mathematical representation. We try to bridge this gap by providing a theoretical framework within which the amplifier theory and device concepts are presented. Our objective is to provide a comprehensive account of light propagation in active media and employ it for describing the signal amplification in different optical amplifier structures. Whenever possible, we obtain approximate analytical results to estimate the operational characteristics of amplifiers, thus enabling the reader to build a detailed intuitive picture about the device operation. We provide sufficient details for numerical implementation of the key algorithms, but we emphasize throughout this text the value and utility of having approximate solutions of different amplification processes. Such an approach not only provides a thorough understanding of the underlying key process of light interaction with matter but also enables one to build a mental picture of the overall operation of the device, without being cluttered with details. However, we have intentionally not included the analysis of noise in amplifiers because a rigorous treatment of noise processes requires sophisticated mathematical machinery beyond the scope of this book. Interested readers may find a thorough discussion of amplifier noise in the book by E. Desurvire, *Erbium-Doped Fiber Amplifiers: Principles and Applications* (Wiley, 1994).

Our presentation style is multi-folded in the sense that we use different descriptions of light, as rays, scalar waves, or vector electromagnetic waves, depending on the sophistication needed to carry out the intended analysis. Our intention is to equip the reader with such sophistication and understanding that he or she gradually develops adequate insight to use a combination of these different descriptions of light to describe the operation of modern optical amplifiers. More specifically, we show that the optical gain of an amplifying medium either can be derived from the first principles using the material susceptibility based on a quantum-mechanical approach or it can be deduced from empirical phenomenological models based on experimental observations. Most importantly, our approach is not to pick a specific model but to show diverse models and techniques to the reader so that reader can make the best possible choice of the method based on the circumstances and the domain of applicability.

In this book, state-of-the-art methods and algorithms suitable for understanding the underlying governing principles and analyzing optical amplifiers are presented in a manner suited to graduate research students and professionals alike. The material in this book may be suitable for scientists and engineers working in the fields of telecommunications, biophotonics, metamaterials, etc., and for those interested in learning and advancing the technology behind modern optical amplifiers. This book is also suitable for individuals opting for self-study as a way for continuing professional development and for professional consultants evaluating technology for industry and government agencies. The readers are assumed to have background in electrical engineering or applied modern physics, including some exposure to electrodynamics and wave motion. Knowledge of quantum mechanics, high-level programming languages such as C++ or Matlab, and numerical software methods is helpful but not essential. We recognize the importance of computer modeling in understanding, analyzing, and designing optical amplifiers. By emphasizing the mathematical and computational issues and illustrating various concepts and techniques with representative examples of modern optical amplifiers, we try to make this book appeal to a wider audience. While any programming language in conjunction with a suitable visualization tool could, in principle, be employed, we feel that reader is best served by learning C++ or a Matlab-type high-level programming language for numerical work associated with this book. We occasionally describe algorithms in this text, but we attempt to avoid programming-specific details.

Our intent has been to provide a self-contained account of the modern theory of optical amplifiers, without dwelling too much on the specific details of each amplifier technology. The book should, therefore, not be considered as a compendium that encompasses applications and design optimizations applicable to a specific amplifying system. Brief derivations of many basic concepts are included to make this book self-contained, to refresh memory of readers who had some prior exposure, or to assist the readers with little or no prior exposure to similar material. The order of the presentation and the level of rigor have been chosen to make the concepts clear and suitable for computer implementation while avoid unnecessary mathematical formalism or abstraction. The mathematical derivations are presented with intermediate steps shown in as much detail as is reasonably possible, without cluttering the presentation or understanding. Recognizing the different degrees of mathematical sophistication of the intended readership, we have provided extensive references to widely available literature and web links to related numerical concepts at the book's website, www.malinp.com.

Malin Premaratne
Govind P. Agrawal

1

Introduction

In this introductory chapter we focus on the interaction of optical fields with matter because it forms the basis of signal amplification in all optical amplifiers. According to our present understanding, optical fields are made of photons with properties precisely described by the laws of quantum field theory [1]. One consequence of this wave–particle duality is that optical fields can be described, in certain cases, as electromagnetic waves using Maxwell's equations and, in other cases, as a stream of massless particles (photons) such that each photon contains an energy $h\nu$, where h is the Planck constant and ν is the frequency of the optical field. In the case of monochromatic light, it is easy to relate the number of photons contained in an electromagnetic field to its associated energy density. However, this becomes difficult for optical fields that have broad spectral features, unless full statistical features of the signal are known [2]. Fortunately, in most cases that we deal with, such a detailed knowledge of photon statistics is not necessary or even required [3, 4]. Both the linear and the nonlinear optical studies carried out during the last century have shown us convincingly that a theoretical understanding of experimental observations can be gained just by using wave features of the optical fields if they are intense enough to contain more than a few photons [5]. It is this semiclassical approach that we adopt in this book. In cases where such a description is not adequate, one could supplement the wave picture with a quantum description.

In Section 1.1 we introduce Maxwell's equations and the Fourier-transform relations in the temporal and spatial domains used to simplify them. This section also establishes the notation used throughout this book. In Section 1.2 we look at widely used dielectric functions describing dispersive optical response of materials. After discussing dispersion relations in Section 1.3, we show in Section 1.4 that they cannot have an arbitrary form because of the constrained imposed by the causality and enforced by the Kramers–Kronig relations. Section 1.5 considers the propagation of plane waves in a dispersive medium because plane waves play a central role in analyzing various amplification schemes.

1.1 Maxwell's equations

In this section we begin with the time-domain Maxwell's equations and introduce their frequency-domain and momentum-domain forms using the Fourier transforms. These forms are then used to classify different optical materials through the constitutive relations.

1.1.1 Maxwell's equations in the time domain

In a semiclassical approach, Maxwell's equations provide the fundamental basis for the propagation of optical fields through any optical medium [6,7]. These four equations can be written in an integral form. Two of them relate the electric field vector E and electric flux-density vector D with the magnetic field vector H and the magnetic flux-density vector B using the line and surface integrals calculated over a closed contour l surrounding a surface S_l shown in Figure 1.1:

$$\text{Faraday's law of induction: } \oint_l E(r,t) \cdot dl = -\frac{d}{dt} \int_{S_l} B(r,t) \cdot dS, \tag{1.1a}$$

$$\text{Ampere's circuital law: } \oint_l H(r,t) \cdot dl = \frac{d}{dt} \int_{S_l} D(r,t) \cdot dS + I(t), \tag{1.1b}$$

where $I(t)$ is the total current flowing across the surface S_l. The pairs E, D and H, B are not independent even in vacuum and are related to each other by

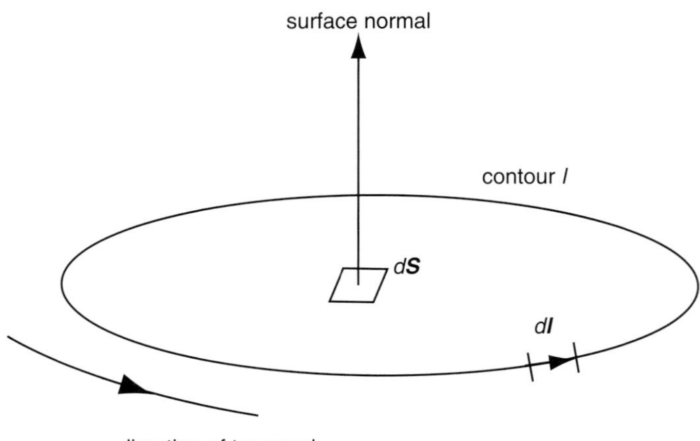

Figure 1.1 Closed contour used for calculating the line and surface integrals appearing in Maxwell's equations.

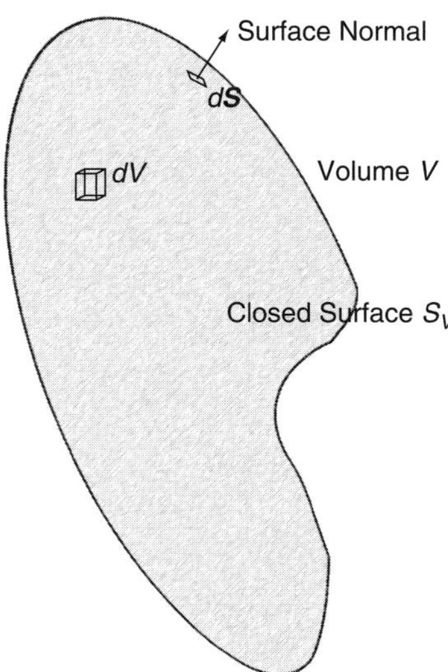

Figure 1.2 Fixed volume used for calculating the surface and volume integrals appearing in Maxwell's equations.

$$\boldsymbol{D} = \varepsilon_0 \boldsymbol{E}, \tag{1.2a}$$

$$\boldsymbol{B} = \mu_0 \boldsymbol{H}, \tag{1.2b}$$

where μ_0 is the permeability and ε_0 is the permittivity in free space.

The other two of Maxwell's equations relate the electric field vector $\boldsymbol{E}$ with the magnetic flux-density vector $\boldsymbol{B}$ using surface integrals calculated over a fixed volume V, bounded by a closed surface S_V as shown in Figure 1.2. Their explicit form is

$$\text{Gauss's law:} \quad \oint_{S_V} \boldsymbol{D}(\boldsymbol{r}, t) \cdot d\boldsymbol{S} = q(t), \tag{1.3a}$$

$$\text{Gauss's law for magnetism:} \quad \oint_{S_V} \boldsymbol{B}(\boldsymbol{r}, t) \cdot d\boldsymbol{S} = 0, \tag{1.3b}$$

where $q(t)$ is the total electric charge contained in the volume V.

In a continuous optical medium, the four integral equations can be recast in an equivalent differential form useful for theoretical analysis and numerical computations [6, 7]. Application of the *Stokes theorem*[1] to Eqs. (1.1) provides us with

---

[1] A continuous vector field $\boldsymbol{F}$ defined on a surface S with a boundary l satisfies $\oint_l \boldsymbol{F} \cdot d\boldsymbol{l} = \int_S \nabla \times \boldsymbol{F} \cdot d\boldsymbol{S}$.

$$\text{Curl equation for electric field:} \quad \nabla \times E(r,t) = -\frac{\partial B(r,t)}{\partial t}, \tag{1.4a}$$

$$\text{Curl equation for magnetic field:} \quad \nabla \times H(r,t) = \frac{\partial D(r,t)}{\partial t} + J(r,t), \tag{1.4b}$$

where $J(r,t)$ is the electric current density. Similarly, application of the *divergence theorem*[2] to Eqs. (1.3) provides us with

$$\text{Divergence equation for electric field:} \quad \nabla \cdot D(r,t) = \rho(r,t), \tag{1.5a}$$

$$\text{Divergence equation for magnetic field:} \quad \nabla \cdot B(r,t) = 0, \tag{1.5b}$$

where $\rho(r,t)$ is the local charge density.

The current density $J(r,t)$ and the charge density $\rho(r,t)$ are related to each other through Maxwell's equations. This relationship can be established by taking the divergence of Eq. (1.4b) and noting the operator identity $\nabla \cdot (\nabla \times F) \equiv 0$ for any vector field F. The result is

$$\nabla \cdot J(r,t) + \frac{\partial \rho(r,t)}{\partial t} = 0. \tag{1.6}$$

This equation is called the charge-continuity equation because it shows that the charge moving out of a differential volume is equal to the rate at which the charge density decreases within that volume. In other words, the continuity equation represents mathematically the principle of charge conservation at each point of space where the electromagnetic field is continuous.

1.1.2 Maxwell's equations in the frequency domain

The differential form of Maxwell's equations can be put into an equivalent format by mapping time variables to the frequency domain [8]. This is done by introducing the Fourier-transform operator $\mathscr{F}_{t+}\{\}(\omega)$ as

$$\widetilde{Y}(\ldots, \omega, \ldots) \triangleq \mathscr{F}_{t+}\{Y(\ldots, t, \ldots)\}(\omega)$$

$$= \int_{-\infty}^{\infty} Y(\ldots, t, \ldots) \exp(+j\omega t)dt, \tag{1.7}$$

where $j = \sqrt{(-1)}$ and ω is the associated frequency-domain variable corresponding to time t; ω can assume any value on the real axis (i.e., $-\infty < \omega < +\infty$). Even though all quantities in the physical world correspond to real variables, their Fourier

---

[2] A vector field F defined on a volume V with a boundary surface S satisfies $\oint_V \nabla \cdot F dV = \int_S F \cdot dS$.

transforms can result in complex numbers. Nonetheless, for physical and mathematical reasons, the Fourier representation can provide a much simpler description of an underlying problem in certain cases.

We consistently use the notation that a tilde over a time-domain variable Y represents its Fourier transform when the Fourier integral is done with the plus sign in the exponential $\exp(+j\omega t)$. The Fourier operator shows this sign convention by displaying a plus sign just after its integration variable t. It is important to note that by adopting this notation, we also explicitly indicate that the Fourier transform takes the real variable t to its Fourier-space variable ω.

A very useful feature of the Fourier transform is that a function can be readily inverted back to its original temporal form using the inverse Fourier-transform operator $\mathscr{F}_{\omega+}^{-1}\{\}\,(t)$, defined as

$$
\begin{aligned}
Y(\ldots,t,\ldots) &\triangleq \mathscr{F}_{\omega+}^{-1}\left\{\widetilde{Y}(\ldots,\omega,\ldots)\right\}(t) \\
&= \frac{1}{(2\pi)^{\dim(\omega)}}\int_{-\infty}^{\infty}\widetilde{Y}(\ldots,\omega,\ldots)\exp(-j\omega t)d\omega,
\end{aligned}
\tag{1.8}
$$

where $\dim(\omega)$ is the dimension of the variable ω. Because $t \to \omega$ is a one-dimensional mapping, we have $\dim(\omega) = 1$ in this instance, but it can assume other positive integer values. For example, the dimension of the Fourier mapping is 3 when we later use spatial Fourier transforms.

To convert Maxwell's equations to the Fourier-transform domain, we need to map the partial differentials in Maxwell's equations to the frequency domain. This can be done by differentiating Eq. (1.8) with respect to t to get

$$
\frac{\partial Y(\ldots,t,\ldots)}{\partial t} = \frac{1}{2\pi}\int_{-\infty}^{\infty}-j\omega\widetilde{Y}(\ldots,\omega,\ldots)\exp(-j\omega t)d\omega.
\tag{1.9}
$$

This equation shows that we can establish the operator identity

$$
\frac{\partial Y(\ldots,t,\ldots)}{\partial t} \equiv \mathscr{F}_{\omega+}^{-1}\left\{-j\omega\widetilde{Y}(\ldots,\omega,\ldots)\right\}(t),
\tag{1.10}
$$

resulting in the mapping $\frac{\partial}{\partial t} \to -j\omega$ from time to frequency domain.

We apply the operator relation (1.10) to the time-domain Maxwell's equations in Eqs. (1.4) and (1.5) to obtain the following frequency-domain Maxwell's equations:

$$
\begin{aligned}
\nabla \times \widetilde{\boldsymbol{E}}(\boldsymbol{r},\omega) &= j\omega\widetilde{\boldsymbol{B}}(\boldsymbol{r},\omega), \\
\nabla \cdot \widetilde{\boldsymbol{D}}(\boldsymbol{r},\omega) &= \widetilde{\rho}(\boldsymbol{r},\omega), \\
\nabla \times \widetilde{\boldsymbol{H}}(\boldsymbol{r},\omega) &= -j\omega\widetilde{\boldsymbol{D}}(\boldsymbol{r},\omega) + \widetilde{\boldsymbol{J}}(\boldsymbol{r},\omega), \\
\nabla \cdot \widetilde{\boldsymbol{B}}(\boldsymbol{r},\omega) &= 0.
\end{aligned}
\tag{1.11}
$$

1.1.3 Maxwell's equations in the momentum domain

Further simplification of Maxwell's equations is possible by mapping the spatial variable r to its equivalent Fourier-domain variable k (also called the momentum domain or the k space). For physical reasons discussed later, the Fourier transform in the spatial domain is defined with a minus sign in the exponential, i.e.,

$$\widehat{Y}(\ldots, k, \ldots) \triangleq \mathscr{F}_{r-}\{Y(\ldots, r, \ldots)\}(k)$$
$$= \int_{-\infty}^{\infty} Y(\ldots, r, \ldots)\exp(-jk \cdot r)dr. \tag{1.12}$$

The Fourier operator shows this sign convention clearly by displaying a minus sign just after its integration variable r. The hat symbol over a field variable Y represents its spatial-domain Fourier transform.

We should point out that a second choice exists for the signs used in the temporal ($t \rightarrow \omega$) and spatial ($r \rightarrow k$) Fourier transforms. The sign convention that we have adopted is often used in physics textbooks. Using Maxwell's equations with this sign convention, one can show that a plane wave of the form $\exp(jk \cdot r - j\omega t)$ moves radially outward from a point source for positive values of r and t [9]. The opposite sign convention, where the temporal variable in Eq. (1.7) carries a negative sign, is widely used in electrical engineering literature.

Similarly to the time-domain Fourier transform, a spatial-domain Fourier transform can be inverted with the following formula:

$$Y(\ldots, r, \ldots) \triangleq \mathscr{F}_{k-}^{-1}\{\widehat{Y}(\ldots, k, \ldots)\}(r)$$
$$= \frac{1}{(2\pi)^{\dim(k)}} \int_{-\infty}^{\infty} \widehat{Y}(\ldots, k, \ldots)\exp(+jk \cdot r)dk, \tag{1.13}$$

where $\dim(k) = 3$ because the mapping is done in three-dimensional space. Using a relation similar to that appearing in Eq. (1.10), we can establish the mapping $\nabla \rightarrow jk$ from the spatial domain to the k space [10]. Applying this mapping to the differential form of Maxwell's equations (1.4) and (1.5) leads to the k-space version of Maxwell's equations:

$$jk \times \widehat{E}(k, t) = -\frac{\partial \widehat{B}(k, t)}{\partial t}, \qquad jk \cdot \widehat{D}(k, t) = \widehat{\rho}(k, t),$$
$$jk \times \widehat{H}(k, t) = \frac{\partial \widehat{D}(k, t)}{\partial t} + \widehat{J}(k, t), \qquad jk \cdot \widehat{B}(k, t) = 0. \tag{1.14}$$

Further simplification occurs if we combine the temporal and spatial mappings to form a $\omega \otimes k$ representation of Maxwell's equations. To achieve this, we define a function that takes us from $t \otimes r$ space to $\omega \otimes k$ space (with a hat over the tilde):

$$\widehat{\widetilde{Y}}(\dots, k, \dots, \omega, \dots) \triangleq \mathscr{F}_{r+,t-}\{Y(\dots, r, \dots, t, \dots)\}(k, \omega)$$

$$= \int_{-\infty}^{\infty} Y(\dots, r, \dots, t, \dots) \exp(jk \cdot r - j\omega t)\, dr\, dt.$$

$$(1.15)$$

Application of this Fourier transform to Maxwell's equations in Eqs. (1.4) and (1.5) gives us their following simple algebraic form:

$$k \times \widehat{\widetilde{E}}(k, \omega) = \omega \widehat{\widetilde{B}}(k, \omega), \qquad k \cdot \widehat{\widetilde{D}}(k, \omega) = \widehat{\widetilde{\rho}}(k, \omega),$$

$$jk \times \widehat{\widetilde{H}}(k, \omega) = -j\omega \widehat{\widetilde{D}}(k, \omega) + \widehat{\widetilde{J}}(k, \omega), \qquad k \cdot \widehat{\widetilde{B}}(k, \omega) = 0.$$

$$(1.16)$$

These vectorial relationships show that, in a linear isotropic medium, the triplets $(\widehat{\widetilde{E}}, \widehat{\widetilde{B}}, k)$ and $(\widehat{\widetilde{D}}, \widehat{\widetilde{H}}, k)$ form two right-handed coordinate systems shown in Figure 1.3, irrespective of material parameters. However, such a general relationship does not exists for the triplets $(\widehat{\widetilde{E}}, \widehat{\widetilde{H}}, k)$ and $(\widehat{\widetilde{D}}, \widehat{\widetilde{B}}, k)$. We shall see later that the signs of the permittivity and permeability associated with a medium play an important role in establishing functional relationships between the four field variables.

The preceding Maxwell's equations need to be modified when they are applied to physical materials because of charge movement (conductance), induced polarization, and induced magnetization within the medium. Charge movement in a material occurs because the electric and magnetic fields exert force on charges. Induced polarization within a material medium is a result of the rearrangement of the bound electrons, and it is responsible for the appearance of additional charge density known as the bound charge density. If these bound charges oscillate as a result of an externally applied electromagnetic field, electric dipoles induce a polarization current within the medium. In addition to the electric dipoles, a magnetization current can also be generated if magnetic dipoles are present in a medium.

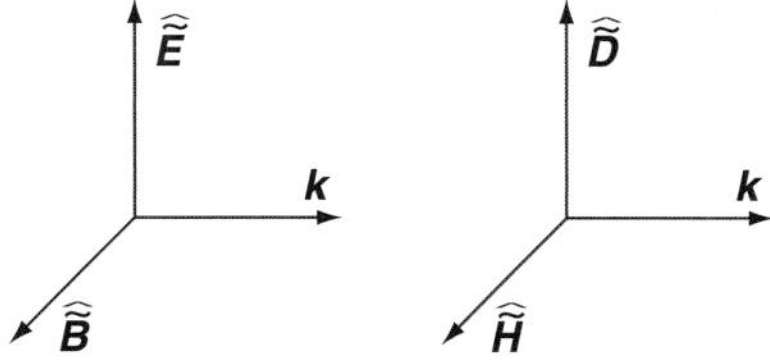

Figure 1.3 An illustration of the orthogonality of the triplets $(\widehat{\widetilde{E}}, \widehat{\widetilde{B}}, k)$ and $(\widehat{\widetilde{D}}, \widehat{\widetilde{H}}, k)$.

Table 1.1. *Constitutive relations for several types of optical media*

Medium type	Functional dependence	Description
Simple	$\widehat{\vec{D}} = \varepsilon \widehat{\vec{E}}$ $\widehat{\vec{B}} = \mu \widehat{\vec{H}}$	ε: constant, scalar μ: constant, scalar
Dispersive	$\widehat{\vec{D}} = \widehat{\varepsilon}(k, \omega) \widehat{\vec{E}}$ $\widehat{\vec{B}} = \widehat{\mu}(k, \omega) \widehat{\vec{H}}$	$\widehat{\varepsilon}(k, \omega)$: complex function $\widehat{\mu}(k, \omega)$: complex function
Anisotropic	$\widehat{\vec{D}} = \varepsilon_0 \widehat{\vec{E}} + \varepsilon_0 \widehat{\chi}_e \cdot \widehat{\vec{E}}$ $\widehat{\vec{B}} = \mu_0 \widehat{\vec{H}} + \mu_0 \widehat{\chi}_m \cdot \widehat{\vec{H}}$	$\widehat{\chi}_e$: 3×3 matrix $\widehat{\chi}_m$: 3×3 matrix
Bi-isotropic	$\widehat{\vec{D}} = \varepsilon \widehat{\vec{E}} + \xi \widehat{\vec{H}}$ $\widehat{\vec{B}} = \zeta \widehat{\vec{E}} + \mu \widehat{\vec{H}}$	ε, ξ: constant, scalars ζ, μ: constant, scalars

All of these phenomena are incorporated within Maxwell's equations through the so-called constitutive relations among the four field vectors.

1.1.4 Constitutive relations for different optical media

The most general linear relationship among $\widehat{\vec{E}}, \widehat{\vec{D}}, \widehat{\vec{H}}$, and $\widehat{\vec{B}}$ occurs in a bi-anisotropic medium, and it can be written as

$$\widehat{\vec{D}}(k, \omega) = \widehat{\varepsilon}(k, \omega) \cdot \widehat{\vec{E}}(k, \omega) + \widehat{\xi}(k, \omega) \cdot \widehat{\vec{H}}(k, \omega), \tag{1.17a}$$

$$\widehat{\vec{B}}(k, \omega) = \widehat{\zeta}(k, \omega) \cdot \widehat{\vec{E}}(k, \omega) + \widehat{\mu}(k, \omega) \cdot \widehat{\vec{H}}(k, \omega), \tag{1.17b}$$

where $\widehat{\varepsilon}(k, \omega)$, $\widehat{\xi}(k, \omega)$, $\widehat{\zeta}(k, \omega)$, and $\widehat{\mu}(k, \omega)$ are in the form of 3×3 matrices. However, some of those matrices are identically zero or have scalar values depending on the nature of material medium. The constitutive relations for several classes of optical media are classified in Table 1.1, based on the dependence of the four parameters on ω and k. A medium is called *frequency dispersive* if these parameters depend explicitly on ω. Similarly, a medium is called *spatially dispersive* if they depend explicitly on k.

Frequency dispersion is an inherent feature of any optical medium because no real medium can respond to an electromagnetic field instantaneously. It is possible to show that the frequency dependence of the permittivity of a medium is fundamentally related to the causality of the response of that medium that leads naturally to the Kramers–Kronig relations [11, 12] discussed later. Using these relations one can also show that a frequency-dispersive medium is naturally lossy. Since the commonly used assumption of a lossless dispersive medium is not generally valid, the notion of an optically thick, lossless medium needs to be handled very cautiously. In certain cases, we may consider the frequency dependence of the refractive index but neglect such dependence for the absorption. Generally, these kinds of assumptions lead to the prediction of noncausal behavior (e.g., a signal traveling faster than the of speed of light in vacuum). However, if the frequency dependence of the refractive index is considered only over a finite frequency range, absorption can be neglected over that range without violating causality [13].

We do not consider spatial dispersion in this text because optical wavelengths are much longer than atomic dimensions and inter-atomic spacing. It is interesting to note that spatial dispersion is implicitly related to the magneto-electric response of a medium. In a magneto-electric medium, the presence of electric fields causes the medium to become magnetized, and the presence of magnetic fields makes the medium polarized. It is evident from Eq. (1.16) that the magnetic field is completely determined by the transverse part of the electric field in the $\omega \otimes k$ space. Thus, it is not possible, in principle, to separate the roles of electric and magnetic fields.

Most physical media are neither linear nor isotropic in the sense that their properties depend on both the strength of the local electric field and its direction. Also, there exist optically active media that can rotate the state of polarization of the electric field either clockwise (dextrorotation) or counterclockwise (levorotation). Although we consider the nonlinear nature of an optical medium whenever relevant, unless otherwise stated explicitly we do not cover this type of optically active medium in this text. Furthermore, unless explicitly stated, we assume that the medium is nonmagnetic. In such a medium, the parameters ξ and ζ in Eqs. (1.1) vanish and the magnetic permeability can be replaced with its vacuum value μ_0. With these simplifications, we only need to know the permittivity ε to study wave propagation in a nonmagnetic medium. It is common to introduce the susceptibility χ of the medium by the relation $\varepsilon = \varepsilon_0(1 + \chi)$.

1.2 Permittivity of isotropic materials

In this section we focus on an isotropic, homogeneous medium and ignore the dependence of the permittivity ε on the propagation vector k. In the frequency domain, this dielectric function is complex because, as mentioned earlier, it must have a nonzero imaginary part to allow for absorption at certain frequencies that correspond to medium resonances. As we show later, the real and imaginary parts of the dielectric function are related to each other through the Kramers–Kronig relations [11].

1.2.1 Debye-type permittivity and its extensions

In some cases one can model an optical medium as a collection of noninteracting dipoles, each of which responds to the optical field independently. The electromagnetic response of such a medium is given by the Debye dielectric function [14],

$$\widetilde{\varepsilon}_{\text{Debye}}(\omega) = \varepsilon_0 \left[\varepsilon_\infty + \frac{\Delta\varepsilon_p}{1 - j\omega\tau_p} \right], \tag{1.18}$$

where ε_∞ is the dielectric constant of the medium in the high-frequency limit and $\Delta\varepsilon_p = \varepsilon_s - \varepsilon_\infty$ is the change in relative permittivity from its static value ε_s owing to the relaxation time τ_p associated with the medium. The parameter ε_∞ represents the deformational electric polarization of positive and negative charge separation due to the presence of an external electric field [15]. The ratio $\varepsilon/\varepsilon_0$ is sometimes referred to as the relative permittivity of the medium.

The permittivity $\varepsilon_{\text{Debye}}(t)$ of the medium in the time domain is obtained by taking the inverse Fourier transform of Eq. (1.18). If we introduce the susceptibility as defined earlier, its time dependence for the Debye dielectric function is exponential and is given by

$$\chi_{\text{Debye}}(t) = \begin{cases} \frac{\Delta\varepsilon_p}{\tau_p} \exp\left(-\frac{t}{\tau_p}\right) & \text{if } t > 0, \\ 0 & \text{otherwise.} \end{cases} \tag{1.19}$$

This equation shows the case of a medium with only one relaxation time. The Debye dielectric function in Eq. (1.18) can be extended readily to a medium with N relaxation times by using

$$\widetilde{\varepsilon}_{\text{Debye}}(\omega) = \varepsilon_0 \left[\varepsilon_\infty + \sum_{m=1}^{N} \frac{\Delta\varepsilon_m}{1 - j\omega\tau_m} \right]. \tag{1.20}$$

To account for the asymmetry and broadness of some experimentally observed dielectric functions, a variant of Eq. (1.18) was suggested in Ref. [16]. This variant, known as $\widetilde{\varepsilon}_{\text{DHN}}(\omega)$, introduces empirically two parameters, α and β:

$$\widetilde{\varepsilon}_{\text{DHN}}(\omega) = \varepsilon_0 \left[\varepsilon_\infty + \frac{\Delta\varepsilon_p}{[1 + (-j\omega\tau_p)^\alpha]^\beta} \right]. \tag{1.21}$$

The value of α is adjusted to match the observed asymmetry in the shape of the permittivity spectrum and β is used to control the broadness of the response. When $\alpha = 1$, this model is known as the Cole–Davidson model [17] and has the form

$$\widetilde{\varepsilon}_{\text{CD}}(\omega) = \varepsilon_0 \left[\varepsilon_\infty + \frac{\Delta\varepsilon_p}{(1 - j\omega\tau_p)^\beta} \right], \quad 0 < \beta \le 1. \tag{1.22}$$

It is interesting to note that the Cole–Davidson form can be seen as a continuous superposition of multiple Debye responses by replacing the sum in Eq. (1.20) with an integral:

$$\tilde{\varepsilon}_{\mathrm{CD}}(\omega) = \varepsilon_0 \left[\varepsilon_\infty + \Delta\varepsilon_p \int_{-\infty}^{\infty} \frac{G_{\mathrm{CD}}(\tau)}{1 - j\omega\tau_p} d\tau \right], \quad 0 < \beta \le 1, \qquad (1.23)$$

where the weighting function has the specific form

$$G_{\mathrm{CD}}(\tau) = \begin{cases} \frac{\sin(\pi\beta)}{\pi} \left(\frac{\tau}{\tau_p - \tau} \right)^\beta & \text{if } -\infty < \tau < \tau_p, \\ 0 & \text{otherwise.} \end{cases} \qquad (1.24)$$

The average relaxation time for the Cole–Davidson model is given by $\beta\tau_p$.

1.2.2 *Lorentz dielectric function*

The classical Lorentz model of dielectric materials is of fundamental importance in optics because it has the ability to describe quite accurately many observed dispersion phenomena such as normal and anomalous types of dispersion [5]. It can also accurately describe the frequency-dependent polarization induced by bound charges. In practice, it provides a simple way to model materials with optical properties dominated by one or more resonance frequencies associated with a specific medium.

The Lorentz dielectric function [14] has the following form:

$$\tilde{\varepsilon}_{\mathrm{Lorentz}}(\omega) = \varepsilon_0 \left[\varepsilon_\infty + \frac{\Delta\varepsilon_p \omega_p^2}{\omega_p^2 - 2j\omega\Gamma_p - \omega^2} \right], \qquad (1.25)$$

where ε_∞, as before, is the permittivity of the medium in the high-frequency limit. The two parameters, ω_p and Γ_p, represent the frequency of the medium resonance and its damping rate. This function has two poles at frequencies

$$\omega_\pm = \sqrt{\omega_p^2 - \Gamma_p^2} \pm j\Gamma_p \qquad (1.26)$$

that form a complex-conjugate pair.

The time-dependent susceptibility associated with the Lorentz model can be found by taking the inverse Fourier transform of Eq. (1.25) and is given by

$$\chi_{\mathrm{Lorentz}}(t) = \frac{\Delta\varepsilon_p \omega_p^2}{\sqrt{\omega_p^2 - \Gamma_p^2}} e^{-\Gamma_p t} \sin\left(\sqrt{\omega_p^2 - \Gamma_p^2}\, t \right), \quad (t > 0). \qquad (1.27)$$

It vanishes for $t < 0$ as required by causality.

Similarly to the Debye case, the Lorentz dielectric function in Eq. (1.25) with a single relaxation time can be easily extended to the case where the medium has multiple resonance frequencies. In the case of N resonance frequencies, it is given by

$$\widetilde{\varepsilon}_{\text{Lorentz}}(\omega) = \varepsilon_0 \left[\varepsilon_\infty + \sum_{m=1}^{N} \frac{\Delta\varepsilon_m \omega_m^2}{\omega_m^2 - 2j\omega\Gamma_m - \omega^2} \right]. \tag{1.28}$$

1.2.3 Drude dielectric function

The Lorentz model is applicable to materials in which bound charges dominate the dielectric response. In contrast, the response of a metallic medium is dominated by the free electrons within the conduction band. For such a medium, the Drude model provides a very good approximation of the dielectric properties over a wide frequency range [18]. It can be obtained by setting $\omega_p = 0$ in the Lorentz model, which corresponds to making the bound electrons free. The simplest Drude response is of the form [14]

$$\widetilde{\varepsilon}_{\text{Drude}}(\omega) = \varepsilon_0 \left[\varepsilon_\infty - \frac{\omega_p^2}{\omega^2 + j\omega\gamma_p} \right], \tag{1.29}$$

where ω_p is the Drude pole frequency and $\gamma_p > 0$ is the relaxation rate. The corresponding time-domain susceptibility is given by

$$\chi_{\text{Drude}}(t) = \begin{cases} \frac{\omega_p^2}{\gamma_p}[1 - \exp(-\gamma_p t)] & \text{if } t > 0, \\ 0 & \text{otherwise.} \end{cases} \tag{1.30}$$

The extension of the Drude dielectric function to a case of multiple relaxation rates is straightforward. As before, we sum over all multiple resonances (or relaxation rates) in Eq. (1.29). For a a metallic medium with N relaxation rates, we obtain

$$\widetilde{\varepsilon}_{\text{Drude}}(\omega) = \varepsilon_0 \left[\varepsilon_\infty - \sum_{m=1}^{N} \frac{\omega_m^2}{\omega^2 + j\omega\gamma_m} \right]. \tag{1.31}$$

1.3 Dispersion relations

In $\omega \otimes k$ space, frequency ω and propagation vector k are independent variables corresponding to time and space variables. When an optical wave propagates through an optical medium, Maxwell's equations establish a functional relation between ω

and k, known as the dispersion relation of the medium. This relationship can be written in the following symbolic form [19]:

$$\mathscr{D}(k, \omega) = 0. \tag{1.32}$$

This equation can be solved to find an explicit relation between ω and k which defines a solution of Maxwell's equations that can be sustained by the medium. Such solutions are known as *modes*. Its physical meaning is that an electromagnetic wave at a specific frequency ω must propagate with a propagation vector k that satisfies Eq. (1.32).

1.3.1 Dispersion relation in free space

As an example, consider wave propagation in free space or vacuum by using Maxwell's equations (1.16) with the "simple" constitutive relations given in Table 1.1. In the case of free space, ε and μ are replaced with their free-space values ε_0 and μ_0. The resulting Maxwell's equations in $\omega \otimes k$ space are

$$k \times \widehat{\widetilde{E}}(k, \omega) = \mu_0 \omega \widehat{\widetilde{H}}(k, \omega), \tag{1.33a}$$

$$k \times \widehat{\widetilde{H}}(k, \omega) = -\omega \varepsilon_0 \widehat{\widetilde{E}}(k, \omega), \tag{1.33b}$$

$$k \cdot \widehat{\widetilde{E}}(k, \omega) = 0, \tag{1.33c}$$

$$k \cdot \widehat{\widetilde{H}}(k, \omega) = 0. \tag{1.33d}$$

Substituting $\widehat{\widetilde{H}}(k, \omega)$ from Eq. (1.33a) into Eq. (1.33b), we obtain

$$k \times k \times \widehat{\widetilde{E}}(k, \omega) + \mu_0 \varepsilon_0 \omega^2 \widehat{\widetilde{E}}(k, \omega) = 0. \tag{1.34}$$

Using the well known identity $k \times k \times F = k(k \cdot F) - (k \cdot k)F$, valid for an arbitrary vector F, together with Eq. (1.33c), we obtain

$$\left[k \cdot k - \mu_0 \varepsilon_0 \omega^2 \right] \widehat{\widetilde{E}}(k, \omega) = 0. \tag{1.35}$$

The term in square brackets has to be identically zero for a finite electric field to exist. Thus, wave propagation in free space must satisfy the dispersion relation

$$k^2 = \mu_0 \varepsilon_0 \omega^2 = \omega^2/c^2, \tag{1.36}$$

where we have used $k^2 = k \cdot k$ and the well-known definition for the speed of light c in vacuum:

$$c = \frac{1}{\sqrt{\mu_0 \varepsilon_0}}. \tag{1.37}$$

The propagation modes can be identified by solving Eq. (1.36). One can use either ω or k as an independent variable. If we use ω as an independent variable and solve for k, we obtain

$$k = \omega/c, \quad \text{or} \quad k = -\omega/c. \tag{1.38}$$

These relations show two possible modes for each frequency ω, corresponding to forward and backward propagating waves in free space. This result is not going to be different if we change the choice of the independent variable. However, we need to keep in mind that ω can be positive or negative in Eqs. (1.33). It is traditional to use positive frequencies when describing dispersion phenomena. This can be justified by noting that the field variables in Eqs. (1.33) are not independent when ω is changed to $-\omega$. The reason is that the electromagnetic field is real-valued in physical space and time. It can be shown using the properties of the Fourier-transform integral that

$$\widehat{\widetilde{E}}^{*}(k, \omega) = \widehat{\widetilde{E}}(-k, -\omega), \qquad \widehat{\widetilde{D}}^{*}(k, \omega) = \widehat{\widetilde{D}}(-k, -\omega), \tag{1.39a}$$

$$\widehat{\widetilde{H}}^{*}(k, \omega) = \widehat{\widetilde{H}}(-k, -\omega), \qquad \widehat{\widetilde{B}}^{*}(k, \omega) = \widehat{\widetilde{B}}(-k, -\omega), \tag{1.39b}$$

where the superscript $*$ denotes the complex-conjugate operation. As long as we are aware of these subtle relations, we could operate anywhere in the frequency domain and arrive at meaningful results.

1.3.2 Dispersion relation in isotropic materials

It is instructive to look at plane-wave propagation in isotropic materials to understand some subtle issues associated with important concepts such as the group and phase velocities and the refractive index. For the moment, we limit our analysis to a passive medium with loss, and discuss the case of active media (relevant for optical amplifiers) later [20, 21]. We also ignore spatial dispersion because it only complicates the following analysis without providing any additional insight.

With the preceding assumptions, the permeability $\varepsilon(\omega)$ and permittivity $\mu(\omega)$ of an isotropic medium can be written in the form

$$\varepsilon(\omega) = \varepsilon_r(\omega) + j\varepsilon_i(\omega), \qquad \varepsilon_i(\omega) > 0, \tag{1.40a}$$

$$\mu(\omega) = \mu_r(\omega) + j\mu_i(\omega), \qquad \mu_i(\omega) > 0, \tag{1.40b}$$

where the subscripts r and i denote the real and imaginary parts of a complex number. The conditions $\varepsilon_i(\omega) > 0$ and $\mu_i(\omega) > 0$ ensure that the material is absorptive. We can understand this by noting that the total electromagnetic energy absorbed in a volume V of an isotropic medium is given by [22]

$$\frac{1}{2\pi} \int_V \int_{-\infty}^{\infty} \left[\omega \varepsilon_i(\omega) |\widetilde{\boldsymbol{E}}(\boldsymbol{r}, \omega)|^2 + \omega \mu_i(\omega) |\widetilde{\boldsymbol{H}}(\boldsymbol{r}, \omega)|^2 \right] d\omega \, d\boldsymbol{r}. \qquad (1.41)$$

Clearly, the positive imaginary parts of the permittivity and permeability ensure that this expression remains positive, in compliance with the thermodynamic requirements for a passive system.

Following an analysis similar to that for the free-space case, the dispersion relation for this material is found to be

$$k^2 = \mu(\omega)\varepsilon(\omega)\omega^2. \qquad (1.42)$$

However, solving this equation to calculate the propagation constant k is not trivial because of some subtle reasons. As we know, light propagates at a slower speed in a material medium compared to its speed in vacuum. To quantify this feature, we can rewrite the preceding equation by using $\mu_0 \varepsilon_0 = 1/c^2$ in the form

$$k^2 = \frac{\mu(\omega)\varepsilon(\omega)}{\mu_0 \varepsilon_0}(\omega^2/c^2) = n^2(\omega)(\omega^2/c^2), \qquad (1.43)$$

where $n^2(\omega)$ is a well-defined complex function of frequency.

To find k, we need to take the square root of a complex function. One has to be careful at this point because such an operation exhibits a branch-cut singularity. Formally, we can introduce two complex parameters n_+ and n_-:

$$n_\pm(\omega) = \pm \sqrt{\frac{\mu(\omega)\varepsilon(\omega)}{\mu_0 \varepsilon_0}}, \qquad (1.44)$$

where both have the same magnitude but opposite signs. However, only one of these parameters makes physical sense, and the one selected is called the complex refractive index of the material.

1.3.3 Refractive index and phase velocity

Physically, the refractive index provides a way to quantify the magnitude and direction of the phase velocity of light in a medium [23]. Recently, it has been argued in the literature that the correct choice of sign for the refractive index in Eq. (1.44) has no bearing on the solution of Maxwell's equations in a finite domain because the concept of refractive index is present in neither the time-domain nor the frequency-domain Maxwell's equations [24]. However, in practice, even the simple problem of wave propagation in a homogeneous isotropic medium, occupying the entire half-space $z > 0$, cannot be successfully resolved without specifying the correct sign of the refractive index. In fact, knowledge of the refractive index provides

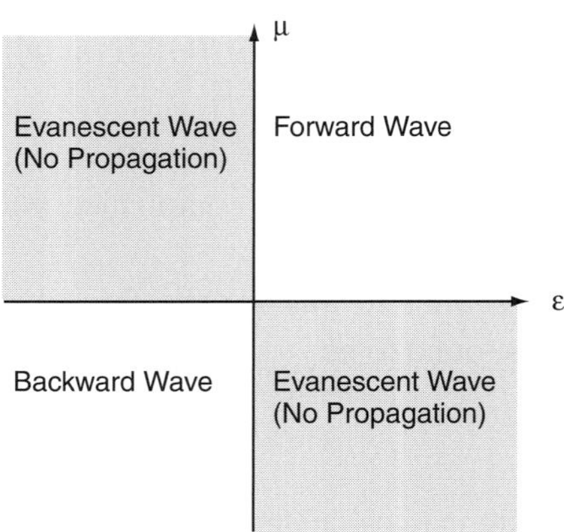

Figure 1.4 A map of wave propagation ability against material parameters for an isotropic material.

the direction of the phase velocity of the medium and dictates whether light will refract positively or negatively at the boundary at $z = 0$. A numerical solution of Maxwell's equations with the associated boundary conditions may resolve this situation, for both active and passive media, without resorting to the concept of the refractive index [20]. Nonetheless, the concept of refractive index is useful for analyzing any amplifier, and we need to investigate it further.

It is clear from Eq. (1.42) that, even when μ and ε are real (no losses), a real solution for the propagation constant k exists only if ε and μ are both positive or both negative. If these quantities have opposite signs, the propagation constant k becomes purely imaginary, leading to evanescent waves that decay exponentially with distance. Figure 1.4 shows the four possible scenarios in ε–μ space. If both $\varepsilon_r(\omega)$ and $\mu_r(\omega)$ are positive, then taking the positive square root in Eq. (1.44) ensures that the phase velocity is correctly represented and is in the direction in which energy flows. This is the case for most optical media. Such a medium is called a right-handed medium since E, B, and k form a right-handed coordinate system as shown in Figure 1.3.

In contrast, a medium in which the phase velocity and power flow are in the opposite directions is an example of a left-handed medium. Such media do not occur naturally but can be designed with artificial structural changes at nanometer scales, and a new field of metamaterials has emerged in recent years. A unique property of left-handed materials is that both $\varepsilon_r(\omega)$ and $\mu_r(\omega)$ are negative. In this case, we may not know the correct sign for the refractive index, but $n_\pm^2$ has a unique magnitude with positive sign. We denoted this quantity by the variable $n^2(\omega)$ in Eq. (1.43). More explicitly, it is defined as [25]

$$n^2(\omega) = n_\pm^2(\omega) = \frac{\mu(\omega)\varepsilon(\omega)}{\mu_0\varepsilon_0} = c^2\mu(\omega)\varepsilon(\omega). \tag{1.45}$$

If the real and imaginary parts of n are n_r and n_i, respectively, these two quantities can be found by equating the real and imaginary parts of Eq. (1.43), resulting in the following two relations:

$$n_r^2(\omega) - n_i^2(\omega) = c^2[\mu_r(\omega)\varepsilon_r(\omega) - \mu_i(\omega)\varepsilon_i(\omega)], \tag{1.46a}$$

$$2n_r(\omega)n_i(\omega) = c^2[\mu_i(\omega)\varepsilon_r(\omega) + \mu_r(\omega)\varepsilon_i(\omega)]. \tag{1.46b}$$

The phase velocity v_p of a plane wave in a medium is given by

$$v_p(\omega) = \frac{c}{n_r(\omega)}. \tag{1.47}$$

It is clear that the sign of n_r determines the direction of phase velocity. It is logical to choose the positive sign in Eq. (1.44) if the medium is right-handed. However, the negative sign should be chosen for a left-handed medium. To prove this, we need to find the direction of power flow using the Poynting vector. One can show [26] that the direction of power flow is proportional to the real part of the ratio $n(\omega)/\mu(\omega)$. Denoting this quantity by $\mathscr{P}_d$, we obtain

$$\mathscr{P}_d = \mathrm{Re}\left[\frac{n(\omega)}{\mu(\omega)}\right] = \frac{n_r(\omega)\mu_r(\omega) + n_i(\omega)\mu_i(\omega)}{\mu_r^2(\omega) + \mu_i^2(\omega)}. \tag{1.48}$$

Therefore, the conditions for obtaining a left-handed medium are

$$n_r(\omega)\mu_r(\omega) + n_i(\omega)\mu_i(\omega) > 0 \quad \text{and} \quad n_r(\omega) < 0. \tag{1.49}$$

Using n_r and n_i in terms of ε_r and ε_i, we find that the preceding two conditions are simultaneously satisfied if

$$\mathscr{H}(\omega) = \frac{\varepsilon_r(\omega)}{|\varepsilon(\omega)|} + \frac{\mu_r(\omega)}{|\mu(\omega)|} < 0. \tag{1.50}$$

We thus arrive at the following definition of the refractive index for a passive medium with $\varepsilon_i(\omega) > 0$ and $\mu_i(\omega) > 0$:

$$n(\omega) = \begin{cases} n_+ & \text{if} \quad \varepsilon_r(\omega) > 0, \ \mu_r(\omega) > 0, \\ n_- & \text{if} \quad \mathscr{H}(\omega) < 0. \end{cases} \tag{1.51}$$

A left-handed material with a negative value of the refractive index is also called a negative-index material and constitutes an example of a metamaterial.

1.3.4 Instabilities associated with a gain medium

Even though we could deduce the sign of the refractive index by noting the handedness of a passive medium, the preceding analysis cannot be extended even to a weakly active medium exhibiting a relatively low gain. It has been argued recently that one needs to know the global characteristics of the permittivity and the permeability functions conclusively to determine the sign and magnitude of the refractive index in active media [27]. However, Eq. (1.51) remains valid for determining the handedness of active media as well [28].

Because our main focus is on analyzing signal propagation in active amplifying media, we need to understand where concepts such as refractive index and wave vector, derived using linear properties of Maxwell's equations, remain meaningful. The main issue is that, if a weak signal launched into a gain medium grows exponentially as it propagates down the medium, does it become unstable at some point? A linear stability analysis can be carried out for this purpose. Such an analysis discards the effect of gain saturation that limits the signal growth and thus allows for unbounded growth indicative of an instability. Instabilities in physical media refer to a blowing up of the field amplitude with time. There are three types of commonly encountered instabilities [20, 29–31]:

Absolute instability refers to a blowing up of the field amplitude with time at a fixed point in space. Figure 1.5(a) shows an illustration of this phenomenon. Absolute instabilities arise from the intrinsic properties of the participating medium, regardless of the boundary conditions. Mathematically, such an instability is associated with the poles or odd-order zeros of the function $\varepsilon(\omega)\mu(\omega)$ in the complex ω plane. As a result, the refractive index $n(\omega)$ is not analytic in the upper half of the complex plane. If an optical wave strikes a semi-infinite medium obliquely, component of the propagation vector perpendicular to the interface may have branch cuts [28].

Convective instability refers to a blowing up of the optical field as it propagates through the medium. The blow-up happens at a point away from the origin (source) of the signal, as illustrated in Figure 1.5(b). Similarly to the case of absolute instability, the convective instability is intrinsic to the medium and occurs regardless of the boundary conditions. An example is provided by the modulation instability occurring in optical fibers [4].

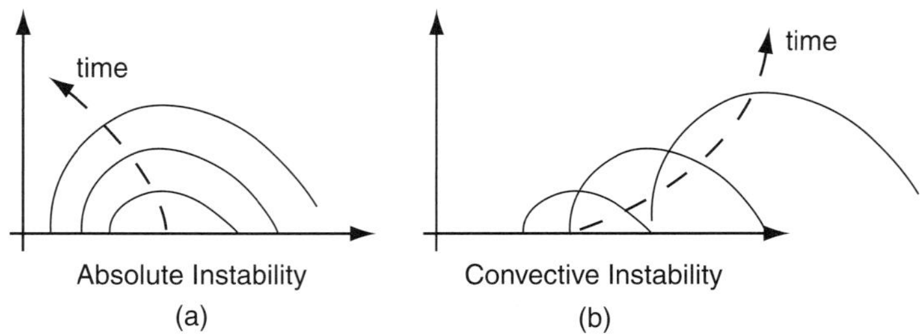

Figure 1.5 Illustration of two common instabilities in causal media. (After Ref. [20]; © APS 2008)

Global instability refers to an instability in a bounded medium such that both the intrinsic properties of the medium and the boundary conditions contribute to blowing up of the field amplitude within the medium. An active medium placed inside a resonator constitutes an example of this scenario. Physically, the active medium tends to amplify the signal without bounds. However, in the presence of gain saturation, such an unlimited growth of the field is hindered beyond the laser threshold [20].

1.4 Causality and its implications

When a dielectric medium is free of instabilities (or the situation can be modified to remove instabilities in the complex frequency plane), it is possible to establish two integral relations between the real and the imaginary parts of the dielectric function. They are known as the Kramers–Kronig relations. The reasons behind the existence of these relations can be traced back to the the fundamental concept of causality, requiring that no physical effect can occur without a cause [11].

1.4.1 Hilbert transform

One can say that the Kramers–Kronig relations restate causality in mathematical terms [19]. To gain an insight into this statement, it is instructive to consider the Fourier transform of a causal function $\zeta(t)$ defined such that it exists only for positive times $t > 0$, i.e., $\zeta(t) = 0$ for $t < 0$. If we use the Heaviside step function[3] $H(t)$, defined as

$$H(t) = \begin{cases} 1 & \text{if } t > 0, \\ 0 & \text{otherwise,} \end{cases} \tag{1.52}$$

then $\zeta(t)$ can be written as $\zeta(t) \equiv H(t)\zeta(t)$.

We now take the Fourier transform of $\zeta(t)$. The Fourier transform of $H(t)$ is well known and is given by

$$\tilde{H}(\omega) = \frac{1}{2}\delta(\omega) + \frac{j}{2\pi\omega}. \tag{1.53}$$

Noting that time-domain multiplication translates to convolution in the frequency domain (denoted by a $*$), we obtain [32]

$$\begin{aligned}
\tilde{\zeta}(\omega) &= \tilde{\zeta}(\omega) * \left(\frac{1}{2}\delta(\omega) + \frac{j}{2\pi\omega} \right) \\
&= \frac{1}{2}\tilde{\zeta}(\omega) + \frac{j}{2\pi}\mathscr{P}\int_{-\infty}^{\infty} \frac{\tilde{\zeta}(\Omega)}{\omega - \Omega}d\Omega,
\end{aligned} \tag{1.54}$$

---

[3] This is also known as a unit step function. It can also be written as $H(t) = \int_{-\infty}^{t} \delta(\tau)\,d\tau$.

where $\mathscr{P}$ in front of the integral sign indicates that the principle value of the integral [33] is taken (in the Cauchy sense). Solving this equation for $\tilde{\zeta}(\omega)$, we obtain the relation

$$\tilde{\zeta}(\omega) = \frac{1}{j\pi}\mathscr{P}\int_{-\infty}^{\infty}\frac{\tilde{\zeta}(\Omega)}{\Omega - \omega}d\Omega. \tag{1.55}$$

This equation forms the basis for the Kramers–Kronig relations. We obtained it by considering only the causality requirement of the function $\zeta(t)$. However, further advances can be made if we employ the well-established machinery of complex variables.

In this approach, the complex function $\tilde{\zeta}(\Omega)$ is a function of the complex variable $\Omega = \Omega_r + j\Omega_i$, where Ω_r and Ω_i are the real and imaginary parts of Ω, respectively. We assume that this function is analytic in the upper half of the complex Ω plane. Then, the application of the Cauchy theorem [33] leads to the result

$$\tilde{\zeta}(\omega) = \oint_C \frac{\tilde{\zeta}(\Omega)}{\Omega - \omega}d\Omega, \tag{1.56}$$

where ω is a real quantity corresponding to a frequency in physical terms and C is a closed contour in the upper half of the complex Ω plane. If we choose this contour as shown in Figure 1.6, assume $\tilde{\zeta}(\Omega) \to 0$ asymptotically as $|\Omega| \to \infty$, and use the residue theorem [33], we recover Eq. (1.55).

We now introduce the real and imaginary parts of $\tilde{\zeta}(\Omega)$ through $\tilde{\zeta} = \tilde{\zeta}_r + j\tilde{\zeta}_i$ and equate the real and imaginary parts of Eq. (1.55). This provides us with the following two relations:

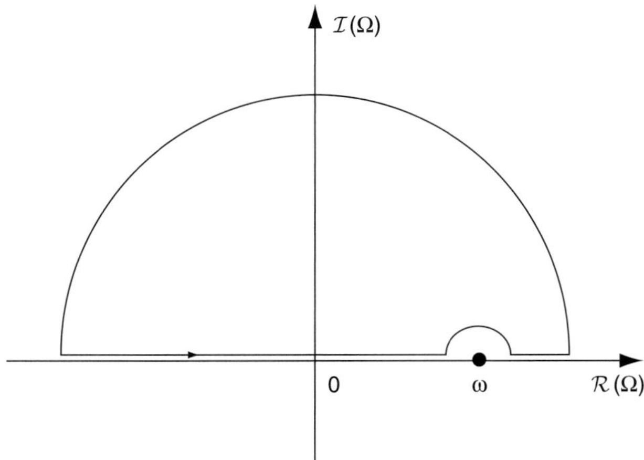

Figure 1.6 Contour in the upper half of the complex plane, used for deriving the Kramers–Kronig relations.

$$\tilde{\zeta}_r(\omega) = \frac{1}{\pi} \mathscr{P} \int_{-\infty}^{\infty} \frac{\tilde{\zeta}_i(\Omega)}{\Omega - \omega} d\Omega,$$

$$\tilde{\zeta}_i(\omega) = -\frac{1}{\pi} \mathscr{P} \int_{-\infty}^{\infty} \frac{\tilde{\zeta}_i(\Omega)}{\Omega - \omega} d\Omega. \tag{1.57}$$

These equations show that $\tilde{\zeta}_r(\omega)$ and $\tilde{\zeta}_r(\omega)$ form a Hilbert-transform pair.

1.4.2 Kramers–Kronig relations

It is useful to write the integrals in Eqs. (1.57) using only positive frequencies. As discussed earlier, negative and positive frequencies carry the same information for a real function $\zeta(t)$ because its Fourier transform satisfies the relation $\tilde{\zeta}^*(\omega) = \tilde{\zeta}(-\omega)$. Applying this to Eq. (1.57), we obtain the Kramers–Kronig relations:

$$\tilde{\zeta}_r(\omega) = \frac{2}{\pi} \mathscr{P} \int_0^{\infty} \frac{\Omega \tilde{\zeta}_i(\Omega)}{\Omega^2 - \omega^2} d\Omega,$$

$$\tilde{\zeta}_i(\omega) = -\frac{2\omega}{\pi} \mathscr{P} \int_0^{\infty} \frac{\tilde{\zeta}_r(\Omega)}{\Omega^2 - \omega^2} d\Omega. \tag{1.58}$$

These relations apply to any causal response function.

In practice, the integrals appearing in the Kramers–Kronig relations converge slowly. A single subtractive Kramers–Kronig relation has been proposed [34] whose use improves accuracy while evaluating it numerically. The idea is to note that Eq. (1.58) holds for any real frequency. Thus, if we pick a suitable frequency ω_0 and subtract $\tilde{\zeta}_r(\omega_0)$, we obtain the relation

$$\tilde{\zeta}_r(\omega) - \tilde{\zeta}_r(\omega_0) = \frac{2}{\pi}(\omega^2 - \omega_0^2) \mathscr{P} \int_0^{\infty} \frac{\Omega \tilde{\zeta}_i(\Omega)}{(\Omega^2 - \omega^2)(\Omega^2 - \omega_0^2)} d\Omega. \tag{1.59}$$

A similar equation can be derived for the imaginary part. This process can be continued to derive multiply-subtractive Kramers–Kronig relations if the convergence still remains an issue.

Let us consider how these relations can be applied to analyze passive dielectric materials. For such materials, $\epsilon(\omega)$ is responsible for the electromagnetic response. As $\omega \to \infty$, the permittivity $\epsilon(\omega) \to \epsilon_0$, where ε_0 is the permittivity of free space. This can be understood intuitively by noting that material cannot respond fast enough to very large frequencies. As a result, an electromagnetic field with an extremely high frequency sees any medium as a vacuum. However, one requirement in deriving the Kramers–Kronig relations was that the function $\zeta(\omega)$ vanish at infinity. Therefore, we must choose $\epsilon(\omega) - \epsilon(\omega_0)$ as an analytical function in the upper half of the complex plane. Substituting $\zeta(\omega) = \epsilon(\omega) - \epsilon_0$ into

Eq. (1.58), we find that the real and imaginary parts of the dielectric constant are related by

$$
\begin{aligned}
\varepsilon_r(\omega) &= \varepsilon_0 + \frac{2}{\pi}\mathscr{P}\int_0^\infty \frac{\Omega\,\varepsilon_i(\Omega)}{\Omega^2 - \omega^2}d\Omega, \\
\varepsilon_i(\omega) &= -\frac{2\omega}{\pi}\mathscr{P}\int_0^\infty \frac{\varepsilon_r(\Omega) - \varepsilon_0}{\Omega^2 - \omega^2}d\Omega.
\end{aligned}
\tag{1.60}
$$

It is important to realize that these two relations are valid only when no singularity of $\varepsilon(\Omega)$ occurs on the real axis. For example, conductors will have a singularity at $\Omega = 0$, and this singularity must be subtracted from the permittivity function before the Kramers–Kronig relations can be established for conductors.

The preceding example shows that causality alone is sufficient to establish the Kramers–Kronig relations in a passive dielectric medium. This is not the case for active, dielectric, and magnetic media because of the presence of instabilities [20]. If instabilities (or singularities) exist in the upper complex plane, the contour in Figure 1.6 needs to be adjusted so that the integral is taken on a line above the singularities.

1.5 Simple solutions of Maxwell's equations

Plane waves form the most fundamental instrument for analyzing electromagnetic wave propagation in optical media because, even though they are an idealization of real propagating waves, they closely resemble waves found far from a point source. For example, light incident on a detector on Earth from distance stars, or radio waves reaching a receiver placed far from an antenna, can be considered plane waves for all practical purposes. In this section, we first focus on the continuous-wave (CW) case and then consider optical pulses in the form of plane waves.

1.5.1 Continuous-wave plane waves

In Section 1.3.1 we found two solutions of Maxwell's equations in free space at a frequency ω in the form of forward and backward propagating plane waves with the propagation constant $k = \pm\omega/c$. We saw in Section 1.3.2 that these solutions apply even in a dispersive medium with frequency-dependent permittivity and permeability, provided we multiply k by its refractive index. Here, we focus on a plane wave at the frequency ω_0 propagating forward in such a medium with the propagation constant $k_0 = n(\omega_0)\omega_0/c$.

A plane wave has a constant phase front that propagates at the phase velocity of the medium. Moreover, the direction of propagation is always perpendicular to this phase front. Figure 1.7 shows a plane wave propagating in a medium with

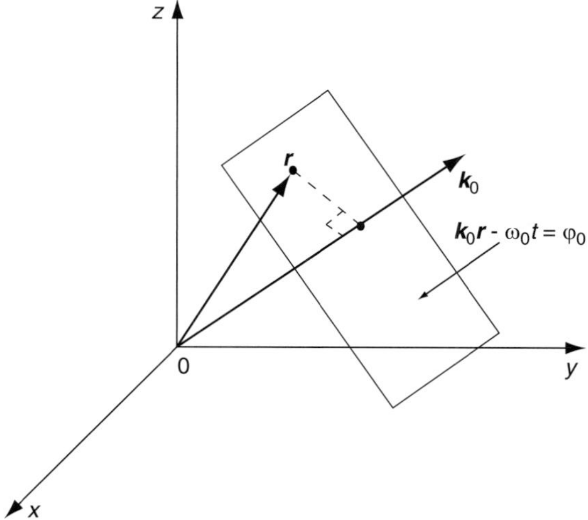

Figure 1.7 A general plane wave. Note that the propagation vector k_0 is perpendicular to the constant-phase wavefront $k_0 \cdot r - \omega_0 t = \varphi_0$.

the propagation vector k_0. Consider a point on the phase front of the plane wave, denoted by the vector r. The electric field of the plane wave at that point can be written as

$$E_p(r, t) = \mathrm{Re}\{E_0 \exp[j(k_0 \cdot r - \omega_0 t)]\} \tag{1.61a}$$

$$= |E_0| \cos(k_0 \cdot r - \omega_0 t + \phi_0), \tag{1.61b}$$

where Re denotes the real part and $E_0 = |E_0| \exp(j\phi_0)$ is the complex amplitude of the plane wave. The constant phase fronts on the plane wave can be tracked by finding spatial and temporal points that satisfy $k_0 \cdot r - \omega_0 t + \phi_0 = \psi_0$, where the value of ψ_0 is chosen to represent a particular wavefront.

The magnetic field associated with the electric field in Eq. (1.61) can be found from Maxwell's equations, given in Eq. (1.33) in $\omega \otimes k$ space. We first transform the electric field to the $\omega \otimes k$ space by taking the Fourier transform with respect to both t and r. The result is given by

$$\widetilde{\widehat{E}}_p(k, \omega) = \frac{1}{2} E_0 \delta(k - k_0) \delta(\omega - \omega_0), \tag{1.62}$$

where we have ignored the negative-frequency component. Using Maxwell's equations (1.33), we then obtain the magnetic field in the form

$$\widetilde{\widehat{H}}_p(k, \omega) = \frac{k \times E_0}{2\omega\mu(\omega)} \delta(k - k_0) \delta(\omega - \omega_0). \tag{1.63}$$

Using the dispersion relation of the medium, given in Eq. (1.42), the preceding equation can be written as

$$\widehat{\widetilde{H}}_p(k, \omega) = \frac{1}{2}\sqrt{\frac{\varepsilon(\omega)}{\mu(\omega)}}\frac{k}{|k|} \times E_0\delta(k - k_0)\delta(\omega - \omega_0). \tag{1.64}$$

However, caution needs to be exercised when determining the sign of the square root in this equation in view of our earlier discussion in Section 1.3.2. We discuss in Chapter 2 how one can determine all field components in an active medium capable of amplifying an electromagnetic field. We can get back to the time domain by using the inverse Fourier-transform relations:

$$H_p(r, t) = \frac{k_0 \times E_0}{2\eta(\omega_0)|k_0|} \exp(jk_0 \cdot r - j\omega_0 t), \tag{1.65}$$

where $\eta(\omega) = \sqrt{\mu(\omega)/\varepsilon(\omega)}$ is the impedance of the dispersive medium.

1.5.2 Pulsed plane waves

Temporal modulation of the amplitude of an electromagnetic wave can create pulses. Although pulses propagating through optical amplifiers are not in the form of plane waves and have spatial beam profiles, in this subsection we consider pulsed plane waves in order to become familiar with the essential features of optical pulses within the framework of Maxwell's equations.

Consider a plane wave whose amplitude is modulated using a time-domain function $s(t)$:

$$E_p(r, t) = \frac{1}{2}s(t)E_0 \exp(jk_0 \cdot r - j\omega_0 t). \tag{1.66}$$

The spectrum of this pulse is found by taking the Fourier transform of this expression. The result is

$$\widetilde{E}_p(r, \omega) = \frac{1}{2}\widetilde{s}(\omega - \omega_0)E_0 \exp(jk_0 \cdot r), \tag{1.67}$$

where $\widetilde{s}(\omega)$ is the temporal Fourier transform of $s(t)$.

The bandwidth of the pulse is governed by the spectral intensity $|\widetilde{s}(\omega)|^2$. However, a unique definition of the pulse bandwidth is not possible in all contexts because it depends to some extent on details of the pulse shape. Figure 1.8 shows three different definitions of bandwidth used in practice. They can be summarized as follows:

Full width at half-maximum (FWHM) bandwidth is equal to the full width of the pulse spectrum measured at half of its maximum amplitude. In mathematical terms, this amounts to finding the largest continuous interval where ω satisfies $\widetilde{s}(\omega) \geq \frac{1}{2}s(\omega_p)$, where ω_p is the frequency at which the pulse spectrum peaks.

Table 1.2. *Three measures of spectral bandwidth for four pulse shapes*

Pulse	$s(t)$	$\widetilde{s}(\omega)$	FWHM	Abs	NtN
Square	$\text{rect}(2t/T)$	$\text{sinc}(\omega T/2)$	$3.8/T$	∞	$4\pi/T$
Gaussian	$\exp(-t^2/T^2)$	$\exp(-\omega^2 T^2/2)$	$1.665/T$	∞	∞
Exponential	$\exp(-|t|/T)$	$1/(1+\omega^2 T^2)$	$2/T$	∞	∞
sech	$\text{sech}(t/T)$	$\text{sech}(\pi\omega T/2)$	$1.122/T$	∞	∞

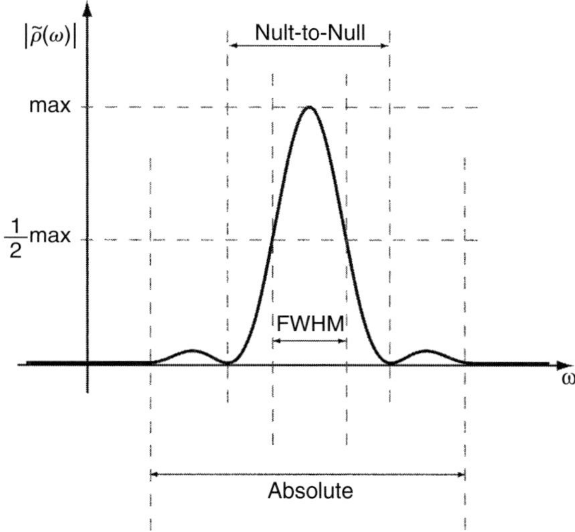

Figure 1.8 Commonly used measures of bandwidth of a pulse.

Null-to-null (NtN) bandwidth is equal to the maximum continuous interval surrounding the peak amplitude of the pulse spectrum between two points of zero amplitude. In mathematical terms this amounts to finding the largest continuous interval near ω_0 where the frequency ω satisfies $\widetilde{s}(\omega) \neq 0$.

Absolute (Abs) bandwidth is equal to the length of the largest continuous interval $[\omega_a, \omega_b]$ such that the spectral amplitude of the pulse vanishes for any frequency lying outside of this interval. In mathematical terms, $\widetilde{s}(\omega) = 0$ for any frequency ω not in the interval $[\omega_a, \omega_b]$.

The shape of pulses launched into an optical amplifier can vary widely. The two most common pulse shapes are Gaussian, with $s(t) = \exp(-t^2/2T^2)$, and hyperbolic secant, with $s(t) = \text{sech}(t/T)$. Other possibilities are a square pulse, with $s(t) = \text{rect}(2t/T)$, and an exponential pulse, with $s(t) = \exp(-|t|/T)$. Table 1.2 lists the FWHM, NtN and Abs bandwidth values for these four types of pulses. The temporal width of a pulse can also be defined using the same three measures. In practice, the FWHM of the intensity profile $|s(t)|^2$ is employed. For a square pulse with $s(t) = \text{rect}(2t/T)$, the FWHM is just T. In contrast, the FWHM of a Gaussian pulse is given by $T_{\text{FWHM}} = 2\sqrt{\ln 2}\,T_0 \approx 1.665\,T_0$. The same relation for a "sech" pulse takes the form $T_{\text{FWHM}} = 2\ln(1 + \sqrt{2})\,T_0 \approx 1.763\,T_0$. We shall encounter Gaussian and sech-shape pulses in later chapters.

As mentioned earlier, plane waves are not realistic because the amplitude of a plane wave has a constant value over the entire plane orthogonal to its direction of propagation. In any amplifier, an optical beam of finite size is launched inside it, and this beam is amplified as it propagates down the amplifier. The spatial profile as well as the temporal profile of the beam may change during amplification, as dictated by the solution of Maxwell's equations in the gain medium. We shall encounter such beams in later chapters.

References

[1] J. M. Jauch and F. Rohrlich, *The Theory of Photons and Electrons*, 2nd ed. Springer, 1976.

[2] L. Mandel and E. Wolf, *Optical Coherence and Quantum Optics*. Cambridge University Press, 1995.

[3] E. Desurvire, *Erbium-Doped Fiber Amplifiers: Principles and Applications*. Wiley, 1994.

[4] G. P. Agrawal, *Nonlinear Fiber Optics*, 4th ed. Academic Press, 2007.

[5] B. E. A. Saleh and M. C. Teich, *Fundamentals of Photonics*, 2nd ed. Wiley InterScience, 2007.

[6] J. D. Jackson, *Classical Electrodynamics*, 3rd ed. Wiley, 1998.

[7] D. J. Griffiths, *Introduction to Electrodynamics*, 3rd ed. Prentice Hall, 1999.

[8] K. E. Oughstun, *Electromagnetic and Optical Pulse Propagation: Spectral Representation in Temporally Dispersive Media*. Springer, 2006.

[9] M. Born and E. Wolf, *Principles of Optics*, 7th ed. Cambridge University Press, 1999.

[10] C. Altman and K. Suchy, *Reciprocity, Spatial Mapping and Time Reversal in Electromagnetics*. Springer, 1991.

[11] H. M. Nussenzveig, *Causality and Dispersion Relationships*. Academic Press, 1972.

[12] J. Raty, K.-E. Peiponen, and T. Asakura, *UV–Visible Reflection Spectroscopy of Liquids*. Springer, 2004.

[13] A. Tip, "Linear dispersive dielectrics as limits of Drude–Lorentz systems," *Phys. Rev. E*, vol. 69, p. 016610 (5 pages), 2004.

[14] A. Taflove and S. C. Hagness, *Computational Electrodynamics: The Finite-Difference Time-Domain Method*, 3rd ed. Artech House, 2005.

[15] Y. A. Lyubimov, "Permittivity at infinite frequency," *Russ. J. Phys. Chem.*, vol. 80, pp. 2033–2040, 2006.

[16] S. Havriliak and S. Negami, "A complex plane representation of dielectric and mechanical relaxation processes in some polymers," *Polymer*, vol. 8, pp. 161–210, 1967.

[17] C. P. Lindsey and G. D. Patterson, "Detailed comparison of the Williams–Watts and Cole–Davidson functions," *J. Chem. Phys.*, vol. 73, pp. 3348–3357, 1980.

[18] N. V. Smith, "Classical generalization of the Drude formula for the optical conductivity," *Phys. Rev. B*, vol. 64, pp. 155106 (6 pages), 2001.

[19] J. S. Toll, "Causality and dispersion relation: Logical foundations," *Phys. Rev.*, vol. 104, pp. 1760–1770, 1956.

[20] B. Nistad and J. Skaar, "Causality and electromagnetic properties of active media," *Phys. Rev. E*, vol. 78, p. 036603 (10 pages), 2008.

[21] J. Skaar, "Fresnel equations and the refractive index of active media," *Phys. Rev. E*, vol. 73, p. 026605 (7 pages), 2006.

[22] L. D. Landau and E. M. Lifshitz, *Electrodynamics of Continuous Media*, 2nd ed. *Course of Theoretical Physics*, vol. 8. Pergamon Press, 1984.

[23] S. A. Ramakrishna, "Physics of negative refractive index materials," *Rep. Prog. Phys.*, vol. 68, pp. 449–521, 2005.

[24] A. Lakhtakia, J. B. Geddes, and T. G. Mackay, "When does the choice of the refractive index of a linear, homogeneous, isotropic, active, dielectric medium matter?" *Opt. Express*, vol. 15, pp. 17709–17714, 2007. Available online: http://www.opticsinfobase.org/abstract.cfm?URI=oe-15-26-17709.

[25] R. A. Depine and A. Lakhtakia, "A new condition to identify isotropic dielectric-magnetic materials displaying negative phase velocity," *Microw. Opt. Technol. Lett.*, vol. 41, pp. 315–316, 2004.

[26] M. W. McCall, A. Lakhtakia, and W. S. Weiglhofer, "The negative index of refraction demystified," *Eur. J. Phys.*, vol. 23, pp. 353–359, 2002.

[27] P. Kinsler and M. W. McCall, "Causality-based criteria for negative refractive index must be used with care," *Phys. Rev. Lett.*, vol. 101, p. 167401 (4 pages), 2008.

[28] J. Skaar, "On resolving the refractive index and the wave vector," *Opt. Lett.*, vol. 31, pp. 3372–3374, 2006.

[29] P. A. Sturrock, "Kinematics of growing waves," *Phys. Rev.*, vol. 112, pp. 1488–1503, 1958.

[30] R. J. Birggs, *Electron-Stream Interactions with Plasmas*. Cambridge University Press, 1964.

[31] A. I. Akhiezer and R. V. Polovin, "Criteria for wave growth," *Sov. Phys. Usp.*, vol. 14, pp. 278–285, 1971.

[32] D. C. Hutchings, M. Sheik-Bahae, D. J. Hagan, and E. W. V. Stryland, "Kramers–Kronig relations in nonlinear optics," *Opt. Quant. Electron.*, vol. 24, pp. 1–30, 1992.

[33] T. W. Gamelin, *Complex Analysis*. Springer, 2001.

[34] R. Z. Bachrach and F. C. Brown, "Exciton-optical properties of TlBr and TlCl," *Phys. Rev. B*, vol. 1, pp. 818–831, 1970.

2

Light propagation through dispersive dielectric slabs

An integral feature of any optical amplifier is the interaction of light with the material used to extract the energy supplied to it by an external pumping source. In nearly all cases, the medium in which such interaction takes place can be classified as a dielectric medium. Therefore, a clear understanding of how light interacts with active and passive dielectric media of finite dimensions is essential for analyzing the operation of optical amplifiers. When light enters such a finite medium, its behavior depends on the global properties of the entire medium because of a discontinuous change in the refractive index at its boundaries. For example, the transmissive and reflective properties of a dielectric slab depend on its thickness and vary remarkably for two slabs of different thicknesses even when their material properties are the same [1].

In this chapter we focus on propagation of light through a dispersive dielectric slab, exhibiting chromatic dispersion through its frequency-dependent refractive index. Even though this situation has been considered in several standard textbooks [2, 3], the results of this chapter are more general than found there. We begin by discussing the state of polarization of optical waves in Section 2.1, followed with the concept of impedance in Section 2.2. We then devote Section 2.3 to a thorough discussion of the transmission and reflection coefficients of a dispersive dielectric slab in the case of a CW plane wave. Propagation of optical pulses through a passive dispersive slab is considered in Section 2.4, where we also provide simple numerical algorithms. The finite-difference time-domain technique used for solving Maxwell's equations directly is presented in Section 2.5. After discussing phase and group velocities in Section 2.6, we focus in Section 2.7 on propagation of pulses through a dispersive dielectric slab. In all cases, special attention is paid to the sign of the imaginary part of the refractive index because this distinguishes active and passive media and can lead to controversial issues.

2.1 State of polarization of optical waves

Electromagnetic waves propagating through a dielectric medium cannot be fully analyzed without knowing the direction of the electric and magnetic fields within the medium. Polarization of an optical wave provides us with a way to quantify these directions. According to Faraday's law, the directions of electric and magnetic fields are intrinsically coupled with each other. As a result, it is sufficient to consider the polarization characteristics of one of them. Traditionally, the electric field has been chosen as the preferred field quantity for describing the polarization of an electromagnetic field.

Most of the fundamental aspects of light polarization can be understood by inspecting the polarization properties of a plane wave propagating in free space. Consider a time-harmonic, monochromatic, plane wave traveling in the $+z$ direction of a right-handed coordinate system. The propagation vector $\mathbf{k}$ has only a z component for such a wave. From the divergence equation for the electric field and the Fourier-transform relations given in Section 1.1, it follows that the electric field lies in the x–y plane perpendicular to the propagation vector $\mathbf{k} = k\mathbf{z}$. Thus, $\mathbf{E}$ can be written in its most general form as

$$\mathbf{E} = E_x\mathbf{x} + E_y\mathbf{y} = E_{ax}\cos(kz - \omega t + \phi_x)\mathbf{x} + E_{ay}\cos(kz - \omega t + \phi_y)\mathbf{y}, \quad (2.1)$$

where $k = |\mathbf{k}| = \omega/c$, ω is the optical frequency, ϕ_x and ϕ_y are the phase angles of the electric field in the x and y directions, and $\mathbf{x}$, $\mathbf{y}$, and $\mathbf{z}$ are unit vectors in the three directions of the coordinate system. The electric field amplitudes, E_{ax} and E_{ay}, are related to the electric field components in the $\mathbf{x}$ and $\mathbf{y}$ directions:

$$E_x = E_{ax}\cos(kz - \omega t + \phi_x), \quad (2.2a)$$

$$E_y = E_{ay}\cos(kz - \omega t + \phi_y). \quad (2.2b)$$

It is possible to eliminate the dependence on $kz - \omega t$ in Eqs. (2.2) and find a relation between the two amplitudes and the corresponding phase angles. To do this, we expand the cosine function in Eqs (2.2) and obtain the following matrix equation:

$$\begin{bmatrix} E_x/E_{ax} \\ E_y/E_{ay} \end{bmatrix} = \begin{bmatrix} \cos(\phi_x) & -\sin(\phi_x) \\ \cos(\phi_y) & -\sin(\phi_y) \end{bmatrix} \begin{bmatrix} \cos(kz - \omega t) \\ \sin(kz - \omega t) \end{bmatrix}. \quad (2.3)$$

Inverting the square matrix on the right side of Eq. (2.3) and noting that the determinant of this matrix is given by $\sin(\phi_x - \phi_y)$, we obtain

$$\frac{E_x}{E_{ax}}\sin(\phi_y) - \frac{E_y}{E_{ay}}\sin(\phi_x) = \sin(kz - \omega t)\sin(\phi_x - \phi_y), \quad (2.4a)$$

$$\frac{E_x}{E_{ax}}\cos(\phi_y) - \frac{E_y}{E_{ay}}\cos(\phi_x) = \cos(kz - \omega t)\sin(\phi_x - \phi_y). \quad (2.4b)$$

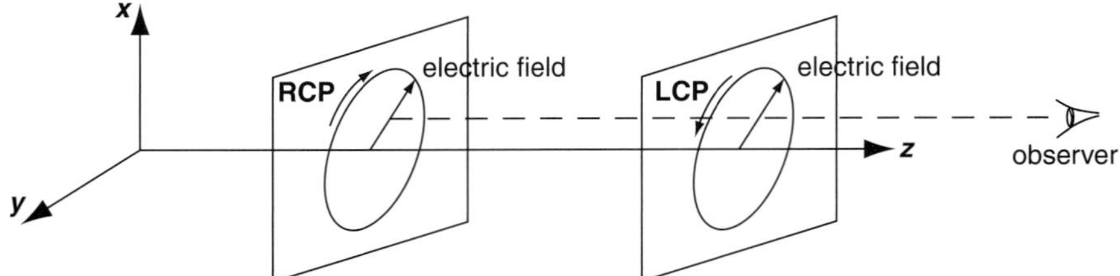

Figure 2.1 Polarization ellipse of a plane wave propagating along the $+z$ axis. The direction of rotation of the electric field along this ellipse corresponds to an observer looking toward the incoming wave. LCP and RCP stand for left and right circular polarizations, respectively.

We can eliminate the time and space dependence from these two equations by squaring and summing them. The resulting equation,

$$\frac{E_x^2}{E_{ax}^2} + \frac{E_y^2}{E_{ay}^2} - \frac{2\cos(\phi_x - \phi_y)}{E_{ax}E_{ay}}E_x E_y = \sin^2(\phi_x - \phi_y), \qquad (2.5)$$

shows that the the trajectory of E_x and E_y is in the form of an ellipse in the x–y plane, often called the polarization ellipse. Figure 2.1 shows two examples of such an ellipse schematically.

We have thus found that, in general, the electric field vector rotates with the frequency ω, in a plane perpendicular to the z axis, tracing out an ellipse. The state of polarization (SOP) of such a field is referred to as being elliptic. Apart from this general case, several special cases are of practical interest.

- **Linear polarization** refers to a scenario where it is possible to establish a linear relationship between the x and y components of the electric field. It is easy to see from Eq. (2.5) that this can occur when the relative phase difference $|\phi_x - \phi_y|$ between the two components is equal to 0 or π. The SOP of the optical wave is also linear when one of the field components, E_{ax} or E_{ay}, is identically equal to zero.
- **Circular polarization** refers to the case in which the polarization ellipse reduces to a circle. It is easy to see from Eq. (2.5) that this occurs when the two field components have the same amplitudes ($E_{ax} = E_{ay}$) and their relative phase difference $\phi_x - \phi_y$ is equal to $\pm\pi/2$. The positive and negative signs of this phase difference determine the direction in which the electric field vector rotates with time. If $\phi_x - \phi_y = -\pi/2$, the rotation is clockwise when viewed by an observer facing the $-z$ direction (see Fig. 2.1). This SOP is referred to as being right circularly polarized (RCP). In the second case, this rotation is counterclockwise, and the SOP represents a left circularly polarized (LCP) state.

- **p and s polarizations** refer to the special choice of a coordinate system in which the electric field components are measured relative to the incident plane of light, defined as a plane that contains the surface normal at the point where light enters the medium and the vector (k) of the incident light ray. Clearly, this type of polarization description makes sense only when interaction of light with a medium is considered. The component of the electric field parallel to the incident plane is termed p-like (parallel) and the component perpendicular to this plane is termed s-like (from *senkrecht*, German for perpendicular) [4]. The corresponding SOPs are called p-polarized and s-polarized, respectively. A similar terminology applies for magnetic fields.

In addition to the preceding classification for plane waves, another scheme is used in practice for light propagating in optical waveguides. This scheme is based on the optical modes supported by such waveguides, and electromagnetic waves are classified according to the nonzero components they have relative to the propagation direction. Modes occurring in optical waveguides are often called transverse electric (TE), transverse magnetic (TM), transverse electromagnetic (TEM), or hybrid.

- **TE modes** correspond to an optical mode whose electric field has no component in the direction of propagation. A subscript may be added to identify explicitly this direction [5]. For example, if a TE mode is propagating in the z direction, it is written as TE_z. In most waveguides, the component of the electric field in the direction normal to the plane of the waveguide layer is so small that TE modes are almost linearly polarized.
- **TM modes** refer to an optical mode whose magnetic field has no component in the direction of propagation. Similarly to the TE case, a subscript may be added to identify explicitly this direction [5]. In most waveguides, the component of the magnetic field in the the plane of the waveguide layer is so small that TM modes are almost linearly polarized.
- **TEM modes** refer to an optical mode for which neither the electric field nor the magnetic field has a component in the direction of propagation. Plane electromagnetic waves discussed in Section 1.5 fall into this category.
- **Hybrid modes** refer to an optical mode for which both the electric and the magnetic fields have components in the direction of propagation. Light propagating in an optical fiber falls in this category.

2.2 Impedance and refractive index

When we consider reflection and refraction of a plane wave at an interface, we need to know its propagation vector in the medium and the impedance of the medium seen by this plane wave. Calculations of these quantities require a selection of the correct sign for the square roots of the permittivity and permeability of the medium. In fact, an incorrect choice of this sign can lead to noncausal results making little physical sense! Moreover, a consistent strategy must be used to select the right sign for both the propagation vector and the impedance of the medium because any incompatibility between them also leads to erroneous results [6, 7].

The impedance η of any medium is defined as

$$\eta = \sqrt{\frac{\mu}{\varepsilon}}, \qquad \mathrm{Re}(\eta) > 0. \tag{2.6}$$

In general, both the permittivity ε and the permeability μ of an optical medium are complex quantities as they are defined in the Fourier-transform domain (see Section 1.3). As a result, the preceding definition is ambiguous because we have not explicitly specified the branch of the square-root function to be used. This choice cannot be arbitrary because causality considerations lead to an additional condition, $\mathrm{Re}(\eta) > 0$ [6, 7]. Moreover, this choice also depends on whether the medium is passive or active because $\mathrm{Im}(\varepsilon) < 0$ for an active dielectric material exhibiting the optical gain needed for optical amplifiers.

The simplest way to select the correct branch of the square-root function in Eq. (2.6) is to express the complex quantities ε and μ in their polar form,

$$\varepsilon = |\varepsilon| \exp[j \arg(\varepsilon) + j2m\pi], \qquad m = 0, \pm 1, \pm, 2, \cdots, \tag{2.7a}$$

$$\mu = |\mu| \exp[j \arg(\mu) + j2p\pi], \qquad p = 0, \pm 1, \pm, 2, \cdots, \tag{2.7b}$$

and select the integers m and p to satisfy the criterion $\mathrm{Re}(\eta) > 0$. The result is

$$\eta = +\sqrt{\frac{|\mu|}{|\varepsilon|}} \exp\left(\frac{j}{2}[\arg(\mu) - \arg(\varepsilon)] + j(p - m)\pi\right), \qquad \mathrm{Re}(\eta) > 0. \tag{2.8}$$

The selected branches of ε and μ are then directly used to calculate the propagation vector $\boldsymbol{k}$:

$$\boldsymbol{k} = \sqrt{\varepsilon\mu} = +\sqrt{|\varepsilon||\mu|} \exp\left(\frac{j}{2}[\arg(\varepsilon) + \arg(\mu)] + j(m + p)\pi\right). \tag{2.9}$$

A nice consequence of the preceding procedure is that the refractive index n of the material can be written with the right sign, for both positive- and negative-index materials, in the following form:

$$n = +\sqrt{\frac{|\varepsilon||\mu|}{\varepsilon_0\mu_0}} \exp\left(\frac{j}{2}[\arg(\varepsilon) + \arg(\mu)] + j(m + p)\pi\right). \tag{2.10}$$

In particular, when both ε and μ are negative real quantities, as may be the case for a negative-index metamaterial, $\arg(\varepsilon) = \arg(\mu) = \pi$, and it is easy to see that n is negative for the choice $m = p = 0$. However, in general, optical loss (or optical gain) exhibited by any material makes ε a complex quantity. In this case, one may have to choose nonzero values of m and p in Eq. (2.8) to ensure $\text{Re}(\eta) > 0$. The same values of m and p should be used to find the the refractive index n.

2.3 Fresnel equations

The passage of an electromagnetic plane wave through a material interface provides considerable insight into the vectorial nature of light. The reflectance and transmission coefficients at the interface depend not only on the optical properties of the two media making up the interface but also on the direction and the polarization of the incoming plane wave. We treat the cases of s and p polarizations separately and refer to them as the s-wave and p-wave, respectively. We neglect medium losses initially but consider the effect of medium loss or gain in the last subsection.

2.3.1 Case of plane s-waves

Figure 2.2 shows a plane s-wave (TE mode) undergoing reflection and refraction at an interface occupying the x–y plane located at $z = 0$. Owing to our assumption of lossless media on both sides of the interface, both the permittivity and the permeability of the two media, (ε_1, μ_1) and (ε_2, μ_2), are real. To sustain propagating waves in these media, we also need to assume that the sign of ε_1 matches the sign of μ_1 and the sign of ε_2 with that of μ_2.

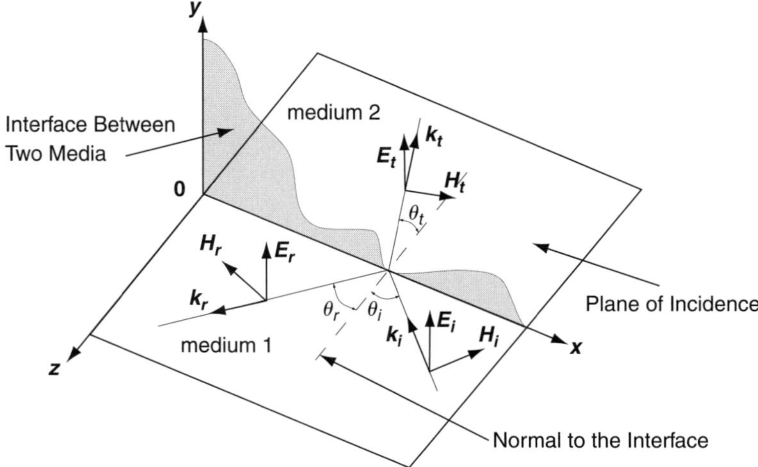

Figure 2.2 Schematic of a plane s-wave (TE mode) reflecting and refracting at an interface located in the $z = 0$ plane between two media with different dielectric properties.

Adopting the notation that subscripts i, r, and t describe incident, reflected, and transmitted quantities, respectively, we can write the following expressions for the corresponding electric fields (in phasor notation):

$$\boldsymbol{E}_i = E_i \, \mathbf{y} \mathrm{e}^{(\boldsymbol{k}_i \cdot \boldsymbol{r} - \omega t)}, \qquad \boldsymbol{k}_i = -k_i \sin(\theta_i) \boldsymbol{x} - k_i \cos(\theta_i) \boldsymbol{z}, \qquad (2.11\text{a})$$

$$\boldsymbol{E}_r = E_r \, \mathbf{y} \mathrm{e}^{(\boldsymbol{k}_r \cdot \boldsymbol{r} - \omega t)}, \qquad \boldsymbol{k}_r = -k_r \sin(\theta_r) \boldsymbol{x} + k_r \cos(\theta_r) \boldsymbol{z}, \qquad (2.11\text{b})$$

$$\boldsymbol{E}_t = E_t \, \mathbf{y} \mathrm{e}^{(\boldsymbol{k}_t \cdot \boldsymbol{r} - \omega t)}, \qquad \boldsymbol{k}_t = -k_t \sin(\theta_t) \boldsymbol{x} - k_t \cos(\theta_t) \boldsymbol{z}, \qquad (2.11\text{c})$$

where $k_i = k_r = \omega \sqrt{\mu_1 \varepsilon_1}$ and $k_t = \omega \sqrt{\mu_2 \varepsilon_2}$. The corresponding magnetic fields can be written as

$$\boldsymbol{H}_i = \frac{\boldsymbol{k}_i}{\eta_i k_i} \times \boldsymbol{E}_i = \frac{E_i}{\eta_i}[-\sin(\theta_i)\boldsymbol{z} + \cos(\theta_i)\boldsymbol{x}]\mathrm{e}^{(\boldsymbol{k}_i \cdot \boldsymbol{r} - \omega t)}, \qquad (2.12\text{a})$$

$$\boldsymbol{H}_r = \frac{\boldsymbol{k}_r}{\eta_r k_r} \times \boldsymbol{E}_r = \frac{E_r}{\eta_r}[-\sin(\theta_r)\boldsymbol{z} - \cos(\theta_r)\boldsymbol{x}]\mathrm{e}^{(\boldsymbol{k}_r \cdot \boldsymbol{r} - \omega t)}, \qquad (2.12\text{b})$$

$$\boldsymbol{H}_t = \frac{\boldsymbol{k}_t}{\eta_t k_t} \times \boldsymbol{E}_t = \frac{E_t}{\eta_t}[-\sin(\theta_t)\boldsymbol{z} + \cos(\theta_t)\boldsymbol{x}]\mathrm{e}^{(\boldsymbol{k}_t \cdot \boldsymbol{r} - \omega t)}. \qquad (2.12\text{c})$$

Because the two optical media are dielectric in nature, there are no surface currents at the interface. In this situation, Maxwell's equations dictate the boundary condition that the tangential components of the electric and magnetic fields should be continuous across the interface. This continuity requirement leads to the relations

$$\boldsymbol{z} \times (\boldsymbol{E}_i + \boldsymbol{E}_r) = \boldsymbol{z} \times \boldsymbol{E}_t, \qquad \boldsymbol{z} \times (\boldsymbol{H}_i + \boldsymbol{H}_r) = \boldsymbol{z} \times \boldsymbol{H}_t. \qquad (2.13)$$

Using Eqs. (2.11) and (2.12) in these two relations, we obtain the following matrix equation:

$$\begin{bmatrix} -1 & 1 \\ \dfrac{\cos(\theta_r)}{\eta_r} & \dfrac{\cos(\theta_t)}{\eta_t} \end{bmatrix} \begin{bmatrix} \Gamma_{\mathrm{TE}} \\ \mathrm{T}_{\mathrm{TE}} \end{bmatrix} = \begin{bmatrix} 1 \\ \dfrac{\cos(\theta_i)}{\eta_i} \end{bmatrix}, \qquad (2.14)$$

where we have defined amplitude reflection and transmission coefficients

$$\Gamma_{\mathrm{TE}} = E_r/E_i, \qquad \mathrm{T}_{\mathrm{TE}} = E_t/E_i. \qquad (2.15)$$

By inverting the coefficient matrix in Eq. (2.14), we obtain the Fresnel equations for a plane s-wave:

$$\Gamma_{\mathrm{TE}} = \frac{\eta_t \cos(\theta_i) - \eta_i \cos(\theta_t)}{\eta_t \cos(\theta_r) + \eta_r \cos(\theta_t)}, \qquad (2.16\text{a})$$

$$\mathrm{T}_{\mathrm{TE}} = \frac{\eta_r \cos(\theta_i) + \eta_i \cos(\theta_r)}{\eta_t \cos(\theta_r) + \eta_r \cos(\theta_t)}. \qquad (2.16\text{b})$$

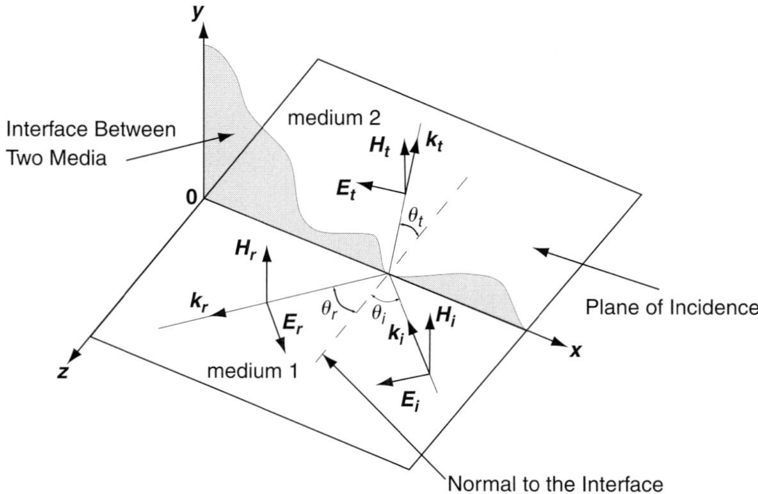

Figure 2.3 Schematic of a plane p-wave (TM mode) reflecting and refracting at an interface located in the $z = 0$ plane between two media with different dielectric properties.

These equations can be simplified by noting that $\theta_r = \theta_i$, $k_r = k_i$, and $\eta_r = \eta_i$. Furthermore, Snell's law, $k_t \sin(\theta_t) = k_i \sin(\theta_i)$, can be used to calculate $\cos(\theta_t)$.

2.3.2 Case of plane p-waves

Figure 2.3 shows a plane p-wave (TM mode) undergoing reflection and refraction at the interface located at $z = 0$. We adopt the notation used in Section 2.3.1 and write the following equations for the magnetic field associated with the incident, reflected, and transmitted plane waves:

$$\boldsymbol{H}_i = H_i\, \boldsymbol{y}\, \mathrm{e}^{(\boldsymbol{k}_i \cdot \boldsymbol{r} - \omega t)}, \qquad \boldsymbol{k}_i = -k_i \sin(\theta_i)\boldsymbol{x} - k_i \cos(\theta_i)\boldsymbol{z}, \qquad (2.17\mathrm{a})$$

$$\boldsymbol{H}_r = H_r\, \boldsymbol{y}\, \mathrm{e}^{(\boldsymbol{k}_r \cdot \boldsymbol{r} - \omega t)}, \qquad \boldsymbol{k}_r = -k_r \sin(\theta_r)\boldsymbol{x} + k_r \cos(\theta_r)\boldsymbol{z}, \qquad (2.17\mathrm{b})$$

$$\boldsymbol{H}_t = H_t\, \boldsymbol{y}\, \mathrm{e}^{(\boldsymbol{k}_t \cdot \boldsymbol{r} - \omega t)}, \qquad \boldsymbol{k}_t = -k_t \sin(\theta_t)\boldsymbol{x} - k_t \cos(\theta_t)\boldsymbol{z}, \qquad (2.17\mathrm{c})$$

where, as before, $k_i = k_r = \omega\sqrt{\mu_1 \varepsilon_1}$ and $k_t = \omega\sqrt{\mu_2 \varepsilon_2}$. The corresponding electric fields are:

$$\boldsymbol{E}_i = -\frac{\eta_i}{k_i} \boldsymbol{k}_i \times \boldsymbol{H}_i = \eta_i H_i [\sin(\theta_i)\boldsymbol{z} - \cos(\theta_i)\boldsymbol{x}] \mathrm{e}^{(\boldsymbol{k}_i \cdot \boldsymbol{r} - \omega t)}, \qquad (2.18\mathrm{a})$$

$$\boldsymbol{E}_r = -\frac{\eta_r}{k_r} \boldsymbol{k}_r \times \boldsymbol{H}_r = \eta_r H_r [\sin(\theta_r)\boldsymbol{z} + \cos(\theta_r)\boldsymbol{x}] \mathrm{e}^{(\boldsymbol{k}_r \cdot \boldsymbol{r} - \omega t)}, \qquad (2.18\mathrm{b})$$

$$\boldsymbol{E}_t = -\frac{\eta_t}{k_t} \boldsymbol{k}_t \times \boldsymbol{H}_t = \eta_t H_t [\sin(\theta_t)\boldsymbol{z} - \cos(\theta_t)\boldsymbol{x}] \mathrm{e}^{(\boldsymbol{k}_t \cdot \boldsymbol{r} - \omega t)}. \qquad (2.18\mathrm{c})$$

Using again the continuity requirements for the tangential components of the electric and magnetic fields, we obtain the matrix equation

$$
\begin{bmatrix} -1/\eta_r & 1/\eta_t \\ \cos(\theta_r) & \cos(\theta_t) \end{bmatrix} \begin{bmatrix} \Gamma_{\mathrm{TM}} \\ \mathrm{T}_{\mathrm{TM}} \end{bmatrix} = \begin{bmatrix} 1/\eta_i \\ \cos(\theta_i) \end{bmatrix},
\tag{2.19}
$$

where we have defined the amplitude reflection and transmission coefficients for the TM wave as

$$
\Gamma_{\mathrm{TM}} = E_r/E_i, \qquad \mathrm{T}_{\mathrm{TM}} = E_t/E_i.
\tag{2.20}
$$

Again, inverting the coefficient matrix, we obtain the Fresnel equations for a plane p-wave in the form

$$
\Gamma_{\mathrm{TM}} = \frac{\eta_r}{\eta_i} \left[\frac{\eta_i \cos(\theta_i) - \eta_t \cos(\theta_t)}{\eta_r \cos(\theta_r) + \eta_t \cos(\theta_t)} \right],
\tag{2.21a}
$$

$$
\mathrm{T}_{\mathrm{TM}} = \frac{\eta_t}{\eta_i} \left[\frac{\eta_r \cos(\theta_r) + \eta_i \cos(\theta_i)}{\eta_r \cos(\theta_r) + \eta_t \cos(\theta_t)} \right].
\tag{2.21b}
$$

These equations can be further simplified by noting that $\theta_r = \theta_i$, $k_r = k_i$, and $\eta_r = \eta_i$. Again, Snell's law, $k_t \sin(\theta_t) = k_i \sin(\theta_i)$, can be used to calculate $\cos(\theta_t)$.

2.3.3 *Fresnel equations in lossy dielectrics*

It is instructive to consider the propagation of a plane wave from a lossless medium to a lossy medium. The scenario where both media are lossy can be handled using a similar strategy, but one cannot avoid sophisticated mathematical machinery because of the appearance of bivectors in the formalism. The interested reader should consult Ref. [8] for details. The functional form of the Fresnel equations does not change when a medium becomes lossy. However, the propagation constant becomes complex in that medium, and care must be exercised in dealing with complex quantities.

When the second medium is lossy, material parameters ε_1 and μ_1 remain real, but ε_2, μ_2, or both become complex. Here we focus on the general case and assume that

$$
\varepsilon_2 = \varepsilon_{2r} + j\varepsilon_{2i}, \qquad \mu_2 = \mu_{2r} + j\mu_{2i}.
\tag{2.22}
$$

The propagation constant and the impedance in the second medium are given by

$$
\eta_2 = \sqrt{\frac{\mu_{2r} + j\mu_{2i}}{\varepsilon_{2r} + j\varepsilon_{2i}}}, \qquad k_2 = \sqrt{(\mu_{2r} + j\mu_{2i})(\varepsilon_{2r} + j\varepsilon_{2i})}.
\tag{2.23}
$$

The correct branch of the square root must be determined, following the discussion in Section 2.2. Assuming this has been done, we write these parameters in their complex form,

$$\eta_2 = \eta_{2r} + j\eta_{2i}, \qquad k_2 = k_{2r} + jk_{2i}. \tag{2.24}$$

Next consider Snell's law, $k_2 \sin(\theta_s) = k_1 \sin(\theta_1)$. Since k_2 is complex, we obtain the relation

$$\sin(\theta_2) = \frac{k_1}{k_{2r} + jk_{2i}} \sin(\theta_1). \tag{2.25}$$

The complex angle θ_t can be calculated by writing the sine function in its exponential form and and solving the resulting quadratic equation for $\exp(j\theta_2)$. Taking the logarithm of the resulting expression, we obtain

$$\theta_2 = -j \ln\left(j\wp + \sqrt{1 - \wp^2} \right), \tag{2.26}$$

where $\wp = k_1 \sin(\theta_1)/k_2$ is a complex number. Subsequent calculations must be carried out in the complex domain to find the correct form of the corresponding Fresnel equations. More specifically, when $\theta = \theta_r + j\theta_i$,

$$\cos(\theta) = \cos(\theta_r + j\theta_i) = \cos(\theta_r)\cosh(\theta_i) - j\sin(\theta_r)\sinh(\theta_i), \tag{2.27a}$$

$$\sin(\theta) = \sin(\theta_r + j\theta_i) = \sin(\theta_r)\cosh(\theta_i) + j\cos(\theta_r)\sinh(\theta_i). \tag{2.27b}$$

The expressions for the reflection and transmission coefficients associated with the TE and TM waves can be obtained by using these relations.

2.4 Propagation of optical pulses

Analysis of optical amplifiers requires a detailed understanding of pulse dynamics in dielectric media. Almost all known amplifier systems make use of a dispersive dielectric medium. In this section we discuss how propagation of optical pulses in such a medium can be treated mathematically. It is important to emphasize that Maxwell's equations can be solved without having any knowledge of the refractive index or the propagation vector because both of them are Fourier-domain concepts. As these concepts have a limited validity under general conditions, they must be used with care. For example, when a Fourier transform does not exist for a certain medium, one could carry out the analysis using Laplace transform. This feature points toward an important theme that recurs in physics and engineering: concepts with limited applicability can often be extended to cover a wider scope with a detailed understanding of the underlying physical principles.

2.4.1 One-dimensional propagation model

Propagation of optical pulses in an amplifying medium can often be studied by using a one-dimensional model involving time and the direction of propagation.[1] The reason is that the gain medium is often in the form of a waveguide that supports a finite number of optical modes whose transverse distributions in the x and y directions do not change much along the z axis. In some cases, such as fiber-based amplifiers, the medium supports a single mode with a definite SOP. As a result, the electric field can be written in the form

$$E(x, y, z, t) = \hat{p} F(x, y) E(z, t), \tag{2.28}$$

where $\hat{p}$ represents a unit vector representing the field SOP and $F(x, y)$ governs the spatial distribution of the optical mode.

The scalar electric field $E(z, t)$ is known in practice at the input end of the amplifier medium, located at $z = 0$, and we want to know how it evolves along the medium length with time. Assuming that nonlinear effects are negligible, it is useful to work in the Fourier domain and use the relation

$$\widetilde{E}(z, \omega) = \int_{-\infty}^{\infty} E(z, t) \exp(+j\omega t) \, dt. \tag{2.29}$$

Since $E(0, t)$ is known, we can find $\widetilde{E}(0, \omega)$ using Eq. (2.29) at $z = 0$. To propagate each Fourier component over a distance z, we only need to multiply $\widetilde{E}(0, \omega)$ by $\exp(jkz)$, where k is the propagation constant of the medium. Using the refractive index, $n(\omega)$, we can write $k = \omega n(\omega)/c$. Taking the inverse Fourier transform of Eq. (2.29), we obtain the electric field at a distance z in the form

$$E(z, t) = \frac{1}{2\pi} \int_{-\infty}^{\infty} \widetilde{E}(0, \omega) \exp\left[j\omega n(\omega)\frac{z}{c} - j\omega t \right] d\omega. \tag{2.30}$$

The preceding expression can be written in an equivalent but intuitively appealing form familiar to engineers by introducing the following transfer function associated with the amplifier medium:

$$H(z, t) = \frac{1}{2\pi} \int_{-\infty}^{\infty} \exp\left(-j\omega[t - n(\omega)z/c] \right) d\omega. \tag{2.31}$$

With this transfer function, it is possible to write Eq. (2.30) as a filtering operation,

$$E(z, t) = \int_{-\infty}^{\infty} H(z, t - \tau) E(0, \tau) d\tau, \tag{2.32}$$

---

[1] We assume that this direction is parallel to the $+z$ axis.

where the integral involves time-domain convolution of the filter function (or the impulse response) with the input electric field at $z = 0$ [9].

Using purely mathematical arguments, it is possible to show that the transfer function $H(z, t)$ is causal. The proof is based on the analytic nature of the complex function $n(\omega)$ in the upper half of the complex ω plane. If $n(\omega) \to 1$ as $\omega \to +\infty$, as required for physical reasons, then $H(z, t)$ must be zero for $t < z/c$ [9]. Note that analyticity in the upper half of the complex plane is a consequence of our $\exp(-j\omega t)$ convention. If $\exp(+j\omega t)$ is used instead, analyticity of $n(\omega)$ in the lower half of the complex plane is required.

2.4.2 Case of a Gaussian pulse

Although the frequency integral in Eq. (2.30) cannot be carried out analytically for an arbitrary input pulse shape, an analytic approach is possible for some specific pulse shapes. A useful example is provided by input pulses whose shape is Gaussian, such that

$$E(0, t) = E_0 \, \exp\left(-\frac{t^2}{2T_0^2} - j\omega_0 t\right), \tag{2.33}$$

where E_0 is the peak amplitude, ω_0 is the carrier frequency, and the parameter T_0 is related to the FWHM of the pulse by $T_{\text{FWHM}} = 2\sqrt{\ln 2}\,T_0 \approx 1.665 T_0$. It is clear that there is no clear starting or ending point for this pulse. Instead, the pulse shape is interpreted based on measurements made by an observer at $z = 0$ who sees a maximum amplitude of the Gaussian pulse at time $t = 0$. Taking the Fourier transform as indicated in Eq. (2.29), we obtain

$$\widetilde{E}(0, \omega) = \frac{E_0 T_0}{\sqrt{2\pi}} \exp\left[-\frac{1}{2}(\omega - \omega_0)^2 T_0^2\right]. \tag{2.34}$$

To find the propagated field at a distance $z > 0$, this expression should be used in Eq. (2.30).

Before proceeding further, we need an analytic form of the refractive index $n(\omega)$ associated with the medium. A common model makes use of the Lorentz response function (see Section 1.2.2) with a single resonance [10]. Assuming that the medium exhibits anomalous dispersion, we use

$$n(\omega) = n_\infty - \frac{\omega_a \omega_p}{\omega(\omega - \omega_a + j\gamma)}, \qquad \left|\frac{\omega_p}{\gamma}\right| \ll n_\infty, \tag{2.35}$$

where ω_a is the atomic resonance frequency, γ governs the bandwidth of the absorption peak, ω_p is a constant (often called the plasma frequency), and n_∞ is the value of $n(\omega)$ at very large frequencies.

To simplify the analysis, we assume that the spectral width of the pulse is substantially smaller than the atomic bandwidth ($\gamma T_0 \gg 1$). Since the dominant contribution to the integral in Eq. (2.30) comes from the neighborhood of $\omega = \omega_0$, we expand $\omega n(\omega)$ in a Taylor series as [10]

$$
\omega n(\omega) = \omega_0 n(\omega_0) + n_\infty(\omega - \omega_0)
$$
$$
- \frac{\omega_a \omega_p(\omega - \omega_0)}{(\omega_0 - \omega_a + j\gamma)^2} + \frac{\omega_a \omega_p(\omega - \omega_0)^2}{(\omega_0 - \omega_a + j\gamma)^3} - \cdots . \tag{2.36}
$$

We truncate this series expansion by assuming that $(\omega - \omega_0)^2 \ll (\omega_0 - \omega_a)^2 + \gamma^2$ for all frequencies that contribute to the integral significantly.

If the distance z is not too large, it is possible to show [10] that frequencies that contribute to the integral are those satisfying the condition $(\omega - \omega_0)^2 T_0^2 \leq 1$. Keeping terms up to second order in $\omega - \omega_0$, and substituting the result back into Eq. (2.30), we obtain

$$
E(z,t) = \exp\left(j(k_0 z - \omega_0 t) + \frac{\zeta_1^2}{4\zeta_2} \right)
$$
$$
\times \frac{E_0 T_0}{(2\pi)^{3/2}} \int_{-\infty}^{\infty} \exp\left[-\zeta_2 \left(u - \frac{\zeta_1}{2\zeta_2} \right)^2 \right] du, \tag{2.37}
$$

where $k_0 = \omega_0 n(\omega_0)/c$, $u = \omega - \omega_0$, and we have introduced

$$
\zeta_1 = j(n_\infty z/c - t) - \frac{j\omega_a \omega_p(z/c)}{(\omega_0 - \omega_a + j\gamma)^2}, \tag{2.38a}
$$

$$
\zeta_2 = \frac{1}{2}T_0^2 - \frac{j\omega_a \omega_p(z/c)}{(\omega_0 - \omega_a + j\gamma)^3}. \tag{2.38b}
$$

The integral in Eq. (2.37) converges if the real part of ζ_2 is positive because the integrand then goes to zero as u tends to $\pm\infty$. Under this condition, an analytical expression for the propagated field can be obtained by using

$$
\int_{-\infty}^{\infty} \exp[-p(t+c)^2]dt = \sqrt{\pi/p}, \qquad \mathrm{Re}(p) > 0, \tag{2.39}
$$

and the final result is given by

$$
E(z,t) = \frac{E_0 T_0}{\sqrt{2\zeta_2}} \exp\left(j(k_0 z - \omega_0 t) + \frac{\zeta_1^2}{4\zeta_2} \right). \tag{2.40}
$$

This expression shows that a Gaussian pulse maintains its Gaussian shape during propagation but its width increases because of dispersion. At the same time, the pulse develops a phase that varies quadratically with time. Such pulses are said to be *linearly chirped*.

2.4.3 Numerical approach

In cases in which the frequency integral in Eq. (2.30) cannot be performed analytically, we must use a numerical approach. Given that this integral is in the form of a a Fourier transform, it is common to employ the fast Fourier transform (FFT) algorithm to evaluate it. The FFT is an efficient way of evaluating the discrete Fourier transform of a periodic signal. So, we must truncate the integral in Eq. (2.30) over a finite time window and assume a periodic extension beyond this window. However, according to Fourier theory, such a truncation can introduce new frequency components in the entire frequency domain [11].

This issue was examined in detail by Slepian [12], who found that, if a signal contains most of its energy in a finite time window, $t_{min} < t < t_{max}$, then it is possible to find a finite window containing that energy in the frequency domain as well. In mathematical terms, given any small positive number ε, the energy contained outside the finite interval can be made arbitrarily low, i.e.,

$$\int_\Psi |E(z, t)|^2 dt < \varepsilon, \tag{2.41}$$

where $\Psi = (-\infty, t_{min}) \bigcup (t_{max}, +\infty)$. Slepian showed that it is possible to find a finite frequency interval $[-\omega_0, +\omega_0]$ such that the following integral holds [12]:

$$\left| \int_\Phi \exp(-j\omega t) \int_\Psi E(z, t) \exp(+j\omega t) dt\, d\omega \right|^2 < \varepsilon, \tag{2.42}$$

where $\Phi = (-\infty, -\omega_0) \bigcup (+\omega_0, +\infty)$. One must use this important result to discretize the Fourier integral in Eq. (2.30).

If a time-domain signal $f(t)$ is defined within the interval $[t_{min}, t_{max}]$, and if the bandwidth W of this signal is known, then we can employ the Shannon sampling theorem well known in communications theory. More specifically, we should sample the signal at intervals Δt such that the sampling frequency $f_s = (\Delta t)^{-1} > 2W$. If the signal is sampled at M points, $\Delta t = (t_{max} - t_{min})/M$. If $F(k)$ denotes the FFT of the function $f(t)$,

$$F(k) = \text{FFT}[f(m)] = \sum_{m=0}^{M-1} f(m) \exp\left(+j\frac{2\pi}{M} km\right), \tag{2.43}$$

with $k = 0, 1, \cdots, M - 1$. It is possible to invert this relation and recover the original sequence using the inverse FFT (IFFT) operation such that

$$f(m) = \text{IFFT}[F(k)] = \frac{1}{M} \sum_{k=0}^{M-1} F(k) \exp\left(-j\frac{2\pi}{M}km\right). \qquad (2.44)$$

Using these definitions, we can evaluate the integral in Eq. (2.30) as

$$E(z, p\Delta t) = \text{IFFT}\left[\text{FFT}[E(0, m)] \exp\left(+j\omega_m n(\omega_m)\frac{z}{c}\right)\right], \qquad (2.45)$$

where $p = 0, 1, \cdots, M - 1$, and $\omega_m = 2\pi m/(M\Delta t)$ are the discrete frequencies employed by the FFT algorithm. Although the FFT technique is widely employed for solving a variety of problems, especially in the context of pulse propagation in optical fibers [13], its use requires adopting the slowly-varying-envelope approximation. Some applications require a more general numerical technique; we discuss this in the next section.

2.5 Finite-difference time-domain (FDTD) method

The finite-difference time-domain (FDTD) method [5] is the most general numerical technique for solving Maxwell's equations directly without recourse to a slowly-varying-envelope approximation [13]. Its use requires adopting spatial and temporal step sizes that are a small fraction of the wavelength and the optical period, respectively, of the incident electromagnetic field, making it quite intense on computational resources. The numerical error is controlled solely by these step sizes. In spite of its general nature and the use of no approximations except those related to finite-difference approximations for various derivatives, the FDTD scheme is numerically dispersive and anisotropic even for a nondispersive, isotropic medium (or a signal propagating in vacuum). If a coarse grid is used to speed up the FDTD code, errors resulting from numerical dispersion and anisotropy accumulate and become unacceptable. If the medium is linear, reasonably accurate results can be obtained with a spatial step size near $\lambda/20$ for an optical wave with wavelength λ. However, if nonlinearities are to be taken into account, even a $\lambda/200$ step size may not provide adequate accuracy or convergence.

2.5.1 FDTD algorithm in one spatial dimension

Although the FDTD method is applicable for optical fields propagating in three spatial dimensions, to present the underlying algorithm as simply as possible, we consider the one-dimensional case by focusing on a plane wave propagating in the

z direction with its electric field polarized along the x axis. Maxwell's equations then reduce to

$$\frac{\partial E_x}{\partial z} = -\mu \frac{\partial H_y}{\partial t}, \tag{2.46a}$$

$$\frac{\partial H_y}{\partial z} = -\varepsilon \frac{\partial E_x}{\partial t}, \tag{2.46b}$$

where ε and μ are assumed to be constant. The case where ε depends on the frequency[2] is handled in Section 2.5.3. It is possible to solve these equations by adopting a central-difference approximation for partial derivatives. However, such discretization is not robust for problems in which electric and magnetic fields are interlinked. It was shown by Yee in 1966 that it is better to use a staggered grid where Faraday's and Ampere's law are coupled via interlinked time and space grids [5]. In physical terms, such an approach implies that a time-varying electric field creates a time-varying magnetic field and vice versa. Owing to the use of central differences, the resulting algorithm is accurate to second order in spatial step size. Moreover, the use of an explicit leapfrog scheme in the time domain also makes it accurate to second order in temporal step size.

We discretize Eqs. (2.46) in both z and t and employ the notation

$$E_x(z, t) = E_x(k\Delta z, n\Delta t) = E_x|_k^n, \tag{2.47a}$$

$$H_y(z, t) = H_y(k\Delta z, n\Delta t) = H_y|_k^n, \tag{2.47b}$$

where Δz and Δt are the step sizes along the spatial and temporal grids, respectively. Figure 2.4 shows how the space and time steps are interleaved by employing the concept of a half-step. More specifically, we evaluate Eq. (2.46a) at the grid point $[(k + 1/2)\Delta z, n\Delta t]$ and Eq. (2.46b) at the grid point $[k\Delta z, (n + 1/2)\Delta t]$ by using

$$\left.\frac{\partial E_x}{\partial z}\right|_{(k+1/2)\Delta z}^{n\Delta t} = -\mu \left.\frac{\partial H_y}{\partial t}\right|_{(k+1/2)\Delta z}^{n\Delta t}, \tag{2.48a}$$

$$\left.\frac{\partial H_y}{\partial z}\right|_{k\Delta z}^{(n+1/2)\Delta t} = -\varepsilon \left.\frac{\partial E_x}{\partial t}\right|_{[k\Delta z}^{(n+1/2)\Delta t}. \tag{2.48b}$$

Employing central differences to approximate partial derivatives on both sides of these equations, they can be written as

$$H_y|_{k+1/2}^{n+1/2} = H_y|_{k+1/2}^{n-1/2} - \frac{1}{\mu}\frac{\Delta t}{\Delta z}\left(E_x|_{k+1}^n - E_x|_k^n\right), \tag{2.49a}$$

---

[2] We do not consider the case where μ depends on frequency.

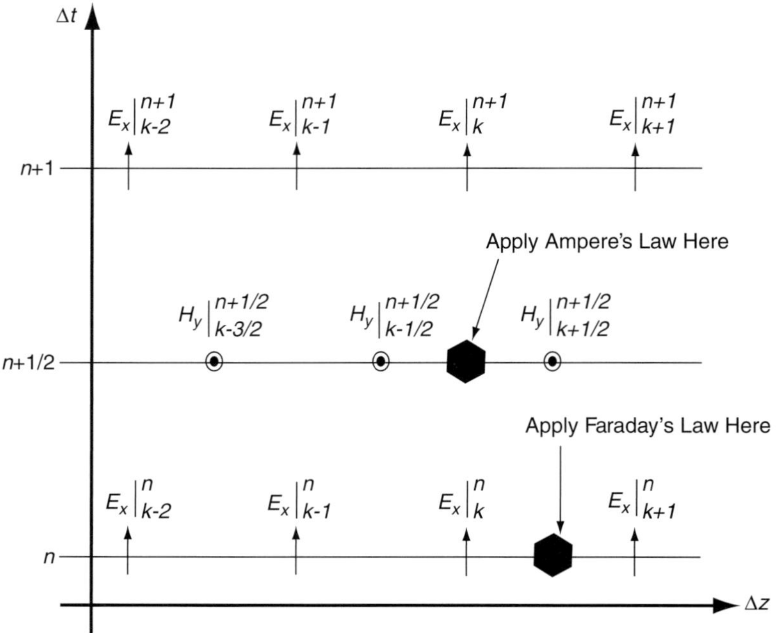

Figure 2.4 Schematic of a staggered grid used for the FDTD algorithm. Arrows mark the direction of electric field and a dot within a circle indicates that the magnetic field is directed normal to the plane of the figure.

$$E_x|_k^{n+1} = E_x|_k^n - \frac{1}{\varepsilon}\frac{\Delta t}{\Delta z}\left(H_y|_{k+1/2}^{n+1/2} - H_y|_{k-1/2}^{n+1/2}\right). \tag{2.49b}$$

When these equations are iterated to find the electric and magnetic fields along the z axis, one may run into problems because of finite reflections at the boundaries of the spatial grid. The remedy is to include artificial boundary conditions at the two ends of this grid such that reflections are nearly eliminated. These boundary conditions are known as the *absorbing boundary conditions*.

Several different absorbing boundary conditions are available for FDTD simulations. In the so-called Mur boundary conditions [5], the electric fields at the two boundary nodes of the z grid are calculated using

$$E_x|_{\text{start}}^{n+1} = E_x|_{\text{start}+1}^n + \left(\frac{1-b}{1+b}\right)\left(E_x|_{\text{start}+1}^{n+1} - E_x|_{\text{start}}^n\right), \tag{2.50a}$$

$$E_x|_{\text{end}}^{n+1} = E_x|_{\text{end}-1}^n + \left(\frac{1-b}{1+b}\right)\left(E_x|_{\text{end}-1}^{n+1} - E_x|_{\text{end}}^n\right), \tag{2.50b}$$

where $b = \Delta z/(c\Delta t)$. The interested reader can find further details in Ref. [5].

2.5.2 Total and scattered electromagnetic fields

One additional concept needs to be introduced to make a plane wave propagate correctly with the FDTD method. The reason is related to the observation that Maxwell's equations support the propagation of both forward and backward propagating waves, regardless of the material properties. Thus, if an electromagnetic wave is generated at some point in the medium, one would see two replicas of that wave propagating in the forward and backward directions. To select the propagation direction in advance, and to stop the backward propagation in the opposite direction, the total field/scattered field (TF/SF) formulation is widely employed [5].

In the TF/SF formulation, the FDTD grid is divided into two regions. The total-field region consists of a superposition of the incident and scattered fields, while the scattered-field region contains only scattered fields from the objects of interest. The point at which the incident field is introduced is called the TF/SF boundary. This boundary can be chosen arbitrarily but it is typically placed such that any scatterers are contained in the total-field region.

Figure 2.5 shows the case where the TF/SF boundary is on the immediate left of the electric field node $z = S\Delta z$. Because the total and scattered fields are "electromagnetic fields" themselves, we can use Eqs. (2.49) without any alterations.

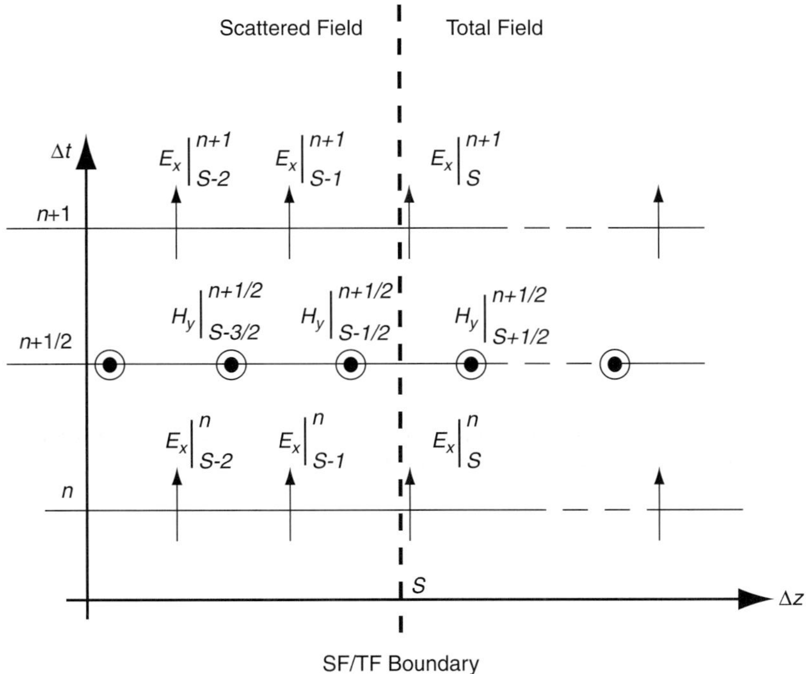

Figure 2.5 Illustration of the boundary separating the total and scattered fields at $z = S\Delta z$.

However, at the vicinity of the boundary, nodes that are immediately to the left or right of the boundary require special attention because the TF/SF boundary feeds the incident field to the FDTD grid. If we know the incident field in the vicinity of the TF/SF boundary, we can update the magnetic and electric fields using

$$H_y^{sc}|_{S-1/2}^{n+1/2} = H_y^{sc}|_{S-1/2}^{n-1/2} - \frac{1}{\mu}\frac{\Delta t}{\Delta z}\left(E_x^{sc}|_S^n - E_x^{sc}|_{S-1}^n\right) + \frac{1}{\mu}\frac{\Delta t}{\Delta z}E_x^{inc}|_S^n, \qquad (2.51a)$$

$$E_x^{tot}|_S^{n+1} = E_x^{tot}|_S^n - \frac{1}{\varepsilon}\frac{\Delta t}{\Delta z}\left(H_y^{tot}|_{S+1/2}^{n+1/2} - H_y^{tot}|_{S-1/2}^{n+1/2}\right) + \frac{1}{\varepsilon}\frac{\Delta t}{\Delta z}H_y^{inc}|_{S-1/2}^{n+1/2}. \qquad (2.51b)$$

Because only the last terms in the preceding equations differ from their normal form, we can apply the update in Eqs. (2.49) first and the correction terms afterwards. The incident fields required for this purpose can be easily calculated by propagating the plane wave to the desired grid point analytically using direct integration in Eq. (2.30), with a similar equation for the magnetic field. However, owing to the discretization of the optical field, the dispersion relation for the medium changes and has the following form:

$$\sin^2\left(\frac{1}{2}k(\omega)\Delta z\right) = \left(\frac{\Delta z}{c\Delta t}\right)^2 \sin^2\left(\frac{1}{2}\omega\Delta t\right). \qquad (2.52)$$

This relation can be solved to find $k(\omega)$, resulting in [14]:

$$k(\omega) = \frac{2}{\Delta z}\sin^{-1}\left[\left(\frac{\Delta z}{c\Delta t}\right)\sin\left(\frac{\omega\Delta t}{2}\right)\right]. \qquad (2.53)$$

There is a possibility that the argument of the $\sin^{-1}$ function acquires a value greater than 1 during numerical calculations, making $k(\omega)$ complex. For this reason, $\sin^{-1}(\zeta) = -j\ln(j\zeta + \sqrt{1-\zeta^2})$ must be used when implementing Eq. (2.53) numerically.

2.5.3 Inclusion of material dispersion

Pulse propagation in a dispersive material requires special care because the standard FDTD algorithm needs to be modified to include variations of the dielectric function with frequency. Many techniques have been proposed for this purpose [5, 15–17]. We discuss here a versatile method that allows the inclusion of any complex linear dispersion profile through appropriate curve fitting. The Debye, Lorentz, and Drude dielectric functions, used widely in practice and discussed in Section 1.2, constitute the special cases of this method.

The basic idea is to employ a pair of complex-conjugate poles in the complex frequency domain to represent an arbitrary dispersion profile, and write the dielectric function in the form [18, 19]

$$\varepsilon(\omega) = \varepsilon_0\varepsilon_\infty + \varepsilon_0 \sum_{p=1}^{M} \left(\frac{c_p}{j\omega - a_p} + \frac{c_p^*}{j\omega - a_p^*} \right), \tag{2.54}$$

where the real part of the pole at a_p is positive to ensure causality of the dielectric response. The complex coefficients c_p and c_p^* are chosen to fit a specific dispersion model of interest. For example, when $a_p = 1/\tau_p$ and $c_p = -\Delta\varepsilon_p/(2\tau_p)$, this formulation reduces to the well-known Debye model [18, 19],

$$\varepsilon_{\text{Debye}}(\omega) = \varepsilon_0\varepsilon_\infty + \varepsilon_0 \sum_{p=1}^{M} \frac{\Delta\varepsilon_p}{1 - j\omega\tau_p}. \tag{2.55}$$

If we choose $a_p = \delta_p - j\sqrt{\omega_p^2 - \delta_p^2}$ and $c_p = j\Delta\varepsilon_p\omega_p^2/(2\sqrt{\omega_p^2 - \delta_p^2})$, we recover the Lorentz model [18, 19],

$$\varepsilon_{\text{Lorentz}}(\omega) = \varepsilon_0\varepsilon_\infty + \varepsilon_0 \sum_{p=1}^{M} \frac{\Delta\varepsilon_p\omega_p^2}{\omega_p^2 - j2\delta_p\omega - \omega^2}. \tag{2.56}$$

In both cases, the model order M needs to be chosen to represent intricate features of the actual material dispersion by using standard curve-fitting techniques. Also, in the case of metals, the Drude model can be seen as a special case of the Lorentz model in which the resonance frequency ω_p is so small compared with ω that ω_p^2 can be neglected in the denominator.

To construct a suitable FDTD model, we introduce two complex quantities $\tilde{J}_p(\omega)$ and $\tilde{I}_p(\omega)$ as follows [18, 19]:

$$\tilde{J}_p(\omega) = \varepsilon_0 \frac{c_p}{j\omega - a_p} j\omega\tilde{E}(\omega), \tag{2.57a}$$

$$\tilde{I}_p(\omega) = \varepsilon_0 \frac{c_p^*}{j\omega - a_p^*} j\omega\tilde{E}(\omega). \tag{2.57b}$$

Taking the inverse Fourier transform of these equations and noting that $j\omega \rightarrow -\frac{\partial}{\partial t}$, we obtain

$$\frac{\partial J_p(t)}{\partial t} + a_p J_p(t) = \varepsilon_0 c_p \frac{\partial E(t)}{\partial t}, \tag{2.58a}$$

$$\frac{\partial I_p(t)}{\partial t} + a_p^* I_p(t) = \varepsilon_0 c_p^* \frac{\partial E(t)}{\partial t}. \tag{2.58b}$$

Because $E(t)$ is a real quantity, these equations show that $I_p(t) = J_p^*(t)$, provided both have the same initial values [18, 19]. Thus, we only need to calculate $J_p(t)$.

The dielectric function enters Maxwell's equations through Ampere's law, which can be written, including the conductivity σ, as

$$\nabla \times \boldsymbol{H} = \frac{\partial \boldsymbol{D}}{\partial t} + \sigma \boldsymbol{E}. \tag{2.59}$$

Noting that $\widetilde{\boldsymbol{D}}(\omega) = \varepsilon(\omega)\widetilde{\boldsymbol{E}}$, we obtain the following time-stepping rule for updating the electric field and $J_p(t)$ as time advances from $n\Delta t$ to $(n+1)\Delta t$:

$$
\begin{aligned}
E^{(n+1)} = {} & \left(\frac{2\varepsilon_0\varepsilon_\infty + 2\sum_{p=1}^{M} \mathrm{Re}(\beta_p) - \sigma\Delta t}{2\varepsilon_0\varepsilon_\infty + 2\sum_{p=1}^{M} \mathrm{Re}(\beta_p) + \sigma\Delta t} \right) E^n \\
& + \frac{2\Delta t \left[\nabla \times H^{(n+1/2)} - \mathrm{Re}\left(\sum_{p=1}^{M}(1 + k_p)J_p^n \right) \right]}{2\varepsilon_0\varepsilon_\infty + 2\sum_{p=1}^{M} \mathrm{Re}(\beta_p) + \sigma\Delta t},
\end{aligned}
\tag{2.60}
$$

$$J_p^{(n+1)} = k_p J_p^n + \frac{\beta_p}{\Delta t}\left(E^{(n+1)} - E^n \right), \tag{2.61}$$

where we have defined k_p and β_p as

$$k_p = \frac{1 + a_p \Delta t/2}{1 - a_p \Delta t/2} \quad \text{and} \quad \beta_p = \frac{\varepsilon_0 c_p \Delta t}{1 - a_p \Delta t/2}. \tag{2.62}$$

These equations permit one to use the FDTD technique for propagation of light in any dispersive medium. However, the adoption of a dispersive model complicates the situation at the boundaries of the grid used for FDTD simulations. To suppress spurious reflections at the grid boundaries, we need to introduce a perfectly matched layer (PML) around the grid.

2.5.4 Perfectly matched layer for dispersive optical media

A perfectly matched layer (PML) is an artificial layer designed to obey Maxwell's equations with superior absorbing properties to those of a real material. A key property of a PML medium is that an electromagnetic wave incident upon the PML from a non-PML medium does not reflect at the interface. Its use eliminates spurious reflections, a source of numerical problems in practice. The concept of PML was first developed by Berenger in 1994 [20] and has gone several reformulations since then. Berenger employed a split form of Maxwell's equations, introducing additional terms for the electric field, but this form still matched Maxwell's equations exactly. However, these additional electric fields cannot be associated with any physical fields. To circumvent this deficiency, two years later Gedney introduced a modified approach called unsplit PML (UPML) that makes use of physical fields [21]. Later, a novel way to construct PML was introduced by Zhao [22]. This approach is called generalized theory-based PML (GTPML) and makes use of a complex coordinate stretching in the frequency domain. More specifically, PMLs were shown to correspond to a coordinate transformation in which one (or more) coordinates are mapped to complex numbers. This transformation essentially constituted an analytic continuation of the wave equation into the complex domain, replacing propagating waves with exponentially decaying waves. We adopt this coordinate-stretching concept in the following discussion.

Let $\widetilde{\mathbf{E}}(\mathbf{r}, \omega)$ and $\widetilde{\mathbf{H}}(\mathbf{r}, \omega)$ be the electric and magnetic fields in the generic dispersive medium described in Section 2.5.3. In a medium where no electromagnetic sources are present, the evolution of these fields (in the frequency domain) is governed by Maxwell's equations (see Section 1.1):

$$\nabla \times \widetilde{\mathbf{E}}(\mathbf{r}, \omega) = j\omega\mu_0 \widetilde{\mathbf{H}}(\mathbf{r}, \omega), \tag{2.63a}$$

$$\nabla \times \widetilde{\mathbf{H}}(\mathbf{r}, \omega) = -j\omega\varepsilon(\omega)\widetilde{\mathbf{E}}(\mathbf{r}, \omega). \tag{2.63b}$$

An absorbing PML, impedance-matched perfectly to its adjacent medium, can be created by transforming coordinates appropriately (see Ref. [21] for details). As a result, in the PML region Eqs. (2.63) take the form

$$\nabla' \times \widetilde{\mathbf{E}}(\mathbf{r}, \omega) = j\omega\mu_0 \widetilde{\Gamma}(\omega)\widetilde{\mathbf{H}}(\mathbf{r}, \omega), \tag{2.64a}$$

$$\nabla' \times \widetilde{\mathbf{H}}(\mathbf{r}, \omega) = -j\omega\varepsilon(\omega)\widetilde{\Gamma}(\omega)\widetilde{\mathbf{E}}(\mathbf{r}, \omega), \tag{2.64b}$$

where ∇' is the operator ∇ in the new coordinate system and $\widetilde{\Gamma}(\omega)$ is the PML material tensor, defined as

$$\widetilde{\Gamma}(\omega) = \begin{pmatrix} \dfrac{s_y(\omega)s_z(\omega)}{s_x(\omega)} & 0 & 0 \\ 0 & \dfrac{s_x(\omega)s_z(\omega)}{s_y(\omega)} & 0 \\ 0 & 0 & \dfrac{s_x(\omega)s_y(\omega)}{s_z(\omega)} \end{pmatrix}. \tag{2.65}$$

The presence of the material tensor $\widetilde{\Gamma}(\omega)$ on the right side of Eqs. (2.64) is equivalent to filling the PML with an artificial absorbing material having anisotropic permittivity $\widetilde{\varepsilon}(\omega)\widetilde{\Gamma}(\omega)$ and anisotropic permeability $\mu_0\widetilde{\Gamma}(\omega)$. Clearly, for any choice of the stretching coefficients $s_r(\omega)$, the impedances of the dispersive medium and the PML are equal with $\eta = \sqrt{\mu_0/\varepsilon(\omega)}$. Attenuation properties of the PML can be controlled by choosing the stretching coefficients $s_r(\omega)$ appropriately. Here, we adopt the form [23]

$$s_r(\omega) = \kappa_r + \frac{\sigma_r}{\gamma - j\omega\varepsilon_0}, \qquad r = x, y, z, \tag{2.66}$$

corresponding to the so-called complex-frequency-shifted (CFS) PML. The impact of parameters $\sigma_r > 0, \kappa_r > 1$, and $\gamma > 0$ on attenuation properties of the CFS-PML is analyzed in [24–26]. From a physical point of view, σ_r resembles the conductivity of the PML and provides attenuation of propagating waves in the rth direction. The parameters κ_r and γ absorb the evanescent waves. To avoid numerical reflections from the PML boundaries, both σ_r and κ_r need to be varied smoothly in space. A common choice of these parameters is [27]

$$\sigma_r = \sigma_{\max}(k/\delta_r)^{m+n}, \tag{2.67a}$$

$$\kappa_r = 1 + (\kappa_{\max} - 1)(k/\delta_r)^n, \tag{2.67b}$$

where $k \leq \delta_r$, δ_r is the PML depth in the rth direction, m and n are two integers taking values in the range $[-3, 3]$ and $[2, 6]$ with the restriction $m + n > 1$, and $\kappa_{\max}$ is a number in the range 1–10. The parameter $\sigma_{\max}$ satisfies $\sigma_{\max} = -c\varepsilon_0(m + n + 1)\ln R_0/(2\delta_r)$, where R_0 is in the range $[10^{-12}, 10^{-2}]$.

The FDTD solution of of Eqs. (2.63) and (2.64) requires discretization of the electric and magnetic fields and the use of a set of update equations on the entire computational grid. It is desirable that update equations do not depend on the exact form of the PML material tensor, as such a dependance would require code modification for each specific choice of the attenuation parameters s_r. This is accomplished by rewriting the update equations in terms of two auxiliary functions,

$$\widetilde{\mathbf{R}}_E = \widetilde{\Gamma}(\omega)\widetilde{\mathbf{E}}(\mathbf{r}, \omega), \tag{2.68a}$$

$$\widetilde{\mathbf{R}}_H = \widetilde{\Gamma}(\omega)\widetilde{\mathbf{H}}(\mathbf{r}, \omega). \tag{2.68b}$$

Once the values of these functions are found in the time domain, they are used to calculate the real fields, $\mathbf{E}(\mathbf{r}, t)$ and $\mathbf{H}(\mathbf{r}, t)$.

The update equations for $\mathbf{R}_E$ and $\mathbf{R}_H$ can be derived as follows. Substituting Eq. (2.54) into Eq. (2.64b), and writing ∇' as ∇ for convenience, we obtain

$$-\nabla \times \widetilde{\mathbf{H}} = j\omega\varepsilon_0\varepsilon_\infty\widetilde{\mathbf{R}}_E + \sum_{p=1}^{M}(\widetilde{\mathbf{J}}_p + \widetilde{\mathbf{K}}_p), \tag{2.69}$$

where (see Section 2.5.3)

$$\widetilde{\mathbf{J}}_p = \varepsilon_0\frac{c_p}{j\omega - a_p}j\omega\widetilde{\mathbf{R}}_E, \tag{2.70a}$$

$$\widetilde{\mathbf{K}}_p = \varepsilon_0 \frac{c_p^*}{j\omega - a_p^*} j\omega \widetilde{\mathbf{R}}_E. \tag{2.70b}$$

Writing Eq. (2.70) in the discrete time-domain form, we obtain

$$\mathbf{J}_p^{n+1} = \alpha_p \mathbf{J}_p^n + \beta_p (\mathbf{R}_E^{n+1} - \mathbf{R}_E^n), \qquad (\mathbf{K}_p^n)^* = \mathbf{J}_p^n, \tag{2.71}$$

where

$$\alpha_p = \frac{1 - a_p \Delta t/2}{1 + a_p \Delta t/2}, \tag{2.72a}$$

$$\beta_p = \frac{\varepsilon_0 c_p}{1 + a_p \Delta t/2}. \tag{2.72b}$$

Using Eq. (2.71), we can write the discrete time-domain version of Eq. (2.69) in the form

$$-\nabla \times \mathbf{H}^n = -\varepsilon_0 \varepsilon_\infty \frac{\mathbf{R}_E^{n+1/2} - \mathbf{R}_E^{n-1/2}}{\Delta t},$$

$$+ \sum_{p=1}^{M} \mathrm{Re}\big[(1 + \alpha_p)\mathbf{J}_p^{n-1/2}\big] + (\mathbf{R}_E^{n+1/2} - \mathbf{R}_E^{n-1/2}) \sum_{p=1}^{M} \mathrm{Re}(\beta_p). \tag{2.73}$$

From this equation, we find the following update equation for $\mathbf{R}_E$:

$$\mathbf{R}_E^{n+1/2} = \mathbf{R}_E^{n-1/2} + \frac{\nabla \times \mathbf{H}^n + \sum \mathrm{Re}\big[(1 + \alpha_p)\mathbf{J}_p^{n-1/2}\big]}{\varepsilon_0 \varepsilon_\infty / \Delta t - \sum \mathrm{Re}(\beta_p)}. \tag{2.74}$$

Similarly, using Eq. (2.64a) one can show that the update equation for $\mathbf{R}_H$ is

$$\mathbf{R}_H^{n+1} = \mathbf{R}_H^n - \frac{\nabla \times \mathbf{E}^{n+1/2}}{\mu_0 / \Delta t}. \tag{2.75}$$

To restore the electric and magnetic fields from auxiliary functions, we need to find the relationships between them in the discrete time domain. This can be done by converting Eqs. (2.68) to continuous time domain with the Fourier correspondence $-j\omega \leftrightarrow \partial/\partial t$ and discretizing the resulting differential equations with appropriate difference operators. This procedure, however, is quite tedious in our case and becomes even more involved for more complicated forms of the coefficients $s_r(\omega)$ (e.g., for negative-index materials [28] or higher-order PMLs [26]). The inconvenience associated with the discretization of high-order differential equations can be avoided by using the Z-transform technique [29, 30] discussed next.

Let us illustrate the Z-transform technique by deriving the difference equation for the x component of the electric field using the relation $\widetilde{R}_{E,x} = \widetilde{\Gamma}_x(\omega)\widetilde{E}_x(\mathbf{r}, \omega)$.

Introducing two new parameters, $\xi_r = \xi_0 + \sigma_r/(\kappa_r \varepsilon_0)$ and $\xi_0 = \gamma/\varepsilon_0$, and utilizing Eq. (2.66), we represent the relation between $\widetilde{R}_{E,x}$ and $\widetilde{E}_x$ in the form

$$\frac{\kappa_x}{\kappa_y \kappa_z} \frac{\widetilde{R}_{E,x}}{(\xi_y - j\omega)(\xi_z - j\omega)} = \frac{\widetilde{E}_x}{(\xi_x - j\omega)(\xi_0 - j\omega)}. \tag{2.76}$$

We recast this equation into the Z domain using the transform pair

$$\frac{1}{\xi - j\omega} \leftrightarrow \frac{1}{1 - Z^{-1}e^{-\xi \Delta t}}, \tag{2.77}$$

where Z is used to represent the Z transform and should not be confused with the spatial variable z. The result is given by

$$\frac{\kappa_x}{\kappa_y \kappa_z} \frac{R_{E,x}(Z)}{1 - Z^{-1}(e^{-\xi_y \Delta t} + e^{-\xi_z \Delta t}) + Z^{-2}e^{-(\xi_y + \xi_z)\Delta t}}$$

$$= \frac{E_x(Z)}{1 - Z^{-1}(e^{-\xi_x \Delta t} + e^{-\xi_0 \Delta t}) + Z^{-2}e^{-(\xi_x + \xi_0)\Delta t}}. \tag{2.78}$$

This equation can be readily inverted to get its equivalent time-domain form suitable for numerical implementation. Using the transform correspondence $Z^{-k}A(Z) \leftrightarrow A^{n-k}$, we obtain the result

$$E_x^{n+1} = E_x^n \left(e^{-\xi_y \Delta t} + e^{-\xi_z \Delta t} \right) - E_x^{n-1} e^{-(\xi_y + \xi_z)\Delta t}$$

$$+ \frac{\kappa_x}{\kappa_y \kappa_z} \left[R_{E,x}^{n+1} - R_{E,x}^n \left(e^{-\xi_x \Delta t} + e^{-\xi_0 \Delta t} \right) \right.$$

$$\left. + R_{E,x}^{n-1} e^{-(\xi_x + \xi_0)\Delta t} \right]. \tag{2.79}$$

Analogous expressions can be obtained to restore the E_y and E_z components by changing the subscript x to y or z in Eq. (2.79). The magnetic field components, H_m^{n+1} ($m = x, y, z$), satisfy equations identical to those for E_m^{n+1} after we replace $R_{E,m}$ with $R_{H,m}$.

The preceding equations allow one to develop a unified FDTD algorithm for simulation of optical phenomena in a wide range of dispersive media truncated by a perfectly matched absorbing boundary. The main steps of the algorithm for updating the electromagnetic field in the PML can be summarized as follows:

(1) Apply the update equation (2.74) and store $\mathbf{R}_E^{n+1/2}$ values. This update requires retrieving of values $\mathbf{H}^n$, $\mathbf{J}_p^{n-1/2}$, and $\mathbf{R}_E^{n-1/2}$ from preceding time steps.

(2) Calculate and store the auxiliary variables $\mathbf{J}_p^{n+1/2}$ for each pole p using Eq. (2.71). This update requires the value $\mathbf{R}_E^{n+1/2}$ calculated in the current time step and requires the retrieving of values $\mathbf{J}_p^{n-1/2}$ and $\mathbf{R}_E^{n-1/2}$ from preceding time steps.

(3) Use Eq. (2.79) with similar equations for y and z components to restore the electric field $\mathbf{E}^{n+1/2}$ from values of $\mathbf{E}^{n-1/2}$, $\mathbf{E}^{n-3/2}$, $\mathbf{R}_E^{n-1/2}$, and $\mathbf{R}_E^{n-3/2}$ retrieved from preceding time steps and the value $\mathbf{R}_E^{n+1/2}$ calculated in the current time step.

(4) Update $\mathbf{R}_H^{n+1}$ values using Eq. (2.75). This update requires values of $\mathbf{R}_H^n$ and $\mathbf{E}^{n+1/2}$ retrieved from preceding time steps.

(5) Employ an analog of Eq. (2.79) to calculate $\mathbf{H}^{n+1}$ using values of $\mathbf{H}^n$, $\mathbf{H}^{n-1}$, $\mathbf{R}_H^n$, and $\mathbf{R}_H^{n-1}$ retrieved from preceding time steps as well as the value of $\mathbf{R}_H^{n+1}$ calculated in the current time step.

It is useful to note that, as expected, this algorithm reduces to the algorithm given in Section 2.5.3 if the PML-specific parameters are set to zero. No restoration of electric and magnetic fields is needed in the absence of the PML since the auxiliary functions are identically equal to these fields. Hence, steps 3 and 5 of the above algorithm should be omitted in this situation and the functions $\mathbf{R}_E^n$ and $\mathbf{R}_H^n$ should be replaced with $\mathbf{E}^n$ and $\mathbf{H}^n$ in the other steps.

2.6 Phase and group velocities

Propagation of pulses in dispersive dielectric media requires a clear understanding of the concepts of phase and group velocities. We discuss both of them in this section, starting from Eq. (2.30). We expand the propagation constant, $k(\omega) = \omega n(\omega)/c$, around the pulse's carrier frequency ω_0, a representative of the mean frequency of the overall signal, as

$$k(\omega) = k(\omega_0) + \frac{(\omega - \omega_0)}{v_g} + \cdots , \tag{2.80}$$

where the group velocity of the pulse is defined as

$$\frac{1}{v_g} = \frac{n_g}{c} = \frac{dk(\omega)}{d\omega}\bigg|_{\omega = \omega_0}, \tag{2.81}$$

n_g being the group index of the medium. Substituting Eq. (2.80) into Eq. (2.30) and rearranging various terms, we obtain

$$E(z, t) = e^{j[k(\omega_0)z - \omega_0 t]} \frac{1}{2\pi} \int_{-\infty}^{\infty} \tilde{E}(0, \omega) e^{-j(\omega - \omega_0)(t - z/v_g)} \, d\omega. \tag{2.82}$$

In this expression, the integral represents the slowly varying part of the electric field associated with a pulse.

The rapidly varying part of the electric field oscillates at the carrier frequency. These oscillations advance with phase velocity v_p given by

$$v_p = \frac{k(\omega_0)}{\omega_0}. \tag{2.83}$$

There is a subtle distinction between the phase and group velocities. The phase of the carrier can be written as

$$\phi(\omega) = \omega n(\omega)z/c - \omega t. \tag{2.84}$$

The phase velocity amounts to the speed with which the phase front of a wave advances. This can be seen by considering a phase front with the constant phase ϕ_0 and propagating it through space such that $\phi(\omega) = \phi_0$. In contrast, if we track the speed where phase is stationary and set the derivative of $\phi(\omega)$ with respect to ω to zero, we obtain the group-velocity expression given in Eq. (2.81).

It is instructive to look at the behavior of $\phi(\omega)$ as ω varies because it gives us better insight into variations of group velocity with frequency. It is clear from Eq. (2.84) that $\phi(0) = 0$. However, if we use $n(\omega) = n_r(\omega) + jn_i(\omega)$ for the complex refractive index, it is not obvious how $\phi(\omega)$ behaves in the neighborhood of the origin at $\omega = 0$. To clarify this, we make use of the Kramers–Kronig relations that relate the real and imaginary parts of the refractive index (see Section 1.4). In particular, the so-called zero-frequency sum rule [31, 32] states that

$$n_r(0) = 1 + \frac{2}{\pi} \mathscr{P} \int_0^\infty \frac{n_i(\omega)}{\omega} d\omega, \tag{2.85}$$

where $\mathscr{P}$ denotes the principle value of the Cauchy integral [33].

The difference between the pulse's propagation in vacuum (i.e., $n_r = 1$) and through a dispersive medium can be gauged by considering the additional phase shift $\Delta\phi(\omega)$ acquired by the pulse in the medium:

$$\Delta\phi(\omega) = \frac{\omega z}{c}[n_r(\omega) - 1]. \tag{2.86}$$

The frequency derivative of this function is the group delay, representing the time taken by the pulse peak to reach a distance z. Focusing on the region near $\omega = 0$ and substituting Eq. (2.85) into Eq. (2.86), we obtain

$$\Delta\phi(\omega \approx 0) = \frac{2\omega z}{\pi c} \mathscr{P} \int_0^\infty \frac{n_i(\omega')}{\omega'} d\omega'. \tag{2.87}$$

This integral behaves differently depending on whether the medium is absorptive or amplifying [31]. In a purely absorptive medium, $n_i(\omega) > 0$, and the integral assumes a positive value. It remains positive even in an amplifying medium with small gain over a relatively small bandwidth. In contrast, in a medium with high gain over a relatively large bandwidth, the integral may become negative. In all cases, the Kramers–Kronig relations show that $n_r(\omega)$ has the following asymptotic form as ω tends to infinity:

$$n_r(\omega) = 1 - \frac{\omega_p^2}{2\omega^2}. \tag{2.88}$$

Therefore, at large frequencies, the phase difference in Eq. (2.86) behaves as

$$\Delta\phi(\omega) = -\frac{\omega_p^2 z}{2c\omega}. \tag{2.89}$$

This quantity is always negative because electrons essentially behave as free particles at very high frequencies [31].

Consider the case for which $\Delta\phi$ in Eq. (2.87) is positive. As seen in Figure 2.6, $\Delta\phi$ is then positive in the vicinity of $\omega = 0$ and becomes negative as $\omega \to \infty$. However, n_r is continuous and differentiable because absorption (or gain) and refractive index are related to each other through the Kramers–Kronig relations [31]. It follows from Eq. (2.86) that the phase difference $\Delta\phi(\omega)$ must also be continuous and differentiable along the entire positive frequency axis. By the mean-value theorem, there is a point where the derivative of $\Delta\phi$ with respect to frequency becomes negative. Since this derivative gives us the group delay of the medium, a negative group delay implies that

$$\frac{d(\Delta\phi)}{\omega} = \left(\frac{1}{v_g} - \frac{1}{c}\right) z < 0. \tag{2.90}$$

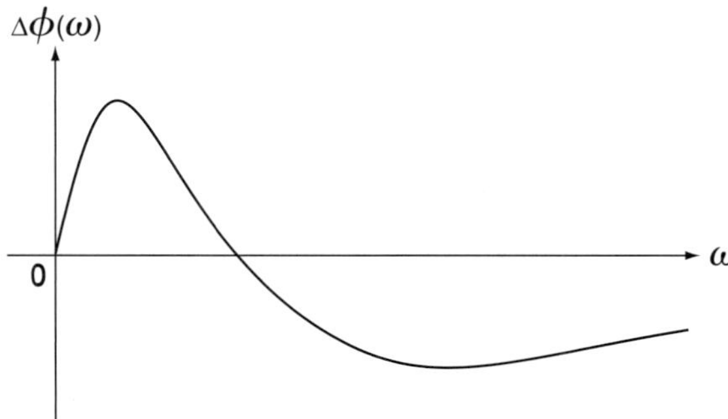

Figure 2.6 Typical frequency dependence of phase difference $\Delta\phi$ in a weakly amplifying or purely absorptive medium. (After Ref. [31]; © APS 1993)

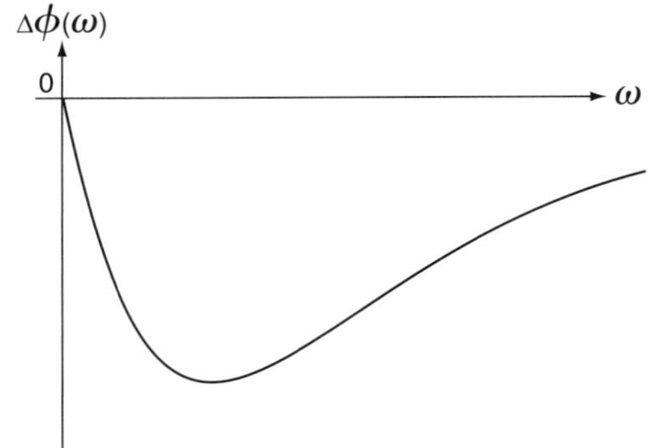

Figure 2.7 Typical frequency dependence of phase difference $\Delta\phi$ in a strongly amplifying medium. (After Ref. [31]; © APS 1993)

There are two ways this expression can become negative: $v_g > c$ (including $v_g = \infty$) or $v_g < 0$. Because we reached this conclusion solely based on the Kramers–Kronig relations, one can definitely say that this conclusion is a consequence of the principle of causality.

The case of a strongly amplifying medium is illustrated in Figure 2.7. For such a medium, $\Delta\phi$ is negative in the vicinity of $\omega = 0$, reaches a minimum value at a specific frequency, and approaches zero when $\omega \to \infty$. Interestingly, the group delay is negative at small frequencies for such a medium. As a result, there exists at least one frequency where $v_g > c$ or $v_g < 0$ holds in a strongly amplifying medium. In the case of an infinite group velocity, the peak of a pulse emerges from a finite-size medium at the same instant as the peak of the pulse enters it. A negative group velocity implies that the output pulse peaks at an earlier time than the peak of the incident pulse. At first sight, these conclusions are hard to understand physically and have been the subject of intense debates in the literature. One conclusion is that the group velocity is not limited by Einstein's relativity theory because it is not the speed with which information can be transmitted through a medium.

It is important to understand that we arrived at the preceding results by considering an unbounded medium, extending to infinity on both sides of the z axis. If a finite slab of the dispersive medium is considered, the phase difference $\Delta\phi(\omega)$ needs to be modified because additional phase shifts occur at the medium boundaries. Even though the mathematical treatment becomes complicated in this case, the main conclusions still remain valid. The interested reader is referred to the appendix of Ref. [31] for details.

2.7 Pulse propagation through a dielectric slab

We next consider pulse propagation through a dielectric slab, a configuration that provides the basic framework needed to understand many kinds of optical

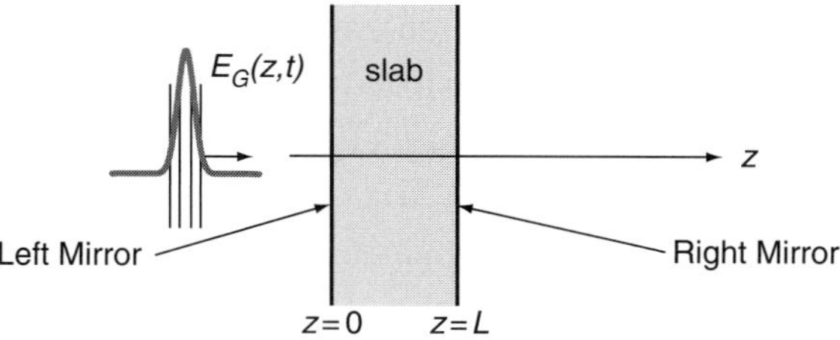

Figure 2.8 A Gaussian pulse transmitted through a Fabry–Perot resonator.

amplifiers. The new feature is that we must take into account the influence of boundaries when solving Maxwell's equations. The transmittance and reflectance at each boundary are quantified through the Fresnel equations given in Section 2.3. However, because of two boundaries, multiple reflections take place and lead to resonance conditions. To understand such resonances, we study the dynamic response of a Fabry–Perot resonator under pulsed conditions.

2.7.1 Fabry–Perot resonators

Figure 2.8 shows a Fabry–Perot resonator made of a dielectric slab with refractive index n_0 and length L. The left facet of this slab, located at $z = 0$, has reflection and transmission coefficients r_1 and t_1, respectively; the corresponding quantities at the right facet, at $z = L$, are r_2 and t_2. A unit-amplitude impulse pulse with a temporal form $\delta(t)$ entering this resonator from the left at time $t = 0$ will undergo multiple reflections at the two facets. A part of it will exit from the right facet with an amplitude $t_1 t_2$ at time $\tau_0 = n_0 L/c$. Another part of the pulse will exit from the resonator after one round-trip time $2\tau_0$ with amplitude $t_1 t_2 r_1 r_2$. This process will continue repeatedly, resulting in the following impulse response function, $T_{\text{FP}}(t)$, for the Fabry–Perot resonator [34]:

$$T_{\text{FP}}(t) = t_1 t_2 \sum_{m=0}^{\infty} r_1^m r_2^m \delta\left(t - (2m + 1)\tau_0\right). \tag{2.91}$$

Because this is a time-invariant expression, the corresponding frequency-domain transfer function can be obtained by taking its temporal Fourier transform, and is given by

$$\tilde{T}_{\text{FP}}(\omega) = t_1 t_2 \sum_{m=0}^{\infty} r_1^m r_2^m \exp[j(2m + 1)\omega\tau_0]$$

$$= \frac{t_1 t_2 \exp(j\omega\tau_0)}{1 - r_1 r_2 \exp(j2\omega\tau_0)}. \tag{2.92}$$

This transfer function shows that a three-component Fabry–Perot resonator (two mirrors and a dielectric medium sandwiched between them) can be mapped to a single optical slab, with no reflections at its boundaries, if we employ the concept of an effective refractive index, $n_{\text{eff}}(\omega)$. The transfer function of this effective slab of length L is $\exp[j\omega n_{\text{eff}}(\omega)L/c]$. Comparing this to Eq. (2.92), we obtain [35]

$$n_{\text{eff}}(\omega) = \frac{c}{j\omega L} \ln\left(\frac{t_1 t_2 \exp(j\omega\tau_0)}{1 - r_1 r_2 \exp(j2\omega\tau_0)} \right). \tag{2.93}$$

Using $n_{\text{eff}}(\omega) = n_e(\omega) + j\kappa_e(\omega)$, we obtain the real and imaginary parts in the form

$$n_e(\omega) = n_0(\omega) + \tan^{-1}\left(\frac{r_1 r_2 \sin(2\omega\tau_0)}{1 - r_1 r_2 \cos(2\omega\tau_0)} \right), \tag{2.94a}$$

$$\kappa_e(\omega) = -\frac{n_0}{\omega\tau_0} \ln\left(\frac{t_1 t_2}{[1 + r_1^2 r_2^2 - 2r_1 r_2 \cos(2\omega\tau_0)]^{1/2}} \right). \tag{2.94b}$$

With this effective formulation, it is possible to employ the techniques developed in the previous sections for analyzing pulse propagation through a Fabry–Perot resonator. Alternatively, realizing that we have a time-invariant transfer function, pulse propagation can be viewed as a filtering operation. This latter approach is taken in this section because it provides more physical insight.

Suppose a chirped Gaussian pulse with an electric field

$$E_g(t) = A \exp\left(-\frac{t^2}{2T_0^2} \right) \exp(-j\omega_0 t + jat^2) \tag{2.95}$$

is incident on a Fabry–Perot resonator from the left [36]. If we view its transmission as a filtering operation, the spectrum of the output pulse can be calculated by multiplying the input spectrum by the Fabry–Perot transfer function. The resulting time-domain output field E_o is then computed by taking the inverse Fourier transform. The result is

$$E_o(t) = \frac{A}{2\pi} \sqrt{\frac{\pi T_0^2}{1 - 2jaT_0^2}} \int_{-\infty}^{\infty} \tilde{T}_{\text{FP}} \exp\left(-\frac{(\omega - \omega_0)^2 T_0^2}{4(1 - 2jaT_0^2)} - j\omega t \right) d\omega. \tag{2.96}$$

If the frequency dependence of the refractive index, $n_0(\omega)$, of the Fabry-Perot medium is important, the preceding integral cannot be performed analytically in general. However, if this dependence is negligible, we can perform the integration and obtain the following analytical result:

$$E_o(t) = t_1 t_2 A e^{-j\omega_0 t}$$

$$\times \sum_{m=0}^{\infty} r_1^m r_2^m \exp\left[j(2m+1)\tau_0 - \frac{(1 - 2jaT_0^2)}{2T_0^2}[t - (2m+1)\tau_0]^2 \right].$$

$$(2.97)$$

This simple result shows how a Gaussian pulse is affected by multiple round trips inside a Fabry–Perot resonator.

2.7.2 Transfer function of a dielectric slab

In this section we consider a dielectric slab of thickness L with a uniform refractive index n_0 surrounded by air on both sides (see Figure 2.9). Such a slab acts as a Fabry–Perot resonator because of an abrupt change in the refractive index at its two interfaces, and the reflection and transmission coefficients can be obtained from the Fresnel relations given in Section 2.3. As shown earlier, reflections at two interfaces of this dielectric slab can be mapped to an effective slab with no reflections, provided we modify the refractive index appropriately.

This approach can be supplemented with a rigorous approach based on Maxwell's equations. We assume that the left and right interfaces of the slab are located at $z = 0$ and $z = L$, respectively, and a plane wave with electric field $\tilde{E}_i(0, \omega)$ is incident on the left facet of the slab. Following the general solution given in Section 2.3 based on Maxwell's equations, the electric field $\tilde{E}(z, \omega)$ at a distance z inside the dielectric slab can be written in the form [37]:

$$\tilde{E}(z, \omega) = \tilde{E}_i(0, \omega) \times \begin{cases} e^{j\omega z/c} + \tilde{R}_{\mathrm{FP}}(\omega)e^{-j\omega z/c}, & \text{if } z < 0, \\ \varrho_f e^{j\omega z/c} + \varrho_b e^{-j\omega z/c}, & \text{if } 0 \le z \le L, \\ \tilde{T}_{\mathrm{FP}}(\omega)e^{-j\omega z/c}, & \text{if } z > L, \end{cases} \quad (2.98)$$

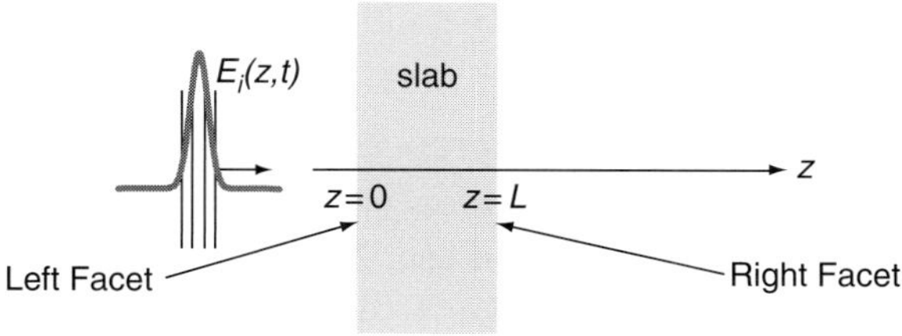

Figure 2.9 An optical pulse transmitted through a dielectric slab with a sudden change in the refractive index at its facets.

where $\widetilde{R}_{\text{FP}}(\omega)$ is the reflectivity of the slab, $\widetilde{T}_{\text{FP}}(\omega)$ is its transfer function, and ϱ_f and ϱ_b are the forward and backward amplitudes inside the slab.

As seen from Eq. (2.98), the incident wave propagates forward in the $z > 0$ region, but it also creates a reflected wave from the left facet of the slab that propagates backward in the $z < 0$ region. The relative amplitude of this reflected wave region is governed by $\widetilde{R}_{\text{FP}}(\omega)$. The forward and backward propagating waves coexist within the slab but only the forward wave exists in the region $z > L$, and its amplitude is governed by $\widetilde{T}_{\text{FP}}(\omega)$.

As discussed in Section 2.3, since no free charges exist within the slab or its facets, the tangential electric field must be continuous at its two interfaces. Matching the electric field at both interfaces, we arrive, after considerable algebra, at the following expressions [37]:

$$\widetilde{T}_{\text{FP}}(\omega) = \frac{4\eta}{(\eta + 1)^2 \exp(-jkL) + (\eta - 1)^2 \exp(jkL)}, \tag{2.99a}$$

$$\widetilde{R}_{\text{FP}}(\omega) = \frac{(\eta^2 - 1)\exp(-jkL) - (\eta^2 - 1)\exp(jkL)}{(\eta + 1)^2 \exp(-jkL) + (\eta - 1)^2 \exp(jkL)}, \tag{2.99b}$$

$$\varrho_f = \frac{2\eta(\eta + 1)}{(\eta + 1)^2 + (\eta - 1)^2 \exp(j2kL)}, \tag{2.99c}$$

$$\varrho_b = \frac{2\eta(\eta - 1)}{(\eta - 1)^2 + (\eta + 1)^2 \exp(-j2kL)}, \tag{2.99d}$$

where $\eta = \sqrt{\mu_0/\varepsilon_0}/n_0$ is the wave impedance inside the slab (see Eq. (2.6)) and $k = n_0\omega/c$ is the propagation constant inside the slab. Even though Eqs. (2.92) and (2.99a) appear quite different, their equivalence can be established by noting that $\exp(jkL) = \exp(j\omega\tau_0)$ and using the Fresnel equations for the reflection and and transmission coefficients at the two facets. More specifically,

$$t_1 = \frac{2}{\eta + 1}, \quad r_1 = \frac{\eta - 1}{\eta + 1}, \quad t_2 = \frac{2\eta}{\eta + 1}, \quad r_2 = \frac{1 - \eta}{\eta + 1}. \tag{2.100}$$

In conclusion, the results of this section show that reflections at the input and output facets of an amplifying medium can be easily taken into account, if necessary, through the concept of effective refractive index.

References

[1] B. Nistad and J. Skaar, "Causality and electromagnetic properties of active media," *Phys. Rev. E*, vol. 78, p. 036603 (10 pages), 2008.
[2] M. Born and E. Wolf, *Principles of Optics*, 7th ed. Cambridge University Press, 1999.
[3] J. D. Jackson, *Classical Electrodynamics*, 3rd ed. Wiley, 1998.
[4] B. E. A. Saleh and M. C. Teich, *Fundamentals of Photonics*, 2nd ed. Wiley InterScience, 2007.

[5] A. Taflove and S. C. Hagness, *Computational Electrodynamics: The Finite-Difference Time-Domain Method*, 3rd ed. Artech House, 2005.

[6] R. W. Ziolkowski and E. Heyman, "Wave propagation in media having negative permittivity and permeability," *Phys. Rev. E*, vol. 64, p. 056625 (15 pages), 2001.

[7] J. Wei and J. Xiao, "Electric and magnetic losses and gains in determining the sign of refractive index," *Opt. Commun.*, vol. 270, pp. 455–464, 2007.

[8] M. A. Dupertius, M. Proctor, and B. Acklin, "Generalization of complex Snell–Descartes and Fresnel laws," *J. Opt. Soc. Am. A*, vol. 11, pp. 1159–1166, 1994.

[9] P. W. Milonni, "Controlling the speed of light pulses," *J. Phys. B: At. Mol. Opt. Phys.*, vol. 35, pp. R31–R56, 2002.

[10] C. G. B. Garrett and D. E. McCumber, "Propagation of a Gaussian light pulse through an anomalous dispersion medium," *Phys. Rev. A*, vol. 1, pp. 305–313, 1970.

[11] C. Gasquet, R. D. Ryan, and P. Witomski, *Fourier Analysis and Applications: Filtering, Numerical Computation, Wavelets*. Springer-Verlag, 1998.

[12] D. Slepian, "On bandwidth," *Proc. IEEE*, vol. 64, pp. 292–300, 1976.

[13] G. P. Agrawal, *Nonlinear Fiber Optics*, 4th ed. Academic Press, 2007.

[14] J. B. Schneider, "FDTD dispersion revisited: Faster-than-light propagation," *IEEE Microw. Guided Wave Lett.*, vol. 9, pp. 54–56, 1999.

[15] D. F. Kelley and R. Luebbers, "Piecewise linear recursive convolution for dispersive media using FDTD," *IEEE Trans. Antennas Propag.*, vol. 44, pp. 792–797, 1996.

[16] M. Okoniewski, M. Mrozowski, and M. Stuchly, "Simple treatment of multi-term dispersion in FDTD," *IEEE Microw. Guided Wave Lett.*, vol. 7, pp. 121–123, 1997.

[17] J. Young and R. Nelson, "A summary and systematic analysis of FDTD algorithms for linearly dispersive media," *IEEE Antennas Propag. Mag.*, vol. 43, pp. 61–77, 2001.

[18] M. Han, R. W. Dutton, and S. Fan, "Model dispersive media in finite-difference time-domain method with complex-conjugate pole-residue pairs," *IEEE Microw. Wireless Compon. Lett.*, vol. 16, pp. 119–121, 2006.

[19] I. Udagedara, M. Premaratne, I. D. Rukhlenko, H. T. Hattori, and G. P. Agrawal, "Unified perfectly matched layer for finite-difference time-domain modeling of dispersive optical materials," *Opt. Express*, vol. 17, pp. 21180–21190, 2009.

[20] J. Berenger, "A perfectly matched layer for the absorption of electromagnetic waves," *J. Comput. Phys.*, vol. 114, pp. 185–200, 1994.

[21] S. Gedney, "An anisotropic perfectly matched layer-absorbing medium for the truncation of FDTD lattices," *IEEE Trans. Antennas Propag.*, vol. 44, pp. 1630–1639, 1996.

[22] L. Zhao and A. C. Cangellaris, "GT-PML: Generalized theory of perfectly matched layers and its application to the reflectionless truncation of finite-difference time-domain grids," *IEEE Trans. Microw. Theory Tech.*, vol. 44, pp. 2555–2563, 1996.

[23] M. Kuzuoglu and R. Mittra, "Frequency dependence of the constitutive parameters of causal perfectly matched anisotropic absorbers," *IEEE Microw. Guided Wave Lett.*, vol. 6, pp. 447–449, 1996.

[24] J.-P. Berenger, "Numerical reflection from FDTD-PMLs: A comparison of the split PML with the unsplit and CFS PMLs," *IEEE Trans. Antennas Propag.*, vol. 50, pp. 258–265, 2002.

[25] ——, "Application of the CFS PML to the absorption of evanescent waves in waveguides," *IEEE Microw. Wireless Compon. Lett.*, vol. 12, pp. 218–220, 2002.

[26] D. Correia and J.-M. Jin, "On the development of a higher-order PML," *IEEE Trans. Antennas Propag.*, vol. 53, pp. 4157–4163, 2005.

[27] Y. Rickard and N. Georgieva, "Problem-independent enhancement of PML ABC for the FDTD method," *IEEE Trans. Antennas Propag.*, vol. 51, pp. 3002–3006, 2003.

[28] S. Cummer, "Perfectly matched layer behavior in negative refractive index materials," *IEEE Antennas Wireless Propag. Lett.*, vol. 3, pp. 172–175, 2004.

[29] A. V. Oppenheim and R. W. Schafer, *Digital Signal Processing.*   Prentice-Hall, 1975.

[30] D. Sullivan, "Z-transform theory and the FDTD method," *IEEE Trans. Antennas Propag.*, vol. 44, pp. 28–34, 1996.

[31] E. L. Bolda, R. Y. Chiao, and J. C. Garrison, "Two theorems for the group velocity in dispersive media," *Phys. Rev. A*, vol. 48, pp. 3890–3894, 1993.

[32] L. D. Landau and E. M. Lifshitz, *Electrodynamics of Continuous Media*, 2nd ed. *Course of Theoretical Physics*, vol. 8. Pergamon Press, 1984.

[33] T. W. Gamelin, *Complex Analysis.* Springer, 2001.

[34] G. Cesini, G. Guattari, G. Lucarini, and C. Palma, "Response of Fabry–Perot interferometers to amplitude-modulated light beams," *Optica Acta*, vol. 24, pp. 1217–1236, 1977.

[35] J. Yu, S. Yuan, J.-Y. Gao, and L. Sun, "Optical pulse propagation in a Fabry–Perot étalon: Analytical discussion," *J. Opt. Soc. Am. A*, vol. 18, pp. 2153–2160, 2001.

[36] X. G. Qiong, W. Z. Mao, and C. J. Guo, "Time delay of a chirped light pulse after transmitting a Fabry–Perot interferometer," *Chinese Phys. Lett.*, vol. 19, pp. pp. 201–202, 2002.

[37] J. Skaar, "Fresnel equations and the refractive index of active media," *Phys. Rev. E*, vol. 73, p. 026605 (7 pages), 2006.

3

Interaction of light with generic active media

The study of optical effects in active media is rich with unexpected consequences. For instance, one may imagine that absorption and amplification in a dielectric medium will exhibit some sort of symmetry because both are related to the same imaginary part of the dielectric constant, except for a sign change. It turns out that such a symmetry does not exist [1]. This issue has also been investigated in detail for random or disordered gain media, with varying viewpoints [2]. Traditionally, much of the research on amplifying media has considered the interaction of light within the entire volume of such a medium. Recent interest in metamaterials and other esoteric structures in which plasmons are used to manipulate optical signals has brought attention to the role of surface waves in active dielectrics [3, 4].

From a fundamental perspective, the main difference between active and passive media is that spontaneous emission cannot be avoided in gain media. As a result, a rigorous analysis of gain media demands a quantum-mechanical treatment. Spontaneous emission in a gain medium depends not only on the material properties of that medium but also on the optical modes supported by the structure containing that material [5]. By a clever design of this structure (e.g., photonic crystals or microdisk resonators with a metallic cladding), it is possible to control the local density of optical modes and the spontaneous emission process itself. In Chapter 4, we discuss how to model a gain medium under such conditions by deploying optical Bloch equations. In this chapter, our focus is on generic properties of a dielectric active medium exhibiting gain, and we disregard the actual mechanism responsible for the gain. We assume that the dielectric gain medium is nonmagnetic and replace the permeability μ in Maxwell's equations with its vacuum value μ_0. In this situation, we can account for the gain by using a complex dielectric permittivity, $\varepsilon(\omega) = \varepsilon_r(\omega) + j\varepsilon_i(\omega)$, with a negative imaginary part ($\varepsilon_i < 0$) in some frequency range. Even though this approach is simplistic, it has been successfully used for describing the operation of lasers and amplifiers [6, 7]. Section 3.1 considers the reflection of light from a gain medium under conditions of total internal reflection.

Section 3.2 focuses on the properties of surface-plasmon polaritons. The remaining sections show that the gain in a dielectric medium can be used to modify the group velocity and other properties of such waves.

3.1 Reflection of light from a gain medium

Before studying light propagation through a gain medium, it is interesting and instructive to consider the behavior at the interface when light is launched into a gain medium from a passive medium. In the case of an interface between two passive media, we can account for the reflection and transmission of light through the interface using the Fresnel equations discussed in Section 2.3.3. This approach can be easily extended to the case where the gain medium has a higher refractive index than the passive medium. It is easy to show from the Fresnel equations that the reflection coefficient has a magnitude less than unity in this situation. In this section, we first discuss the phenomenon of total internal reflection at an interface between two passive media and then study what happens when one of the media exhibits gain.

3.1.1 Total internal reflection and the Goos–Hänchen effect

Figure 3.1 shows a light beam incident on an interface from the side of a medium with higher refractive index. If the incident medium has refractive index of n_i, and the beam gets refracted into a medium with refractive index n_r, then Snell's law, $n_i \sin \theta_i = n_r \sin \theta_r$, provides a way to calculate the refraction angle θ_r from the

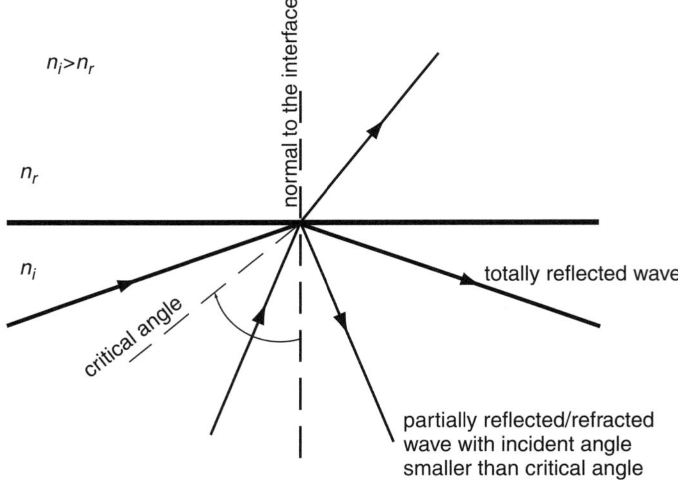

Figure 3.1 Illustration of total internal reflection at an interface.

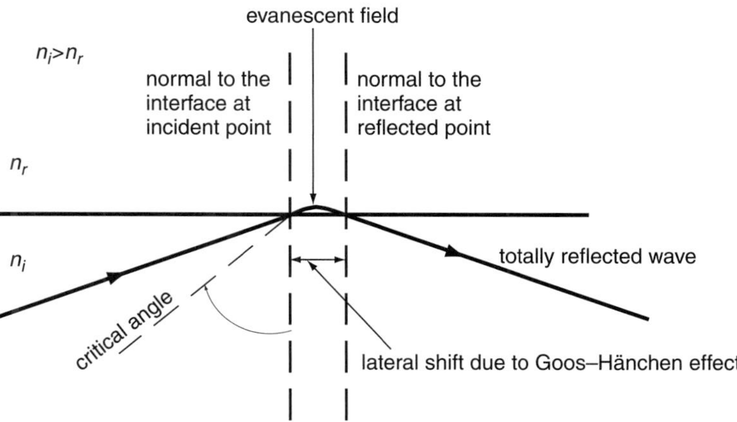

Figure 3.2 Schematic illustration of the Goos–Hänchen shift occurring at an interface during total internal reflection.

incident angle θ_i. Because of the condition $n_i > n_r$, θ_r is larger than θ_i. Therefore, as the incident angle θ_i is increased, a situation is reached in which $\theta_r = 90°$, i.e., the refracted ray becomes parallel to the interface. The incident angle corresponding to this condition is called the *critical angle* and can be obtained from the well-known expression

$$\theta_c = \sin^{-1}\left(\frac{n_r}{n_i}\right). \tag{3.1}$$

If the incident angle exceed this critical angle, the incident light is totally reflected, obeying normal specular reflection laws (the incident and reflected angles are equal). Such a reflection from an interface is known as *total internal reflection*.

One peculiarity of total internal reflection is that, in the case of a linearly polarized input, the reflected light undergoes a lateral shift from the position predicted by geometrical optics, as shown in Figure 3.2. This phenomenon is known as the Goos–Hänchen effect. The lateral shift happens in the incident plane, and the magnitude of this shift is comparable to the wavelength of light for incident angles slightly greater than the critical angle. The analogous effect for circularly or elliptically polarized light is known as the Imbert–Fedorov effect. If the geometrical-optics description is replaced with wave optics appropriate for a beam of light, in addition to the lateral shift the reflected beam also experiences a focal shift, an angular shift, and a change in its beam waist [8, 9].

3.1.2 *Total internal reflection from an active cladding*

The ability to amplify light in a passive medium while undergoing total internal reflection at the interface of an active material was first noted by Koester in a

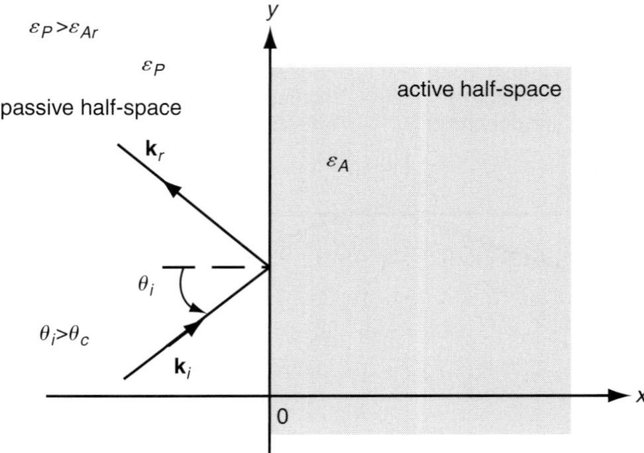

Figure 3.3 Schematic illustration of total internal reflection at an interface when light enters from a lossless passive medium into an active cladding material.

1966 experiment [10]. In this experiment, an index-guided mode of a dielectric waveguide, surrounded with an active cladding, experienced amplification along the waveguide length. Recent classical and quantum studies [11–13] have shown conclusively that light in a completely passive medium can be amplified using evanescent-gain techniques based on total internal reflection.

To get a quantitative understanding of this process, we consider total internal reflection with the geometry shown in Figure 3.3. The interface coinciding with the plane $x = 0$ separates a passive lossless medium with real permittivity $\varepsilon_P > 0$, on its left, from an active gain medium, on its right, whose complex permittivity has the form $\varepsilon_A = \varepsilon_{Ar} + j\varepsilon_{Ai}$. Both media are assumed to be nonmagnetic with a magnetic permeability equal to the free-space value μ_0. To enable total internal reflection, the refractive index of the active medium has to be lower than that of the passive medium, leading to the condition $\varepsilon_P > \varepsilon_{Ar} > 0$. The critical angle in this case is found from Eq. (3.1) to be

$$\theta_c = \sin^{-1}\left(+\sqrt{\frac{\varepsilon_{Ar}}{\varepsilon_P}}\right), \qquad (3.2)$$

where we have taken the positive value of the square root because of the condition $\varepsilon_P > \varepsilon_{Ar} > 0$.

For a general incident angle θ_i, the propagation vector of the incident field in the passive medium can be written as

$$\boldsymbol{k}_{Pi} = \sqrt{\frac{\varepsilon_P}{\varepsilon_0}}k_0 \cos(\theta_i)\boldsymbol{x} + \sqrt{\frac{\varepsilon_P}{\varepsilon_0}}k_0 \sin(\theta_i)\mathbf{y}, \qquad (3.3)$$

where $k_0 = \omega/c$ for an optical field with carrier frequency ω. The corresponding reflected wave has a propagation vector $\boldsymbol{k}_{Pr}$ whose x component is reversed and has a negative sign, i.e.,

$$\boldsymbol{k}_{Pr} = -\sqrt{\frac{\varepsilon_P}{\varepsilon_0}}\, k_0 \cos(\theta_i)\boldsymbol{x} + \sqrt{\frac{\varepsilon_P}{\varepsilon_0}}\, k_0 \sin(\theta_i)\boldsymbol{y}. \tag{3.4}$$

Owing to the continuity of the tangential component of the electric field, the y components of all propagation constants coincide (this amounts to obeying Snell's law at the interface). Therefore, the propagation vector in the active medium can be written as

$$\boldsymbol{k}_A = k_{Ax}\boldsymbol{x} + \sqrt{\frac{\varepsilon_P}{\varepsilon_0}}\, k_0 \sin(\theta_i)\boldsymbol{y}, \tag{3.5}$$

where the complex quantity k_{Ax} needs to be determined using the known medium and incident-wave properties. If this medium is passive and the incident angle is less than the critical angle, this quantity can be written as

$$k_{Ax} = \sqrt{\frac{\varepsilon_{Ar}}{\varepsilon_0}}\, k_0 \sqrt{1 - \frac{\varepsilon_{Ar}}{\varepsilon_P}\sin^2(\theta_i)}, \qquad \varepsilon_{Ai} = 0, \ 0 < \theta_i < \theta_c. \tag{3.6}$$

In the case of an active medium, the dispersion relation for the active medium allows us to find k_{Ax}. Suppose σ_A is the conductivity of the active medium. It is related to the imaginary part of the dielectric constant through the relation $\sigma_A = \omega \varepsilon_{Ai}$. Using the following relation obtained from Maxwell's equations,

$$\nabla \times \nabla \times \boldsymbol{E} = -\nabla^2 \boldsymbol{E} = -\frac{\partial \nabla \times \boldsymbol{B}}{\partial t} = -\frac{\partial}{\partial t}\left(\mu_0 \sigma_A \boldsymbol{E} + \mu_0 \varepsilon_{Ar}\frac{\partial \boldsymbol{E}}{\partial t}\right), \tag{3.7}$$

and noting that $\boldsymbol{E} = E_A \exp[j\boldsymbol{k}_A \cdot (x\boldsymbol{x} + y\boldsymbol{y}) - j\omega t]$, we obtain the relation

$$k_{Ax}^2 + \frac{\varepsilon_P}{\varepsilon_0}k_0^2 \sin^2(\theta_i) = \mu_0 \omega(\varepsilon_{Ar}\omega + j\sigma_A). \tag{3.8}$$

Using $k_0 = \omega\sqrt{\mu_0\varepsilon_0}$ and $\sigma_A = \omega\varepsilon_{Ai}$, this equation can be solved to obtain

$$k_{Ax} = k_0 \sqrt{\frac{\varepsilon_A - \varepsilon_P \sin^2(\theta_i)}{\varepsilon_0}}. \tag{3.9}$$

It is useful to introduce a polar form for the complex argument of the square root in the preceding equation [13]:

$$\frac{\varepsilon_A - \varepsilon_P \sin^2(\theta_i)}{\varepsilon_0} \equiv \xi = |\xi|\exp(j\theta_\xi), \qquad -\frac{\pi}{2} \le \theta_\xi < \frac{3\pi}{2}. \tag{3.10}$$

It is clear from this definition that we have introduced a branch cut along the imaginary axis of the complex ξ plane. The x component of the propagation constant in the active medium can now be written as

$$k_{Ax} = k_0\sqrt{|\xi|}\exp\left(j\frac{\theta_\xi}{2}\right). \tag{3.11}$$

It follows from the definition of θ_c that when the incident angle θ_i exceeds this value, the real part of ξ becomes negative, and thus θ should be in the range $\pi/2 < \theta_\xi < 3\pi/2$. However, if propagation needs to happen towards the interface and the signal has to grow because of gain, the permissible range of this angle is $\pi < \theta_\xi < 3\pi/2$.

Using our treatment of Fresnel reflection coefficients in Section 2.3, we obtain the following expressions for the reflection coefficients Γ_{TE} and Γ_{TM} for the TE and TM polarizations, respectively:

$$\Gamma_{TE} = \frac{\sqrt{\varepsilon_P}\cos(\theta_i) - \sqrt{\varepsilon_A - \varepsilon_P\sin^2(\theta_i)}}{\sqrt{\varepsilon_P}\cos(\theta_i) + \sqrt{\varepsilon_A - \varepsilon_P\sin^2(\theta_i)}}, \tag{3.12a}$$

$$\Gamma_{TM} = \frac{\sqrt{\varepsilon_A}\cos(\theta_i) - \sqrt{\varepsilon_A - \varepsilon_P\sin^2(\theta_i)}}{\sqrt{\varepsilon_A}\cos(\theta_i) + \sqrt{\varepsilon_A - \varepsilon_P\sin^2(\theta_i)}}. \tag{3.12b}$$

Using the definition in Eq. (3.10), we can write these reflectivity coefficients in a more compact form as

$$\Gamma_{TE} = \frac{\sqrt{\varepsilon_P/\varepsilon_0}\cos(\theta_i) - \xi}{\sqrt{\varepsilon_P/\varepsilon_0}\cos(\theta_i) + \xi}, \tag{3.13a}$$

$$\Gamma_{TM} = \frac{\sqrt{\varepsilon_A/\varepsilon_0}\cos(\theta_i) - \xi}{\sqrt{\varepsilon_A/\varepsilon_0}\cos(\theta_i) + \xi}. \tag{3.13b}$$

There is no need to investigate these two quantities separately because it is possible to establish the following unique relation between them by eliminating the variable ξ:

$$\Gamma_{TM} = \left(\frac{\Gamma_{TE} - \cos(2\theta_i)}{1 - \Gamma_{TE}\cos(2\theta_i)}\right)\Gamma_{TE}. \tag{3.14}$$

We thus focus our attention on Γ_{TE}. Because we are mainly interested in the TE-mode reflectivity, we calculate $|\Gamma_{TE}|^2$ by multiplying Eq. (3.14) by its complex conjugate. The result is given by

$$|\Gamma_{TE}|^2 = \frac{[\sqrt{\varepsilon_P/\varepsilon_0}\cos(\theta_i) - |\xi|\cos(\theta_\xi)]^2 + |\xi|^2\sin^2(\theta_\xi)}{[\sqrt{\varepsilon_P/\varepsilon_0}\cos(\theta_i) + |\xi|\cos(\theta_\xi)]^2 + |\xi|^2\sin^2(\theta_\xi)}. \tag{3.15}$$

This quantity exceeds 1 for θ_ξ in the range $\pi < \theta_\xi < 3\pi/2$. We saw previously that this range corresponds to the case in which an evanescent wave propagates in the active medium towards the interface, and the TE mode propagating in the passive medium grows because of the gain experienced during total internal reflection. From the relation (3.14), the TM mode will also be amplified through this mechanism.

3.2 Surface-plasmon polaritons

Surface-plasmon polaritons (SPPs) represent TM-polarized waves that are localized at the interface between two materials whose refractive indices (related to the real part of the permittivity) have opposite signs. They can also be viewed as optical waves trapped close to the interface because of interactions between photons and electrons [14]. The ability of an electromagnetic field to propagate as a surface wave on an interface was first studied by Ritchie [15]. At optical frequencies, a commonly used configuration employs a metal–dielectric interface, as shown schematically in Figure 3.4. At the oscillation frequency of the SPP, the metal has a negative real permittivity and the dielectric has a positive real permittivity [16].

3.2.1 SPP dispersion relation and other properties

The dispersion relation for SPPs at a metal–dielectric interface was first derived by Teng and Stern [17]. Figure 3.5 shows a typical dispersion relation for SPPs at such an interface. Clearly, at a given frequency ω, SPPs have a longer wave vector than that of an electromagnetic wave propagating in vacuum (i.e., the SPP dispersion curve lies beyond the light line $\omega = ck$). As a result, the momentum of an SPP exceeds that of a photon in vacuum at the same frequency ω [14]. However, the coupling of light to SPPs requires matching of the momentum at the coupling point. This is achieved in practice by making light travel through a transparent

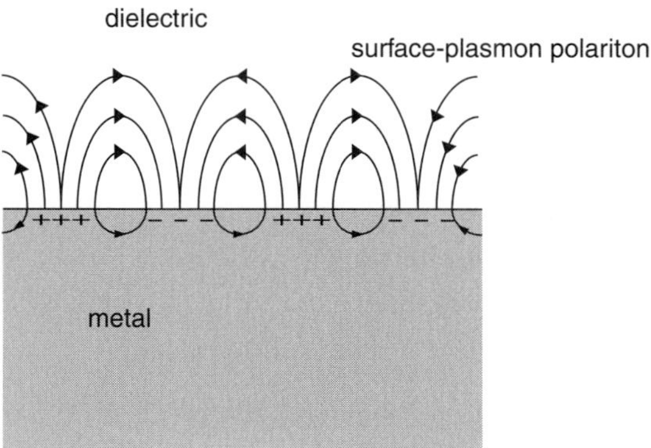

Figure 3.4 Illustration of surface-plasmon polaritons propagating along a metal–dielectric interface as a surface wave.

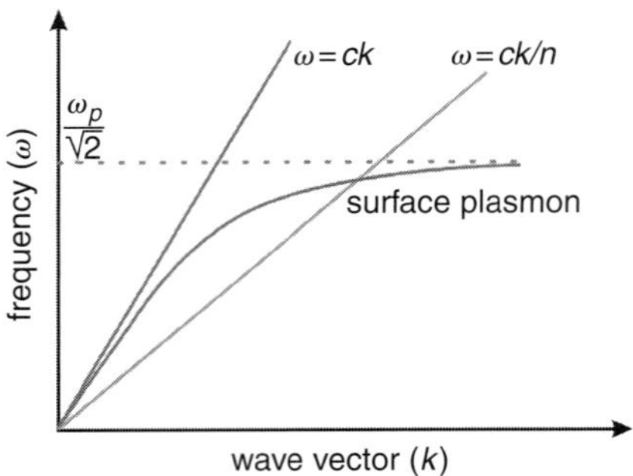

Figure 3.5 Dispersion curve of surface-plasmon polaritons at a metal–dielectric interface. Two straight lines show the dispersion curves of light in vacuum and in a substrate of refractive index $n > 1$.

medium with high refractive index and varying the incident angle until momentum or phase matching is achieved in the SPP propagation direction (see Figure 3.5). Because such phase matching requires a propagation vector whose magnitude is larger than that obtained in vacuum, evanescent fields are generally required in practice. Further details can be found in Refs. [16] and [18].

The word "plasmon" in SPP emphasizes the underlying physical process. Negative permittivity in metals results from oscillations in the density of electrons within it. To a very good approximation, electrons in a conductor can be viewed as a free ideal gas with properties resembling that of a plasma (consisting of plasmons in a quantum treatment). The word "polariton" in SPPs represents the phonon oscillations associated with surface modes in polar dielectrics. SPPs represent a hybrid wave associated with the interaction between these two distinctive processes [14]. Within the framework of Maxwell's equations, once the material properties of the dielectric and the metal are specified, SPPs refer to a solution that decays exponentially in both directions from the interface. Figure 3.6 shows this behavior schematically.

Recent advances in metamaterials and artificial materials have widened the frequency range over which an SPP can be sustained [19, 20]. For example, in the microwave region, metals are not very effective for initiating SPPs because their response becomes dominated by their high conductivity. However, properly designed metamaterials with a negative refractive index can be interfaced with conventional dielectrics to form composite structures in which SPPs can be sustained on the interface at microwave and terahertz frequencies.

3.2.2 Propagation loss and its control

Even though it is relatively easy to find dielectrics that exhibit very low losses in the frequency range of SPPs, it is not possible to avoid a significant energy loss resulting

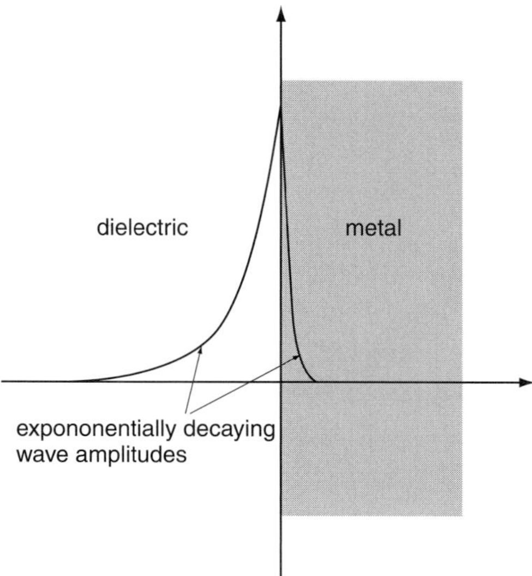

Figure 3.6 Exponential decay of an SPP surface wave strength in both directions from the metal–dielectric interface.

from the positive imaginary part of the permittivity of metals. The overall effect is a rather lossy propagation of SPPs on a metal–dielectric interface. Often, the effective propagation length of SPPs is limited to distances $\sim 10\lambda$ when radiation of wavelength λ is used to excite such a surface wave [14]. This lossy nature of SPPs is the main obstacle to their use in channeling light using sub-wavelength size structures, because it effectively limits the maximum size of plasmonic devices that can be made. In addition, if SPPs propagate through sharp bends, there is a chance of losing additional energy to radiation losses [21].

Two fundamentally different strategies have been proposed to reduce the propagation losses of SPPs [22, 23]. One scheme utilizes gain in the dielectric region to compensate for the losses occurring within the metal [22]. The other scheme is known by the term *spaser*, an acronym similar to laser and standing for *surface plasmon amplification by stimulated emission of radiation*. The spaser is indeed the nano-plasmonic counterpart of a conventional laser, but it amplifies plasmon resonances rather than photons, and employs a plasmonic resonator (i.e., nanoparticles) instead of a mirror-based cavity [24]. A spaser receives energy from a conventional gain medium, pumped externally, using a mechanism similar to Forster's dipole–dipole energy transfer [25], transferring excitation energy nonradiatively to the resonant nanosystem.

Figure 3.7 shows one possible design of a spaser based on a metal–dielectric nanoshell [24]. More specifically, a silver nanoshell is surrounded by active quantum dots that are pumped to provide gain. When a quantum dot is not in the immediate vicinity of the nanoshell, it relaxes to a lower energy state by emitting

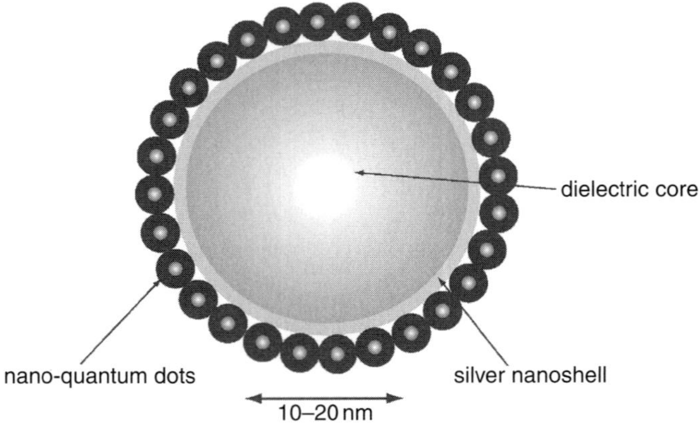

Figure 3.7 Illustration of the spaser scheme in which gain in quantum dots surrounding a metal–dielectric nanoshell compensates for propagation losses within the metal.

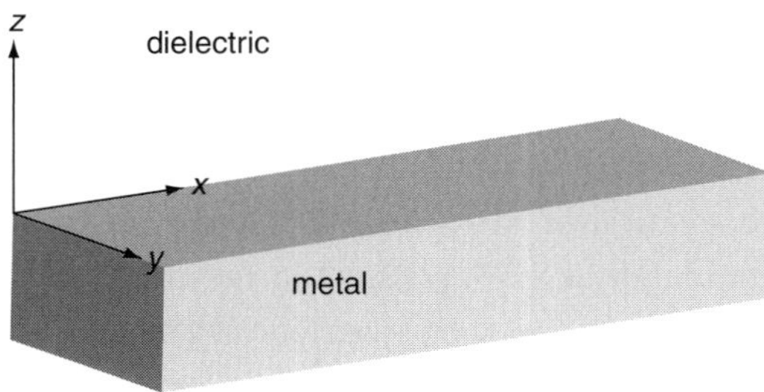

Figure 3.8 Configuration used for analyzing SPPs confined to a metal–dielectric interface.

photons. However, when there is tighter coupling between the nanoshell and the quantum dots, the relaxation process does not lead to emission of new photons but transfers energy to the resonant surface plasmons of the nanoshell. Moreover, surface plasmons on the nanoshell can stimulate further such transitions, leading to stimulated emission of identical surface plasmons [24].

3.2.3 Gain-assisted propagation of SPPs

The basic idea of using an active dielectric medium to compensate for losses in the metallic region of SPPs on an interface [22] has been developed further in recent years [3,26]. We follow the analysis of Ref. [3] to gain a fundamental understanding of the analytical reasoning behind the physical process.

Figure 3.8 shows the geometry used. We assume that an SPP surface wave is propagating along the positive x direction on the metal–dielectric boundary, which lies in the x–y plane, and that the electromagnetic fields associated with this surface wave decays in the positive (dielectric) and negative (metal) z directions. The permittivities of the dielectric and metal media are assumed to be constant within each medium at a given frequency, but we allow for the possibility that they may vary with frequency. The dielectric has a complex permittivity, $\varepsilon_D = \varepsilon_{Dr} + j\varepsilon_{Di}$ with $\varepsilon_{Dr} > 0$. When the dielectric is pumped externally to provide gain, the permittivity also satisfies $\varepsilon_{Di} < 0$. The excitation of an SPP wave at the metal–dielectric interface requires that the real parts of the permittivities of these two media have opposite signs. Therefore, the metal permittivity, $\varepsilon_M = \varepsilon_{Mr} + j\varepsilon_{Mi}$, must satisfy $\varepsilon_{Mr} < 0$. Also, the lossy nature of the metal implies that $\varepsilon_{Mi} > 0$.

Owing to our assumption of homogeneity in the y direction, the global[1] electric and magnetic fields, $\boldsymbol{E}_G$ and $\boldsymbol{H}_G$, can be written in the following general form [16]:

$$\boldsymbol{E}_G(x, y, z) = \boldsymbol{E}(z)\exp[j(k_x x - \omega t)], \tag{3.16a}$$

$$\boldsymbol{H}_G(x, y, z) = \boldsymbol{H}(z)\exp[j(k_x x - \omega t)], \tag{3.16b}$$

where k_x is the magnitude of the wave vector of the SPP wave pointing in the x direction and ω is its frequency. Substitution of these fields in the Maxwell equation $\nabla \times \boldsymbol{E}_G = -\mu_0\frac{\partial \boldsymbol{H}_G}{\partial t}$ gives us the relations

$$\frac{\partial E_y}{\partial z} = -j\omega\mu_0 H_x, \tag{3.17a}$$

$$\frac{\partial E_x}{\partial z} = j\omega\mu_0 H_y + jk_x E_z, \tag{3.17b}$$

$$jk_x E_y = j\omega\mu_0 H_z. \tag{3.17c}$$

Since the permittivity is discontinuous across the metal–dielectric interface, we introduce a single permittivity function $\varepsilon(z)$ of the form

$$\varepsilon(z) = \begin{cases} \varepsilon_D & \text{if } z > 0, \\ \varepsilon_M & \text{if } z < 0. \end{cases} \tag{3.18}$$

Using the Maxwell equation for the magnetic field, $\nabla \times \boldsymbol{H}_G = -\varepsilon\frac{\partial \boldsymbol{E}_G}{\partial t} + \boldsymbol{J}$, we then obtain

$$\frac{\partial H_y}{\partial z} = j\omega\varepsilon(z)E_x, \tag{3.19a}$$

$$\frac{\partial H_x}{\partial z} = -j\omega\varepsilon(z)E_y + jk_x H_z, \tag{3.19b}$$

$$jk_x H_y = -j\omega\varepsilon(z)E_z, \tag{3.19c}$$

---

[1] We use the qualifier "global" to designate the whole system in Figure 3.8.

where we have assumed $J = 0$ because there are no currents present in the system under study.

Two self-consistent solutions exist for the preceding six field equations (corresponding to the TE and TM modes) when the metal in Figure 3.8 is replaced with a dielectric. However, it is known that there is no nonzero TE solution in the case of a metal, and SPP waves are necessarily in the form of TM modes [16]. For a TM solution, only the field components E_x, E_z, and H_y are nonzero, and we need to solve the following set of three equations:

$$\frac{\partial E_x}{\partial z} = j\omega\mu_0 H_y + jk_x E_z, \tag{3.20a}$$

$$E_x = -j\frac{1}{\omega\varepsilon}\frac{\partial H_y}{\partial z}, \tag{3.20b}$$

$$E_z = -\frac{k_x}{\omega\varepsilon}H_y. \tag{3.20c}$$

Substitution of Eqs. (3.20b) and (3.20c) into Eq. (3.20a) gives the wave equation for H_y,

$$\frac{\partial^2 H_y}{\partial z^2} + (\varepsilon\mu_0\omega^2 - k_x^2)H_y = 0. \tag{3.21}$$

Because we are interested in finding a solution confined to the interface and because the two sides of the interface require distinct z components of the propagation vector, it is useful to assign different labels to the electric and magnetic field components in each medium. We use the subscripts D and M to indicate the dielectric and metal regions, respectively, and write the TM-mode solution in the form

$$\boldsymbol{E}_D(x, z, t) = (E_{xD}\boldsymbol{x} + E_{zD}\boldsymbol{z})\exp[j(k_x x + k_{zD}z - \omega t)], \tag{3.22a}$$

$$\boldsymbol{E}_M(x, z, t) = (E_{xM}\boldsymbol{x} + E_{zM}\boldsymbol{z})\exp[j(k_x x + k_{zM}z - \omega t)]), \tag{3.22b}$$

$$\boldsymbol{H}_D(x, z, t) = H_{yD}\boldsymbol{y}\exp[j(k_x x + k_{zD}z - \omega t)], \tag{3.22c}$$

$$\boldsymbol{H}_M(x, z, t) = H_{yM}\boldsymbol{y}\exp[j(k_x x + k_{zM}z - \omega t)]. \tag{3.22d}$$

The continuity of the tangential magnetic field H_y at the interface requires

$$H_{yD} = H_{yM}, \tag{3.23}$$

and the continuity of the tangential electric field E_x gives us the relation

$$\frac{k_{zD}}{\varepsilon_D} = \frac{k_{zM}}{\varepsilon_M}, \tag{3.24}$$

where we have used Eq. (3.20b). Since H_y satisfies the wave equation (3.21) in both media, we also obtain the relations

$$k_x^2 + k_{zD}^2 = \varepsilon_D \mu_0 \omega^2, \tag{3.25a}$$

$$k_x^2 + k_{zM}^2 = \varepsilon_M \mu_0 \omega^2. \tag{3.25b}$$

The preceding three equations can be used to find the three unknowns, k_x, k_{zD}, and k_{zM}. First, we solve for the squares of these variables because it avoids any ambiguity in choosing the appropriate branch cut in the complex plane. The result is given by

$$k_x^2 = \mu_0 \omega^2 \frac{(\varepsilon_{Dr} + j\varepsilon_{Di})(\varepsilon_{Mr} + j\varepsilon_{Mi})}{(\varepsilon_{Dr} + \varepsilon_{Mr}) + j(\varepsilon_{Di} + \varepsilon_{Mi})}, \tag{3.26a}$$

$$k_{zD}^2 = \mu_0 \omega^2 \frac{(\varepsilon_{Dr} + j\varepsilon_{Di})^2}{(\varepsilon_{Dr} + \varepsilon_{Mr}) + j(\varepsilon_{Di} + \varepsilon_{Mi})}, \tag{3.26b}$$

$$k_{zM}^2 = \mu_0 \omega^2 \frac{(\varepsilon_{Mr} + j\varepsilon_{Mi})^2}{(\varepsilon_{Dr} + \varepsilon_{Mr}) + j(\varepsilon_{Di} + \varepsilon_{Mi})}, \tag{3.26c}$$

where we have made the substitutions $\varepsilon_D = \varepsilon_{Dr} + j\varepsilon_{Di}$ and $\varepsilon_M = \varepsilon_{Mr} + j\varepsilon_{Mi}$.

For a general complex number $a + jb$ with $b \neq 0$, the square root $\sqrt{a + jb}$ can be written as

$$\sqrt{a + jb} = \frac{1}{\sqrt{2}} \sqrt{a + \sqrt{a^2 + b^2}} + j \frac{\mathrm{sgn}(b)}{\sqrt{2}} \sqrt{-a + \sqrt{a^2 + b^2}}. \tag{3.27}$$

This equation shows clearly that the imaginary part of the square root of a complex number has a branch cut along the negative real axis. To have bound SPP modes on the interface, both $\mathrm{Im}(k_{zD})$ and $\mathrm{Im}(k_{zM})$ must be positive to ensure exponentially decaying fields in the z direction. From Eq. (3.27), this amounts to requiring $\mathrm{sgn}(b) > 0$. Applying this criterion to k_{zD} and k_{zM} gives the following two conditions:

$$(\varepsilon_{Di}^2 - \varepsilon_{Dr}^2)(\varepsilon_{Di} + \varepsilon_{Mi}) + (\varepsilon_{Dr} + \varepsilon_{Mr})\varepsilon_{Dr}\varepsilon_{Di} > 0, \tag{3.28a}$$

$$(\varepsilon_{Mi}^2 - \varepsilon_{Mr}^2)(\varepsilon_{Di} + \varepsilon_{Mi}) + (\varepsilon_{Dr} + \varepsilon_{Mr})\varepsilon_{Mr}\varepsilon_{Mi} > 0. \tag{3.28b}$$

Similarly, we can find the threshold gain at which lossless propagation of the SPP mode happens by setting $\mathrm{Im}(k_x) = 0$. The result is given by

$$(\varepsilon_{Di}\varepsilon_{Mi} - \varepsilon_{Dr}\varepsilon_{Mr})(\varepsilon_{Di} + \varepsilon_{Mi}) + (\varepsilon_{Di}\varepsilon_{Mr} + \varepsilon_{Dr}\varepsilon_{Mi})(\varepsilon_{Dr} + \varepsilon_{Mr}) = 0. \tag{3.29}$$

3.3 Gain-assisted management of group velocity

As discussed in Section 2.6, three different velocities are associated with an electromagnetic field, namely the phase velocity, the group velocity, and the energy

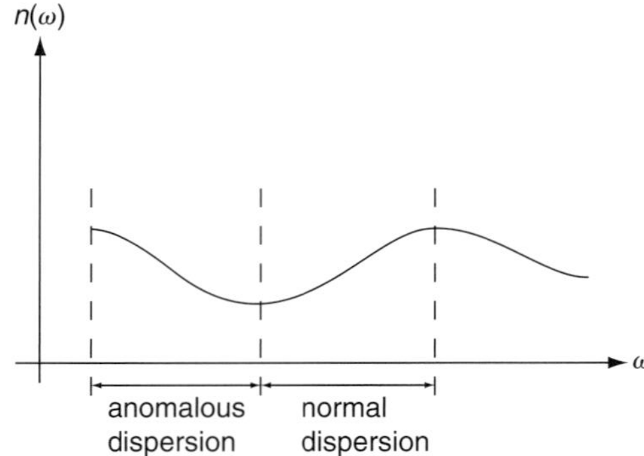

Figure 3.9 Frequency dependence of the refractive index. The normal- and anomalous-dispersion regions of an optical medium are related to the change in slope of this curve.

velocity. In this section we look at ways of controlling the group velocity of an optical pulse by clever use of gain at particular frequencies to create anomalous-dispersion regions. Before going into detail, it is instructive to consider how the anomalous nature of dispersion characteristics in a medium can drastically alter the group velocity of light in that medium.

3.3.1 Dispersion and group velocity

Figure 3.9 shows an example of the frequency dependence of the refractive index (dispersion) in a particular medium. Normal dispersion occurs when the refractive index $n(\omega)$ increases with increasing ω, i.e., when $dn/d\omega > 0$. Anomalous dispersion refers to the opposite situation, in which the refractive index decreases when the frequency increase locally, i.e., when $dn/d\omega < 0$. The group index $n_g(\omega)$ of a medium, introduced earlier in Section 2.6, can be obtained from $k(\omega) = n(\omega)\omega/c$ in the form

$$n_g(\omega) = n(\omega) + \omega\frac{d}{d\omega}n(\omega). \tag{3.30}$$

As a result, the group index exceeds $n(\omega)$ in the normal-dispersion region of a medium, but it becomes smaller than $n(\omega)$ in the anomalous-dispersion region. The group velocity is related to the group index by the relation $v_g(\omega) = c/n_g(\omega)$ given earlier in Section 2.6.

The frequency dependence of the group velocity becomes important for short optical pulses with a large bandwidth. If the group velocity of a medium remains nearly constant over the entire bandwidth of the pulse, the pulse retains its shape when propagating through such a medium [27]. However, if that is not the case, the pulse may broaden or shorten in duration within the medium, depending on whether it is chirped or unchirped and on whether the medium exhibits normal or anomalous dispersion [28].

If the material is anomalously dispersive within the frequency range of the pulse bandwidth, the term $\omega(dn/d\omega)$ can be negative and large, a situation that makes the group index in Eq. (3.30) also large and negative. Also, there may be frequencies where the group index in Eq. (3.30) is positive but close to zero. The former case leads to negative velocities and the latter case leads to group velocities greater than the speed of light in vacuum. Negative group velocities seem to contradict the causality constraints, whereas group velocities larger than the vacuum speed of light seem to contradict the foundations of the special theory of relativity. Therefore a closer look at these phenomena is essential for a proper understanding of the underlying concepts.

3.3.2 Gain-assisted superluminal propagation of light

Consider the propagation of light over a distance L in a medium having a group velocity v_g. The transit time of light is given by $T = L/v_g = Ln_g/c$. If the medium is free space, the transit time is given by L/c. Therefore the delay seen by the light while propagating through this medium, compared with the same distance in free space, is given by $\Delta T = (n_g - 1)L/c$. Interestingly, if $n_g < 1$, the delay becomes negative, and the pulse seems to travel faster than the speed of light in vacuum [27]. This is referred to as the superluminal propagation of light, and this phenomenon has attracted wide attention because of its controversial nature. By now, it is well understood that such a behavior does not represent a noncausal effect and occurs as a natural consequence of the wave nature of light [29].

When the optical frequency of the incident light is far from an atomic resonance of a passive medium, the medium exhibits weak absorption and appears nearly transparent. Such a medium also exhibits normal dispersion ($dn/d\omega > 0$), and it is hard to observe superluminal effects [30]. Specifically, if a medium is absorptive (i.e. $\text{Im}[\chi(\omega)] \geq 0$), the use of the Kramers–Kronig relations shows that [30]

$$n_g(\omega) > \max\left\{n(\omega), \frac{1}{n(\omega)}\right\}, \qquad \text{if} \quad \text{Im}[\chi(\omega)] \geq 0. \qquad (3.31)$$

Because transparent dielectrics have a refractive index $n(\omega) > 1$ at frequencies below the bandgap of the material, it follows that $n_g(\omega)$ will also exceed 1 [29]. Similarly, a metal becomes transparent at frequencies above its plasma frequency. The refractive index of such a metal satisfies the condition $n(\omega) < 1$ but the group index n_g exceeds 1 again [29]. Therefore, both the dielectric and the metallic media exhibit a group velocity smaller than the vacuum speed of light because their group index satisfies $n_g(\omega) > 1$.

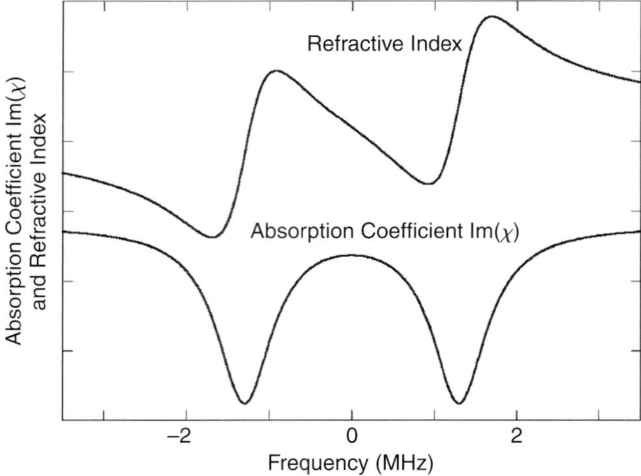

Figure 3.10 Gain and refractive index of the medium described in the text. (After Ref. [29]; © APS 2001)

In active media exhibiting optical gain, $\text{Im}[\chi(\omega)] < 0$, and the relation (3.31) does not hold. Interestingly, close to the gain peak, the medium can even become anomalously dispersive and may support superluminal propagation of light [29, 31]. Consider an active medium with two closely spaced gain peaks as shown in Figure 3.10. The susceptibility of such a medium has the form

$$\chi(\omega) = \frac{M}{(\omega - \omega_0 - \Delta\omega) + j\Gamma} + \frac{M}{(\omega - \omega_0 + \Delta\omega) + j\Gamma}, \qquad (3.32)$$

where M depends on the pumping level of the medium and $2\Delta\omega$ represents the spacing between the two gain peaks. Numerical values of these parameters corresponding to a real gain medium can be found in Ref. [29].

The refractive index of any medium is related to its susceptibility by

$$n(\omega) = \text{Re}[\sqrt{1 + \chi(\omega)}]. \qquad (3.33)$$

Using this expression, the group index is found to be

$$n_g(\omega) = \frac{1}{n(\omega)} \left(1 + \frac{2M\Delta\omega}{D^2(\omega)}[\omega(\omega - \omega_0 + j) + D^2(\omega)] \right), \qquad (3.34)$$

where $D(\omega) = [(\omega - \omega_0 + \Delta\omega) + j\Gamma][(\omega - \omega_0 - \Delta\omega) + j\Gamma]$. Figure 3.11 shows the delay or advance in the arrival time of a pulse as its carrier frequency is tuned across the two gain peaks for the specific set of parameters given in Ref. [29]. The negative values in the region between the two gain peaks indicate that the pulse arrives earlier because of an increase in its group velocity.

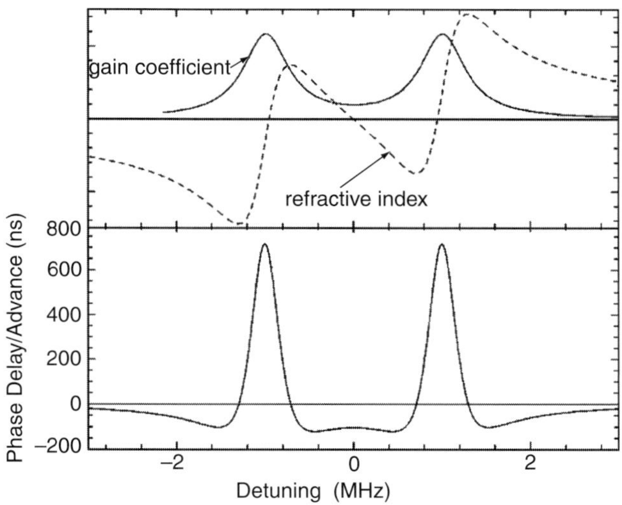

Figure 3.11 Pulse delay/advance as a function of the carrier frequency of the light pulse. (After Ref. [29]; © APS 2001)

3.4 Gain-assisted dispersion control

Even though active materials are primarily used to compensate for propagation losses or to boost the signal strength before detection, they can also be used for manipulating dispersion in nanowaveguides [32]. If an active medium is used for this purpose, the group velocity in a nanowaveguide can be controlled drastically by varying the distributed gain along the propagation path, because both the material properties and the waveguide geometry play important roles in setting the group velocity (as compared with larger replicas of such waveguides such as optical fibers). What is even more striking is that dispersive properties can be controlled using gain alone, even in nanowaveguides made of materials that are weakly dispersive (i.e., materials for which $\partial\varepsilon/\partial\omega \approx 0$).

3.4.1 Group velocity of a multilayer cylindrical nanowaveguide

Consider the multilayer cylindrical nanowaveguide shown in Figure 3.12, made with N layers of different refractive indices. For simplicity, we assume that the waveguide is homogeneous along the the propagation direction z with a propagation constant k_z in this direction. Application of Maxwell's equations to this composite structure results in a dispersion relation for the nanowaveguide that predicts how k_z varies with ω [32]. This dispersion relation is found by solving Maxwell's equations in each layer and relating field components at layer interfaces using the appropriate boundary conditions [33]. Since the contribution of each layer to the dispersion

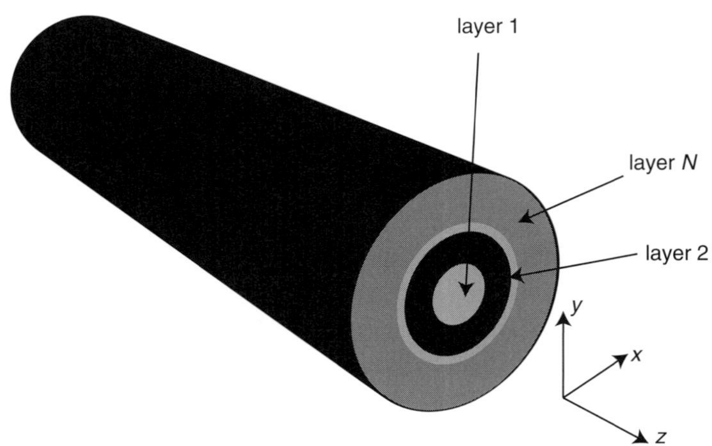

Figure 3.12 Schematic geometry of a multi-layered nanowaveguide.

relation $\mathcal{D}$ is governed by a structural parameter κ_s ($s = 1$ to N), we can write this relation in the symbolic form

$$\mathcal{D}(k_z, \omega, \kappa_1, \kappa_2, \ldots, \kappa_N) = 0. \tag{3.35}$$

In this equation the contribution of each layer appears through a single parameter κ_s. Such a one-to-one mapping may not hold if the waveguide is not cylindrically symmetric or is made of materials that are anisotropic [32].

In addition to this global relationship, each layer has its own dispersion relation $\mathcal{D}_s$, when considered in isolation, giving us N individual dispersion relations of the form

$$\mathcal{D}_s(k_z, \omega, \kappa_s) = 0, \qquad s = 1, 2, \ldots, N. \tag{3.36}$$

The solution of these $N + 1$ transcendental equations provides sufficient information to completely characterize any electromagnetic wave propagating through the structure depicted in Figure 3.12. In particular, the phase and group velocities can be obtained from the relations

$$v_p = \frac{\omega}{k_z}, \qquad v_g = \frac{\partial \omega}{\partial k_z}. \tag{3.37}$$

To derive an expression for the group velocity while maintaining the constraints imposed on ω and k_z by the dispersion relations, we calculate the total derivatives of Eqs. (3.35) and (3.36) with respect to k_z and obtain

$$\frac{d\mathscr{D}}{dk_z} = \frac{\partial\mathscr{D}}{\partial k_z} + \frac{\partial\mathscr{D}}{\partial\omega}v_g + \sum_{s=1}^{N}\frac{\partial\mathscr{D}}{\partial\kappa_s}\frac{\partial\kappa_s}{\partial k_z} = 0 \tag{3.38a}$$

$$\frac{d\mathscr{D}_s}{dk_z} = \frac{\partial\mathscr{D}_s}{\partial k_z} + \frac{\partial\mathscr{D}_s}{\partial\omega}v_g + \frac{\partial\mathscr{D}_s}{\partial\kappa_s}\frac{\partial\kappa_s}{\partial k_z} = 0. \tag{3.38b}$$

To find v_g, we first solve Eq. (3.38b) for $\partial\kappa_s/\partial k_z$ and obtain

$$\frac{\partial\kappa_s}{\partial k_z} = -\left(\frac{\partial\mathscr{D}_s}{\partial\kappa_s}\right)^{-1}\frac{\partial\mathscr{D}_s}{\partial k_z} - \left(\frac{\partial\mathscr{D}_s}{\partial\kappa_s}\right)^{-1}\frac{\partial\mathscr{D}_s}{\partial\omega}v_g. \tag{3.39}$$

Substitution of this result into Eq. (3.38a) then provides us with the following expression for the group velocity:

$$v_g = -\frac{\sum_{s=1}^{N}\frac{\partial\mathscr{D}}{\partial\kappa_s}\left(\frac{\partial\mathscr{D}_s}{\partial\kappa_s}\right)^{-1}\frac{\partial\mathscr{D}_s}{\partial k_z} - \frac{\partial\mathscr{D}}{\partial k_z}}{\sum_{s=1}^{N}\frac{\partial\mathscr{D}}{\partial\kappa_s}\left(\frac{\partial\mathscr{D}_s}{\partial\kappa_s}\right)^{-1}\frac{\partial\mathscr{D}_s}{\partial\omega} - \frac{\partial\mathscr{D}}{\partial\omega}}. \tag{3.40}$$

It is clear from this expression that the group velocity can be controlled pretty much independently of the phase velocity by controlling the composite and individual dispersion relationships. This amounts to controlling the geometry and material properties of the indicated nanowaveguide, giving a design flexibility unheard of for conventional waveguides.

3.4.2 Silver nanorod immersed in an active medium

To illustrate how optical gain be used to control the group velocity, we analyze the nanowaveguide structure shown in Figure 3.13, where a silver nanorod is immersed

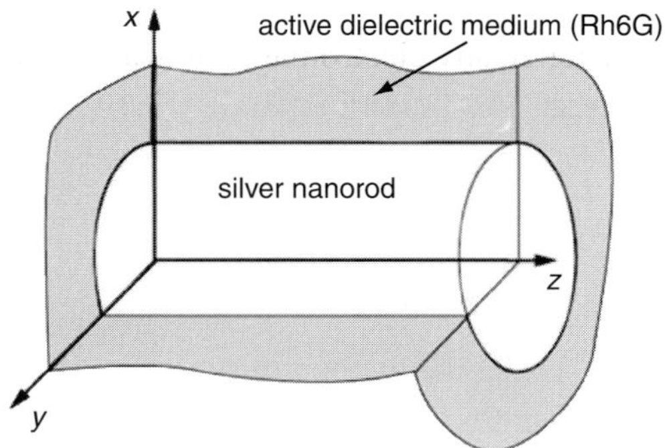

Figure 3.13 Silver nanorod in an active dielectric medium.

in an active dielectric medium (Rh6G dye). We assume that the silver's permittivity is adequately described by the Drude model:

$$\varepsilon_1(\omega) = \varepsilon_\infty^{\mathrm{Ag}}\varepsilon_0 - \frac{\omega_p^2\varepsilon_0}{\omega(\omega + j\gamma_p)}. \tag{3.41}$$

Typically, $\varepsilon_\infty^{\mathrm{Ag}} = 5$, $\omega_p = 13.88 \times 10^{15}$ s^{-1}, and $\gamma_p = 3.3 \times 10^{13}$ s^{-1} [32, 34].

The active medium surrounding the silver nanorod is assumed to be a 1% solution of rhodamine 6G in methanol (Rh6G). The following permittivity model can be used for this solution in the range of optical frequencies [32, 35]:

$$\varepsilon_2(\omega) = \varepsilon_\infty^{\mathrm{Rh6G}}\varepsilon_0 - \frac{G_1\varepsilon_0}{\omega^2 - \omega_1^2 + j\gamma_1\omega} - \frac{G_2\varepsilon_0}{\omega^2 - \omega_2^2 + j\gamma_2\omega}. \tag{3.42}$$

Typical parameter values are $\varepsilon_\infty^{\mathrm{Rh6G}} = 1.81$, $\gamma_1 = 1.2 \times 10^{14}$ s^{-1}, $\omega_1 = 3.85 \times 10^{15}$ s^{-1}, $\gamma_2 = 6 \times 10^{13}$ s^{-1}, and $\omega_2 = 3.52 \times 10^{15}$ s^{-1}. The gain of the dye solution can be adjusted by fixing G_1 and varying G_2, which amounts to increasing the excited-state population of Rh6G with an external optical pumping scheme. In the following analysis, we fix $G_1 = 0.001$ but vary G_2 such that the excited-state population relative to the ground state remains below 50% [32]. Because the Rh6G medium is not inverted under these conditions, the imaginary part of $\varepsilon_2(\omega)$ remains positive, making the medium effectively lossy. However, this loss is much less than that of the unpumped medium.

Unlike waveguides with dimensions larger than the wavelength of light, a nanowaveguide cannot support propagation of light as a bound optical mode because all bound modes are typically below the standard cut-off limit. Indeed, one can show by using the field-continuity requirements at the metal–dielectric interface that only SPP modes are supported. For a TM-polarized SPP mode propagating along the $+z$ direction, $H_z = 0$. Since all nonzero field components can be calculated in terms of E_z, we only need to find this field component.

Within the silver nanorod, the E_z component satisfies Maxwell's wave equation, which takes the following form when written in the frequency domain using cylindrical coordinates (r, ϕ, z):

$$\frac{\partial^2 E_z}{\partial r^2} + \frac{1}{r}\frac{\partial E_z}{\partial r} + \frac{1}{r^2}\frac{\partial^2 E_z}{\partial \phi^2} + \frac{\partial^2 E_z}{\partial z^2} + \varepsilon_1(\omega)\frac{\omega^2}{c^2}E_z = 0. \tag{3.43}$$

The same equation can be used in the active medium surrounding the nanorod, after replacing ε_1 with ε_2. Using the method of separation of variables, the general solution of the preceding equations takes the form

$$E_z(r, \phi, z) = F(r)e^{im\phi}\exp(jk_z z - j\omega t), \tag{3.44}$$

where the radial function $F(r)$ satisfies

$$\frac{d^2 F}{dr^2} + \frac{1}{r}\frac{dF}{dr} - \left(\kappa_s^2 + \frac{m^2}{r^2}\right) F = 0, \tag{3.45}$$

with

$$\kappa_s^2 = k_z^2 - \varepsilon_s \frac{\omega^2}{c^2}, \quad s = 1, 2. \tag{3.46}$$

The radial equation has two independent solutions of the form $I_m(r)$ and $K_m(r)$, where I_m and K_m are the modified Bessel functions of the first and second kinds, respectively. Therefore, the most general solution should be a linear combination of I_m and K_m. The choice $m = 0$ leads to the cylindrically symmetric solution that is of primary interest here. In this case, the radial part is given by

$$F(\rho) = \begin{cases} A I_0(\kappa_1 r) + A' K_0(\kappa_1 r), & r \le R, \\ B I_0(\kappa_2 r) + B' K_0(\kappa_2 r), & r > R. \end{cases} \tag{3.47}$$

If we use the boundary conditions that the field must be finite at $r = 0$ and should vanish at $r = \infty$ (the radiation condition), we find that we must set $A' = 0$ and $B = 0$ in Eq. (3.47). This can be understood by noting that, when $\zeta \to 0$, the functions I_0 and K_0 behave approximately as [36]

$$I_0(\zeta) \sim 1 + \frac{1}{4}\zeta^2, \qquad K_0(\zeta) \sim -\ln\left(\frac{\zeta}{2}\right) + \gamma, \tag{3.48}$$

where γ is Euler's constant. Similarly, when $\zeta \to \infty$, I_0 and K_0 behave approximately as [36]

$$I_0(\zeta) \sim \frac{\exp(\zeta)}{\sqrt{2\pi\zeta}}, \qquad K_0(\zeta) \sim \sqrt{\frac{\pi}{2\zeta}}\exp(-\zeta). \tag{3.49}$$

Since K_0 blows up at the origin and I_0 blows up at ∞, they cannot be part of field solutions and the corresponding terms should be removed from Eq. (3.47).

The requirement that the E_z component be continuous everywhere gives us the following expression for the E_z component of the SPP mode supported by the nanorod (for $m = 0$):

$$E_z(r, z, t) = \begin{cases} E_{z0}[I_0(\kappa_1 r)/I_0(\kappa_1 R)]\exp(jk_z z - j\omega t), & r \le R, \\ E_{z0}[K_0(\kappa_2 r)/K_0(\kappa_2 R)]\exp(jk_z z - j\omega t), & r > R. \end{cases} \tag{3.50}$$

The continuity of the ϕ component of the magnetic field provides us with the dispersion relation for the waveguide. To calculate it, we note that

$$H_{\phi,s} = j\frac{\varepsilon_s}{\mu_0 \kappa_s^2 c}\frac{\partial E_z}{\partial r}, \quad s = 1, 2. \tag{3.51}$$

To calculate $H_{\varphi,s}$, we need to know the derivatives of I_0 and K_0. These can be obtained from the well-known relations [37]

$$\frac{dI_0(\zeta)}{d\zeta} = I_1(\zeta), \qquad \frac{dK_0(\zeta)}{d\zeta} = -K_1(\zeta), \tag{3.52}$$

where I_1 and K_1 are the modified Bessel functions of order 1. Matching $H_{\varphi,s}$ at the interface ($r = R$), we obtain the following dispersion relation for the silver nanorod:

$$\mathscr{D} \equiv \frac{\varepsilon_1}{\kappa_1}I_1(\kappa_1 R) + \frac{\varepsilon_2}{\kappa_2}K_1(\kappa_2 R) = 0. \tag{3.53}$$

The individual dispersion relations can be obtained from Eq. (3.47) and are given by

$$\mathscr{D}_1 = \varepsilon_1(\omega)\frac{\omega^2}{c^2} - k_z^2 + \kappa_1^2 = 0, \tag{3.54a}$$

$$\mathscr{D}_2 = \varepsilon_2(\omega)\frac{\omega^2}{c^2} - k_z^2 + \kappa_2^2 = 0. \tag{3.54b}$$

We can now obtain the group velocity in terms of the derivatives indicated in Eq. (3.40). The result is given by

$$v_g = \frac{k_z\left(\dfrac{1}{\kappa_1}\dfrac{\partial\mathscr{D}}{\partial\kappa_1} + \dfrac{1}{\kappa_2}\dfrac{\partial\mathscr{D}}{\partial\kappa_2}\right)}{\dfrac{\omega}{c^2}\displaystyle\sum_{s=1}^{2}\dfrac{1}{\kappa_s}\dfrac{\partial\mathscr{D}}{\partial\kappa_s}\left(\varepsilon_s + \dfrac{\omega}{2}\dfrac{d\varepsilon_s}{d\omega}\right) - \dfrac{\partial\mathscr{D}}{\partial\omega}}. \tag{3.55}$$

In the absence of material dispersion, the derivatives of $\mathscr{D}$, $\mathscr{D}_1$, and $\mathscr{D}_2$ with respect to ω vanish and Eq. (3.55) leads to the simple relation [32]

$$v_g v_p = \text{constant}. \tag{3.56}$$

In the presence of material dispersion, the derivative $\partial\mathscr{D}/\partial\omega$ in Eq. (3.55) can be controlled by adjusting the gain in the Rh6G medium. As an example, Figure 3.14 shows how the phase and group velocities vary with the gain (left column) over a wavelength range extending from 480 to 560 nm for a 35-nm-radius nanorod. Their dependence on the nanorod radius is shown in the right column at a wavelength of 534 nm. One can see that both superluminal and subluminal speeds can be achieved for a silver nanorod under varying gain conditions.

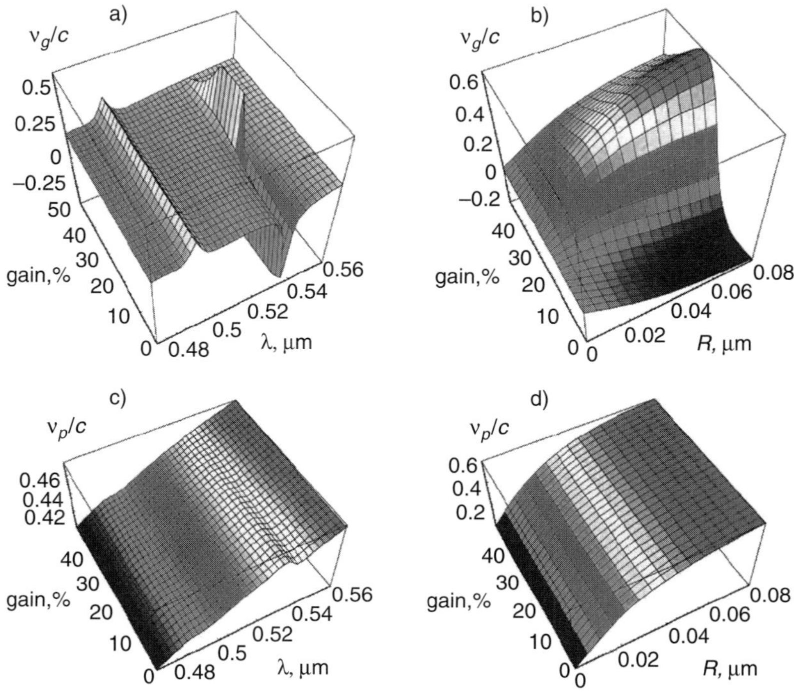

Figure 3.14 Group and phase velocities of the TM SPP mode supported by a metal nanorod in Rh6G methanol solution as a function of gain, wavelength, and the radius of the nanorod. The radial dependence is given at $\lambda = 534$ nm, and the the wavelength dependence is given for $R = 35$ nm. (After Ref. [32]; © APS 2006)

References

[1] X. Jiang, Q. Li, and C. M. Soukoulis, "Symmetry between absorption and amplification in disordered media," *Phys. Rev. B*, vol. 59, no. 14, pp. R9007–R9010, 1999.

[2] S. A. Ramakrishna, "On the dual symmetry between absorbing and amplifying random media," *Pramana - J. Phys.*, vol. 62, pp. 1273–1279, 2004.

[3] M. P. Nezhad, K. Tetz, and Y. Fainman, "Gain assisted propagation of surface plasmon polaritons on planar metallic waveguides," *Opt. Express*, vol. 12, pp. 4072–4079, 2004.

[4] M. A. Noginov, V. A. Podolskiy, G. Zhu, *et al.* "Compensation of loss in propagating surface plasmon polariton by gain in adjacent dielectric medium," *Opt. Express*, vol. 16, pp. 1385–1392, 2008.

[5] C. H. Henry and R. F. Kazarinov, "Quantum noise in photonics," *Rev. Mod. Phys.*, vol. 68, pp. 801–853, 1996.

[6] C. H. Henry, "Theory of spontaneous emission noise in open resonators and its application to lasers and optical amplifiers," *J. Lightw. Technol.*, vol. LT-4, pp. 288–297, 1986.

[7] T. Makino and J. Glinski, "Transfer matrix analysis of the amplified spontaneous emission of DFB semiconductor laser amplifiers," *IEEE J. Quantum Electron.*, vol. 24, pp. 1507–1518, 1988.

[8] T. Tamir, "Nonspecular phenomena in beam fields reflected by multilayered media," *J. Opt. Soc. Am. A*, vol. 3, pp. 558–565, 1986.

[9] M. A. Porras, "Nonspecular reflection of general light beams at a dielectric interface," *Opt. Commun.*, vol. 135, pp. 369–377, 1997.

[10] C. J. Koester, "Laser action by enhanced total internal reflection," *IEEE J. Quantum Electron.*, vol. 2, pp. 580–584, 1966.

[11] C. W. Lee, K. Kim, J. Noh, and W. Jhe, "Quantum theory of amplified total internal reflection due to evanescent mode coupling," *Phys. Rev. A*, vol. 62, p. 053805 (7 pages), 2000.

[12] J. Fan, A. Dogariu, and L. J. Wang, "Amplified total internal reflection," *Opt. Express*, vol. 11, pp. 299–308, 2003.

[13] K. J. Willis, J. B. Schneider, and S. C. Hagness, "Amplified total internal reflection: Theory, analysis, and demonstration of existence via FDTD," *Opt. Express*, vol. 16, pp. 1903–1914, 2008.

[14] W. L. Barnes, A. Dereux, and T. W. Ebbesen, "Surface plasmon subwavelength optics," *Nature*, vol. 424, pp. 824–830, 2003.

[15] R. H. Ritchie, "Plasma losses by fast electrons in thin films," *Phys. Rev.*, vol. 106, pp. 874–881, 1957.

[16] S. A. Maier, *Plasmonics Fundamentals and Applications*. Springer, 2007.

[17] Y. Teng and E. A. Stern, "Plasma radiation from metal grating surfaces," *Phys. Rev. Lett.*, vol. 19, pp. 511–514, 1967.

[18] H. Raether, *Surface Plasmons on Smooth and Rough Surfaces and on Gratings*. Springer, 1988.

[19] J. B. Pendry, L. Martin-Moreno, and F. J. Garcia-Vidal, "Mimicking surface plasmons with structured surfaces," *Science*, vol. 305, pp. 847–848, 2004.

[20] W. Barnes and R. Sambles, "Only skin deep," *Science*, vol. 305, pp. 785–786, 2004.

[21] P. Berini and J. Lu, "Curved long-range surface plasmon–polariton waveguides," *Opt. Express*, vol. 14, pp. 2365–2371, 2006.

[22] A. N. Sudarkin and P. A. Demkovich, "Excitation of surface electromagnetic waves on the boundary of a metal with an amplifying medium," *Sov. Phys. Tech. Phys.*, vol. 34, pp. 764–766, 1989.

[23] D. J. Bergman and M. I. Stockman, "Surface plasmon amplification by stimulated emission of radiation: Quantum generation of coherent surface plasmons in nanosystems," *Phys. Rev. Lett.*, vol. 90, p. 027402 (4 pages), 2003.

[24] M. I. Stockman, "Spasers explained," *Nat. Photonics*, vol. 2, pp. 327–329, 2008.

[25] D. L. Dexter, "A theory of sensitized luminescence in solids," *J. Chem. Phys.*, vol. 21, pp. 836–850, 1953.

[26] I. Avrutsky, "Surface plasmons at nanoscale relief gratings between a metal and a dielectric medium with optical gain," *Phys. Rev. B*, vol. 70, p. 155416 (6 pages), 2004.

[27] L. J. Wang, A. Kuzmich, and A. Dogariu, "Gain-assisted superluminal light propagation," *Nature*, vol. 406, pp. 277–279, 2000.

[28] G. P. Agrawal, *Nonlinear Fiber Optics*, 4th ed. Academic Press, 2007.

[29] A. Dogariu, A. Kuzmich, and L. J. Wang, "Transparent anomalous dispersion and superliminal light-pulse propagation at a negative group velocity," *Phys. Rev. A*, vol. 63, p. 053806 (12 pages), 2001.

[30] L. D. Landau and E. M. Lifshitz, *Electrodynamics of Continuous Media*, 2nd ed. *Course of Theoretical Physics*, vol. 8. Pergamon Press, 1984.

[31] E. Bolda, J. C. Garrison, and R. Y. Chiao, "Optical pulse propagation at negative group velocities due to a nearby gain line," *Phys. Rev. A*, vol. 49, pp. 2938–2947, 1994.

[32] A. A. Govyadinov and V. A. Podolskiy, "Gain-assisted slow to superluminal group velocity manipulation in nanowaveguides," *Phys. Rev. Lett.*, vol. 97, p. 223902 (4 pages), 2006.

[33] ——, "Sub-diffraction light propagation in fibers with anisotropic dielectric cores," *J. Mod. Opt.*, vol. 53, pp. 2315–2324, 2006.

[34] E. Palik, ed., *The Handbook of Optical Constants of Solids*. Academic Press, 1997.

[35] W. Leupacher and A. Penzkofer, "Refractive-index measurement of absorbing condensed media," *Appl. Opt.*, vol. 23, pp. 1554–1558, 1984.

[36] G. N. Watson, *A Treatise on the Theory of Bessel Functions*, 2nd ed. Cambridge University Press, 1996.

[37] M. Abramowitz and I. A. Stegun, eds., *Handbook of Mathematical Functions: With Formulas, Graphs, and Mathematical Tables*. Dover Publications, 1965.

4

Optical Bloch equations

The analysis in Chapter 3 used a phenomenological form of the permittivity to describe active materials. A proper understanding of optical amplification requires a quantum-mechanical approach for describing the interaction of light with atoms of an active medium [1]. However, even a relatively simple atom such as hydrogen or helium allows so many energy transitions that its full description is intractable even with modern computing machinery [2, 3]. The only solution is to look for idealized models that contain the most essential features of a realistic system. The semiclassical two-level-atom model has proven to be quite successful in this respect [4]. Even though a real atom has infinitely many energy levels, two energy levels whose energy difference nearly matches the photon energy suffice to understand the interaction dynamics when the atom interacts with nearly monochromatic radiation. Moreover, if the optical field contains a sufficiently large number of photons (>100), it can be treated classically using a set of optical Bloch equations. In this chapter, we learn the underlying physical concepts behind the optical Bloch equations. We apply these equations in subsequent chapters to actual optical amplifiers and show that they can be solved analytically under certain conditions to provide a realistic description of optical amplifiers.

It is essential to have a thorough understanding of the concept of a quantum state [5]. To effectively use the modern machinery of quantum mechanics, physical states need to be represented as vectors in so-called Hilbert space [6]. However, a vector in Hilbert space can be represented in different basis states that amount to having different coordinate systems for standard vectors (such as velocity and momentum). To emphasize the independence of a vector from its representation in a chosen basis, Dirac introduced the bra–ket notation in quantum mechanics [7, 8]. This notation is not only a nice analytical tool but also has the power to aid the thought process [9]. For this reason, we provide in Section 4.1 an overview of the bra–ket notation. However, we should emphasize that our presentation ignores subtle differences between concepts such as a

Hilbert space and a rigged Hilbert space (see Ref. [8]). We introduce the concept of a density operator in Section 4.2 and use it in Section 4.3 to derive the optical Bloch equations. The rotating-wave approximation is also discussed in Section 4.3. We introduce in Section 4.4 the damping terms in the optical Bloch equations that are essential for describing active media. Section 4.5 provides the full set of Maxwell–Bloch equations, and their numerical integration is discussed in Section 4.6.

4.1 The bra and ket vectors

A complete set of *ket* vectors constitutes a linear vector space $\mathbb{V}$. The description of such a vector space involves the following properties [5]:

- If we denote different ket vectors as $|v_k\rangle$, $(k = 1, 2, \ldots)$, they obey the following properties:
  Associativity property: $|v_1\rangle + (|v_2\rangle + |v_3\rangle) = (|v_1\rangle + |v_2\rangle) + |v_3\rangle$.
  Commutativity property: $|v_1\rangle + |v_2\rangle = |v_2\rangle + |v_1\rangle$.
  Zero element: There exists a zero vector such that $|v\rangle + |0\rangle = |v\rangle$.
  Inverse element: for each vector $|v\rangle \in \mathbb{V}$, there exists an inverse element $-|v\rangle$ such that $|v\rangle + (-|v\rangle) = |0\rangle$.
- Addition and multiplication operations with respect to a scalar field consisting of complex numbers $(c \to \mathbb{C})$ satisfy the following relations:
  Distributivity property for vector addition: $c(|v_1\rangle + |v_2\rangle) = c\,|v_1\rangle + c\,|v_2\rangle$.
  Distributivity property for the scalar field: $(c_1 + c_2)\,|v\rangle = c_1\,|v\rangle + c_2\,|v\rangle$.
  Associative property for the scalar field: $c_1(c_2\,|v\rangle) = (c_1 c_2)\,|v\rangle$.
  Scalar unit element: the unit element $1 \in \mathbb{C}$ obeys $1\,|v\rangle = |v\rangle$.

In quantum mechanics, ket vectors $|v\rangle$ and $c\,|v\rangle$, where c is an arbitrary complex number, represent the same physical state. A normalization condition is used in practice to provide a unique representation of any physical state.

It is possible to define a mapping between two arbitrary elements in the ket-vector space $\mathbb{V}$ and the complex-number space $\mathbb{C}$. This mapping, called the *scalar product* and denoted by $\langle \cdot, \cdot \rangle$, obeys the following relations [5]:

Vector addition in the first ket: $\langle |v_1\rangle + |v_2\rangle, |v_3\rangle \rangle = \langle |v_1\rangle, |v_3\rangle \rangle + \langle |v_2\rangle, |v_3\rangle \rangle$.
Vector addition in the second ket: $\langle |v_1\rangle, |v_2\rangle + |v_3\rangle \rangle = \langle |v_1\rangle, |v_2\rangle \rangle + \langle |v_1\rangle, |v_3\rangle \rangle$.
Scalar multiplication of the first ket: $\langle c\,|v_1\rangle, |v_2\rangle \rangle = c^*\,\langle |v_1\rangle, |v_2\rangle \rangle$.
Scalar multiplication of the second ket: $\langle |v_1\rangle, c\,|v_2\rangle \rangle = c\,\langle |v_1\rangle, |v_2\rangle \rangle$.
Conjugate symmetry: $\langle |v_1\rangle, |v_2\rangle \rangle = \langle |v_2\rangle, |v_1\rangle \rangle^*$.
Positive definiteness: $\langle |v\rangle, |v\rangle \rangle > 0$, $\quad$ if $|v\rangle \neq |0\rangle$.

Owing to the last property, one could naturally introduce another measure called the "norm" of the ket-space and defined as

$$\| \, |v\rangle \, \| \equiv \sqrt{\langle |v\rangle , |v\rangle \rangle}, \tag{4.1}$$

where the positive sign of the square root is always chosen. Because of this definition, the norm of a ket vector is never negative and provides a way to quantify the magnitude of that ket. In the case of a three-dimensional vector in Euclidian space, its norm is equal to the length of that vector. Thus, the norm carries many of the familiar properties of lengths in Euclidian space to the abstract linear spaces found in quantum mechanics [5]. Two examples are provided by the triangle inequality, $\| \, |v_1\rangle + |v_2\rangle \, \| \leq \| \, |v_1\rangle \, \| + \| \, |v_2\rangle \, \|$, and the Cauchy–Schwarz inequality, $| \langle |v_1\rangle , |v_2\rangle \rangle | \leq \| \, |v_1\rangle \, \| \cdot \| \, |v_2\rangle \, \|$.

We are now in a position to introduce *bra* vectors in the associated dual space of a linear vector space using the concept of dual correspondence [5, 10]. Dual correspondence refers to a mapping (i.e., an operator, which we denote by $\dagger$) that takes a ket $|v\rangle$ to its corresponding bra $\langle v|$. The vector space in which bra $\langle v|$ vectors live is called the dual space and is denoted as $\mathbb{V}^\dagger$ to emphasize the one-to-one relationship it has with the ket-space $\mathbb{V}$. Mathematically, this mapping is denoted as

$$\dagger : \qquad \mathbb{V} \to \mathbb{V}^\dagger. \tag{4.2}$$

The action of bra, $\langle v| \in \mathbb{V}^\dagger$, on a ket, $|u\rangle \in \mathbb{V}$, is defined as

$$\langle v| \, |u\rangle \equiv \langle v| (|u\rangle) \equiv \langle v|u\rangle = \langle |v\rangle , |u\rangle \rangle . \tag{4.3}$$

The operation $\dagger$ which maps a vector space to its dual space is known as *Hermitian conjugation* [5]. It can be used to compactly write the following equality between a bra and its corresponding ket:

$$\langle v| = (|v\rangle)^\dagger . \tag{4.4}$$

For all practical purposes, this is a unique relationship between a vector space and its dual space.[1] Because of this unique relationship, the inverse mapping from the dual space should carry the bra to its original ket:

$$|v\rangle = (\langle v|)^\dagger = (|v\rangle)^{\dagger\dagger} . \tag{4.5}$$

Combining these properties with the conjugate-symmetry property of the scalar product gives us

$$(c \, |v\rangle)^\dagger = c^* \langle v| \tag{4.6}$$

---

[1] It is one-to-one and onto for all finite-dimensional and infinite-dimensional Hilbert spaces.

for an arbitrary complex number c.

We now consider representation of the bra and ket vectors in a fixed basis. Suppose the vector space $\mathbb{V}$ has an orthonormal basis given by $\{\,|e_i\rangle\,\}$, where the index i may be finite, countably infinite, or a continuous set. It could even be a combination of discrete and continuous elements. Being an orthonormal basis, its elements satisfy

$$\|\,|e_i\rangle\,\| = \sqrt{\langle e_i|e_i\rangle} = 1, \tag{4.7a}$$

$$\langle\,|e_i\rangle\,,\,|e_j\rangle\,\rangle = \langle e_i|e_j\rangle = 0, \quad \text{if } i \neq j. \tag{4.7b}$$

Using this basis, any ket vector $|\psi\rangle$ can be written as

$$|\psi\rangle = \sum_i \langle\,|e_i\rangle\,,\,|\psi\rangle\,\rangle\,|e_i\rangle = \sum_i \langle e_i|\psi\rangle\,|e_i\rangle. \tag{4.8}$$

It is convenient and instructive to introduce the unit operator as

$$\mathbb{I} = \sum_i |e_i\rangle\langle e_i|. \tag{4.9}$$

This operator sends any ket to itself in the vector space in which $\{\,|e_i\rangle\,\}$ forms an orthonormal basis. Therefore, $|\psi\rangle$ can be written as

$$|\psi\rangle = \mathbb{I}\,|\psi\rangle = \left(\sum_i |e_i\rangle\langle e_i|\right)|\psi\rangle = \sum_i \langle e_i|\psi\rangle\,|e_i\rangle. \tag{4.10}$$

It is clear from the definition of the unit operator in Eq. (4.9) that the term $|e_i\rangle\langle e_i|$ is responsible for projecting ψ along the base ket $|e_i\rangle$. This idea can be generalized by introducing a projection operator $\mathbb{P}$ along the direction of $|e\rangle$ (assuming that $|e\rangle$ is a normalized ket, i.e., $\|\,|e\rangle\,\| = 1$):

$$\mathbb{P} = |e\rangle\langle e|. \tag{4.11}$$

If this operator is applied to an arbitrary ket $|\psi\rangle$, we get

$$\mathbb{P}\,|\psi\rangle = |e\rangle\langle e|\,|\psi\rangle = \langle e|\psi\rangle\,|e\rangle, \tag{4.12}$$

which is clearly the projection of $|\psi\rangle$ along the direction $|e\rangle$.

4.2 Density operator

In classical mechanics the state of a particle at any time is completely described by its position and momentum at that time [11], and Newton's equations of motion specify how this state evolves with time. In quantum mechanics, a ket vector $|\psi\rangle$ in the relevant Hilbert space specifies the particle state completely, and the temporal evolution of this state is governed by the Schrödinger equation [5],

$$j\hbar\frac{\partial\,|\psi\rangle}{\partial t} = \mathbb{H}\,|\psi\rangle\,, \tag{4.13}$$

where $\mathbb{H}$ is the Hamiltonian operator and $\hbar = h/(2\pi)$ is related to the Planck constant h.

How is the solution of Eq. (4.13) related to measurements of a particle's position or momentum? In quantum mechanics any measurable quantity is represented by a Hermitian operator [6]. This is a fundamental difference from classical mechanics. As an example, the momentum in quantum mechanics is represented by the operator $\frac{\hbar}{j}\frac{\partial}{\partial x}$, where x is the position coordinate. In any measurement of the operator $\mathbb{O}$, the only values that can be observed are the eigenvalues o_λ obtained by solving the eigenvalue equation

$$\mathbb{O}\,|\psi_\lambda\rangle = o_\lambda\,|\psi_\lambda\rangle\,, \tag{4.14}$$

where $|\psi_\lambda\rangle$ is called an eigenstate of the operator $\mathbb{O}$. Owing to the Hermitian nature of this operator, all of its eigenvalues are real. This ensures that measurements on a quantum system always yield real results, in agreement with our everyday experience. Of course, measurements on identical systems in the same quantum state $|\psi\rangle$ may yield different results for a given operator. The mean value of the operator $\mathbb{O}$ is then given by the expectation value $\langle\psi|\mathbb{O}|\psi\rangle$.

We assume that all eigenstates of the operator $\mathbb{O}$ are normalized such that $\|\,|\psi_\lambda\rangle\,\| = 1$. Suppose we do not have definitive information about any of the observable states but are aware that there is a probability P_λ of finding the quantum system in the state $|\psi_\lambda\rangle$. Since the system must be in one of these states at any given instant, the relation $\sum_\lambda P_\lambda = 1$ must hold. We note that the index λ may take finite, countably infinite, or a continuous range of values without affecting the following discussion.

Similarly to the statistical description of a classical system, adopted when one does not know the initial state of the system with certainty, it is common to adopt the concept of the density operator ρ for a quantum system using the definition

$$\rho = \sum_\lambda P_\lambda\,|\psi_\lambda\rangle\,\langle\psi_\lambda|\,. \tag{4.15}$$

This operator can be used to calculate the expectation value of the operator $\mathbb{O}$ using the relation $\langle\mathbb{O}\rangle = \mathrm{Tr}(\rho\mathbb{O})$, where Tr represents the trace operation that corresponds the sum of the eigenvalues of the operator in the parenthesis.

To prove the preceding relation, we first note that $\langle \mathbb{O} \rangle = \sum_{\lambda \in \Lambda} P_\lambda o_\lambda$. Next, noting that $o_\lambda = \langle \psi_\lambda | \mathbb{O} | \psi_\lambda \rangle$, we obtain

$$
\begin{aligned}
\langle \mathbb{O} \rangle &= \sum_{\lambda \in \Lambda} P_\lambda o_\lambda = \sum_\lambda P_\lambda \langle \psi_\lambda | \mathbb{O} | \psi_\lambda \rangle \\
&= \sum_\lambda P_\lambda \langle \psi_\lambda | \left(\sum_\mu | \psi_\mu \rangle \langle \psi_\mu | \right) \mathbb{O} | \psi_\lambda \rangle ,
\end{aligned}
\tag{4.16}
$$

where the bracketed term is an identity operator (see Eq. (4.9)). Using the orthonormal property of the eigenvectors, we obtain

$$
\langle \mathbb{O} \rangle = \sum_\mu \langle \psi_\mu | \left(\sum_\lambda P_\lambda | \psi_\lambda \rangle \langle \psi_\lambda | \right) \mathbb{O} | \psi_\mu \rangle .
\tag{4.17}
$$

Using the definition of the density operator in Eq. (4.15), we finally obtain

$$
\langle \mathbb{O} \rangle = \sum_\mu \langle \psi_\mu | \rho \mathbb{O} | \psi_\mu \rangle \equiv \mathrm{Tr}(\rho \mathbb{O}).
\tag{4.18}
$$

The time evolution of the density operator is governed by the Schrödinger equation (4.13). Taking the partial derivative of ρ with respect to time, we obtain

$$
\begin{aligned}
\frac{\partial \rho}{\partial t} &= \sum_\lambda P_\lambda \left(\frac{\partial | \psi_\lambda \rangle}{\partial t} \langle \psi_\lambda | + | \psi_\lambda \rangle \frac{\partial \langle \psi_\lambda |}{\partial t} \right) \\
&= \frac{1}{j\hbar} \sum_\lambda P_\lambda \left(\mathbb{H} | \psi_\lambda \rangle \langle \psi_\lambda | - | \psi_\lambda \rangle \langle \psi_\lambda | \mathbb{H} \right).
\end{aligned}
\tag{4.19}
$$

Defining the commutator of two operators $\mathbb{A}$ and $\mathbb{B}$ as

$$
[\mathbb{A}, \mathbb{B}] \equiv \mathbb{A}\mathbb{B} - \mathbb{B}\mathbb{A},
\tag{4.20}
$$

we finally obtain the evolution equation for the density operator in the form

$$
\frac{\partial \rho}{\partial t} = \frac{1}{j\hbar} [\mathbb{H}, \rho].
\tag{4.21}
$$

4.3 Density-matrix equations for two-level atoms

As mentioned earlier, even though an atom may have many energy levels, as shown schematically in Figure 4.1, if the optical frequency of incident light is close to an atomic resonance, one can approximate the atom with an effective two-level system. This is possible because interaction between atoms and electromagnetic radiation

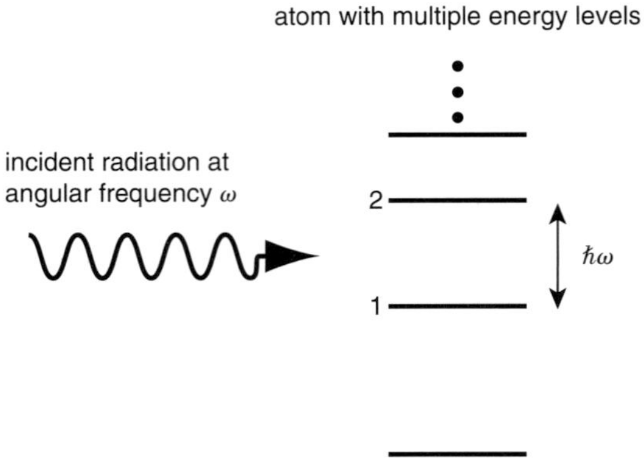

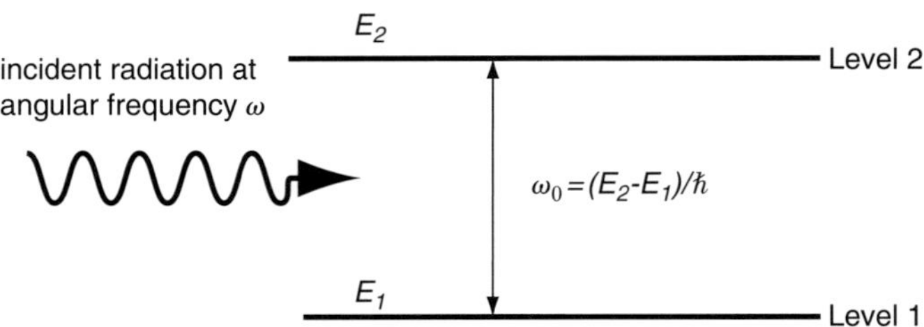

Figure 4.1 Schematic of the interaction of a multilevel atom with radiation whose photon energy $\hbar\omega$ nearly matches the energy difference between two specific atomic levels.

Figure 4.2 Schematic of an effective two-level system when $\omega \approx \omega_0$.

occurs through dipole moments. If the energy $\hbar\omega$ of the incoming radiation closely matches the energy difference $E_2 - E_1$ between two specific levels, the strength of the corresponding dipole moment dominates over all other dipoles involving nonresonant energy levels of the same atom.

4.3.1 Atomic states with even or odd parity

Figure 4.2 shows a two-level atom interacting with an optical field E whose frequency ω is close to the atomic transition frequency $\omega_0 = (E_2 - E_1)/\hbar$. We assume that the two eigenstates of the atom, $|1\rangle$ and $|2\rangle$, possess opposite parity. To understand the exact meaning of this statement, we need to consider the parity operator Π. Consider a wave function, $\psi(x) \equiv \langle x|\psi\rangle$, associated with the ket vector $|\psi\rangle$. Applying the parity operator to this wave function, we get

$$\Pi\psi(x) = \langle x|\Pi|\psi\rangle = \langle -x|\psi\rangle = \psi(-x). \tag{4.22}$$

It is evident that two consecutive parity operations on a wave function produce no change at all, i.e., $\Pi^2 \psi(x) = \psi(x)$. Therefore, the eigenvalues of the parity operator are ± 1. Even parity corresponds to a $+1$ eigenvalue ($\Pi |\psi\rangle = |\psi\rangle$) and odd parity corresponds to a -1 eigenvalue ($\Pi |\psi\rangle = -|\psi\rangle$). The opposite parity of states $|1\rangle$ and $|2\rangle$ implies that if level 1 has even parity, level 2 then has odd parity (or vice versa).

One consequence of the opposing parities is that diagonal elements of the position operator $\mathbb{X}$ are identically zero if the Hamiltonian is invariant under space inversion (i.e., $\Pi \mathbb{H} \Pi^\dagger = \mathbb{H}$). To show this, consider the state $|1\rangle$ satisfying the relation $\mathbb{H} |1\rangle = E_1 |1\rangle$. It follows that

$$\Pi \mathbb{H} |1\rangle = E_1 \Pi |1\rangle . \tag{4.23}$$

Noting that $\Pi \Pi^\dagger = \mathbb{I}$, we obtain

$$\Pi \mathbb{H} \mathbb{I} |1\rangle = \Pi \mathbb{H} \Pi^\dagger \Pi |1\rangle = \mathbb{H} \left(\Pi |1\rangle \right) = E_1 \left(\Pi |1\rangle \right) , \tag{4.24}$$

i.e., $\Pi |1\rangle$ is another eigenstate of this two-level atom with the same energy. However, since our atom has only two distinct energy levels, the state $\Pi |1\rangle$ must differ from $|1\rangle$ by a constant that we denote by ς. Multiplying by the parity operator twice, it is easy to show that $|\varsigma^2| = 1$. Armed with this knowledge, we calculate the diagonal matrix element of $\mathbb{X}$ in the basis state $|1\rangle$:

$$\langle 1 |\mathbb{X}| 1 \rangle = \left\langle 1 \left| \Pi^\dagger \Pi \mathbb{X} \Pi^\dagger \Pi \right| 1 \right\rangle = -\left\langle 1 \left| \Pi^\dagger \mathbb{X} \Pi \right| 1 \right\rangle = -|\varsigma^2| \langle 1 |\mathbb{X}| 1 \rangle , \tag{4.25}$$

where we have used the definition of the parity operator, $\Pi \mathbb{X} = -\mathbb{X} \Pi$, together with $\Pi \Pi^\dagger = 1$. Since $|\varsigma^2| = 1$, this relation implies that $\langle 1 |\mathbb{X}| 1 \rangle$ must be zero. Similarly, we can show that $\langle 2 |\mathbb{X}| 2 \rangle = 0$.

4.3.2 Interaction of a two-level atom with an electromagnetic field

We now focus on the interaction of a two-level atom with an electromagnetic field taking place through the dipole moment of the atom [4]. This notion of dipole moment comes from a classical analysis according to which an electric dipole with dipole moment $\boldsymbol{d}$ acquires an interaction energy $-\boldsymbol{d} \cdot \boldsymbol{E}$ in the presence of an electric field $\boldsymbol{E}$, where the minus sign has its origin in the negative charge of electrons. In quantum mechanics, $\boldsymbol{d}$ is replaced with the corresponding operator $\mathbf{d}$ and this interaction energy is added to the atomic Hamiltonian, $\mathbb{H}$, resulting in the total Hamiltonian

$$\mathbb{H}_T = \mathbb{H} - \mathbf{d} \cdot \boldsymbol{E}, \tag{4.26}$$

where $\mathbb{H}$ is the Hamiltonian associated with the isolated two-level atom. Because the dipole operator $\mathbf{d}$ has the same parity as the position operator $\mathbb{X}$, its diagonal matrix elements vanish. As a result, the dipole operator can be expanded in the two-level basis as

$$\mathsf{d} = \sum_{m=1}^{2} \sum_{n=1}^{2} \boldsymbol{d}_{mn} |m\rangle \langle n| = \boldsymbol{d} \left(|1\rangle \langle 2| + |2\rangle \langle 1| \right), \tag{4.27}$$

where $\boldsymbol{d}_{mn} = \langle m|\mathsf{d}|n\rangle$. We have also used $\boldsymbol{d}_{21} = \boldsymbol{d}_{12} \equiv \boldsymbol{d}$, assuming $\boldsymbol{d}_{21}$ is a vector with real components.

Since the incident optical field $\boldsymbol{E} = \boldsymbol{E}_0 \cos(\omega t)$ is oscillating at an angular frequency ω with amplitude $\boldsymbol{E}_0$, we can write the total Hamiltonian (4.26) in the form

$$\mathbb{H}_T = \mathbb{H} + \hbar\Omega \cos(\omega t) \left(|1\rangle \langle 2| + |2\rangle \langle 1| \right) \equiv \mathbb{H} + \mathbb{H}_I, \tag{4.28}$$

where we have introduced the *Rabi frequency* Ω using its standard definition,

$$\Omega = -\frac{\boldsymbol{d} \cdot \boldsymbol{E}_0}{\hbar}. \tag{4.29}$$

As will be seen later, the Rabi frequency characterizes the strength of the coupling between a two-level atom and the incident electromagnetic field.

We are now ready to apply the density-matrix formalism to a medium in the form of an ensemble of such atoms. Using Eqs. (4.21) and (4.28), the density operator for this ensemble of two-level atoms evolves as [4]:

$$\frac{\partial \rho}{\partial t} = \frac{1}{j\hbar} [\mathbb{H} + \mathbb{H}_I, \rho]. \tag{4.30}$$

Evolution of the density-matrix elements $\rho_{mn} \equiv \langle m|\rho|n\rangle$ is obtained using Eq. (4.26) and leads to the following four equations (with m, $n = 1$ and 2):

$$\frac{\partial \rho_{11}}{\partial t} = \frac{1}{j\hbar} \langle 1|[\mathbb{H} + \mathbb{H}_I, \rho]|1\rangle = j\Omega \cos(\omega t) \left(\rho_{12} - \rho_{21} \right), \tag{4.31a}$$

$$\frac{\partial \rho_{12}}{\partial t} = \frac{1}{j\hbar} \langle 1|[\mathbb{H} + \mathbb{H}_I, \rho]|2\rangle = j\omega_0 \rho_{12} + j\Omega \cos(\omega t) \left(\rho_{11} - \rho_{22} \right), \tag{4.31b}$$

$$\frac{\partial \rho_{21}}{\partial t} = \frac{1}{j\hbar} \langle 2|[\mathbb{H} + \mathbb{H}_I, \rho]|1\rangle = -j\omega_0 \rho_{21} + j\Omega \cos(\omega t) \left(\rho_{22} - \rho_{11} \right), \tag{4.31c}$$

$$\frac{\partial \rho_{22}}{\partial t} = \frac{1}{j\hbar} \langle 2|[\mathbb{H} + \mathbb{H}_I, \rho]|2\rangle = j\Omega \cos(\omega t) \left(\rho_{21} - \rho_{12} \right). \tag{4.31d}$$

These equations are not independent because the elements of the density matrix satisfy the following two relations:

$$\rho_{11} + \rho_{22} = 1, \qquad \rho_{12} = \rho_{21}^*. \tag{4.32}$$

Indeed, one can easily verify that

$$\frac{\partial \rho_{11}}{\partial t} + \frac{\partial \rho_{22}}{\partial t} = 0, \tag{4.33a}$$

$$\frac{\partial \rho_{12}}{\partial t} - \frac{\partial \rho_{21}^*}{\partial t} = 0. \tag{4.33b}$$

The first relation is a consequence of the normalization condition $\text{Tr}\rho = 1$. The second relation follows from the Hermitian property of the density operator.

4.3.3 Feynman–Bloch vector

Since some of the density-matrix equations are redundant, it is common to recast them using a quantity called the Feynman–Bloch vector, whose three components are defined as

$$\textbf{Dipole moment:} \quad \beta_1 = \rho_{12} + \rho_{21}, \tag{4.34a}$$

$$\textbf{Dipole current:} \quad \beta_2 = j(\rho_{21} - \rho_{12}), \tag{4.34b}$$

$$\textbf{Population inversion:} \quad \beta_3 = \rho_{22} - \rho_{11}. \tag{4.34c}$$

Owing to the Hermitian nature of the density operator, all three elements are real. As mentioned before, the off-diagonal elements of the dipole moment are responsible for the atomic transitions. For this reason, the off-diagonal density-matrix components give rise to a dipole moment β_1 and a dipole current β_2. The difference in the diagonal components, β_3, indicates the relative probability that the atom is in the excited state and is thus a measure of population inversion. Substituting Eq. (4.31) into Eq. (4.34), we get

$$\frac{\partial \beta_1}{\partial t} = -\omega_0 \beta_2, \tag{4.35a}$$

$$\frac{\partial \beta_2}{\partial t} = \omega_0 \beta_1 - 2\Omega \cos(\omega t)\beta_3, \tag{4.35b}$$

$$\frac{\partial \beta_3}{\partial t} = 2\Omega \cos(\omega t)\beta_2. \tag{4.35c}$$

These equations clearly show the underlying physics of the two-level system when interpreted using the Feynman–Bloch vector defined in Eq. (4.34). The other main advantage of introducing the Feynman–Boch vector is that we can write the preceding Eqs. (4.35) (and associated four density-matrix equations in Eq. (4.31)) in the very compact vector form

$$\frac{d\boldsymbol{\beta}}{dt} = \boldsymbol{\Omega}(t) \times \boldsymbol{\beta}, \tag{4.36}$$

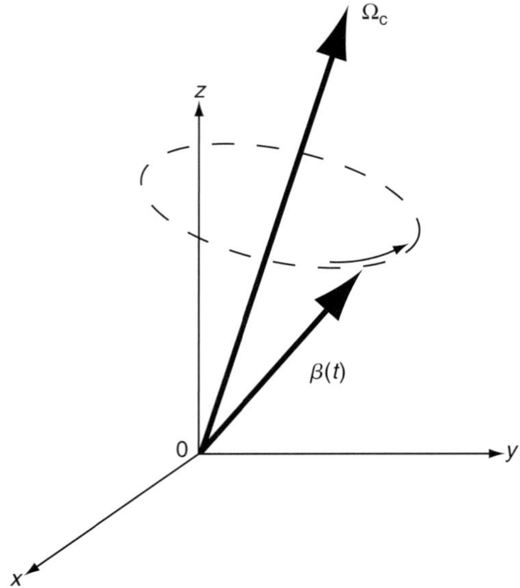

Figure 4.3 Precession of vector $\boldsymbol{\beta}$ around a fixed vector $\boldsymbol{\Omega}(t) = \boldsymbol{\Omega}_c$.

where the vector $\boldsymbol{\Omega}(t)$ is defined as

$$\boldsymbol{\Omega}(t) = [2\Omega\cos(\omega t), 0, \omega_0]. \tag{4.37}$$

When $\boldsymbol{\Omega}(t)$ is a constant, time-independent vector, denoted by $\boldsymbol{\Omega}_c$, it is well-known from classical mechanics that Eq. (4.36) describes the precession of the vector $\boldsymbol{\beta}$ around the fixed vector $\boldsymbol{\Omega}_c$, as shown in Figure 4.3. The precession trajectory in this case is a circle because the vector $\boldsymbol{\beta}$ is normalized to unity at all times.

We can show the fixed unit length of the Feynman–Boch vector $\boldsymbol{\beta}$ using the following simple argument. At any time, the state $|\psi\rangle$ of a two-level atom can be written as

$$|\psi\rangle = c_1(t)\,|1\rangle + c_2(t)\,|2\rangle, \tag{4.38}$$

where normalization of $|\psi\rangle$ demands $|c_1(t)|^2 + |c_2(t)|^2 = 1$. The density operator of this atom can then be written as

$$\rho \equiv |\psi\rangle\,\langle\psi| = |c_1(t)|^2\,|1\rangle\,\langle 1| + |c_2(t)|^2\,|2\rangle\,\langle 2|$$
$$+ c_1(t)c_2^*(t)\,|1\rangle\,\langle 2| + c_2(t)c_1^*(t)\,|2\rangle\,\langle 1|. \tag{4.39}$$

This relation immediately provides the following expressions for the elements of the density matrix:

$$\rho_{11} = |c_1(t)|^2, \qquad \rho_{12} = c_1(t)c_2^*(t),$$
$$\rho_{22} = |c_2(t)|^2, \qquad \rho_{21} = c_2(t)c_1^*(t). \tag{4.40}$$

Using these relations with Eq. (4.34), we can calculate the magnitude of the Feynman–Boch vector $\boldsymbol{\beta}$ as

$$\begin{aligned}
|\boldsymbol{\beta}|^2 &= \beta_1^2 + \beta_2^2 + \beta_3^2 \\
&= \rho_{12}^2 + 2\rho_{12}\rho_{21} + \rho_{21}^2 - \rho_{12}^2 + 2\rho_{12}\rho_{21} - \rho_{21}^2 + \rho_{22}^2 + \rho_{11}^2 - 2\rho_{22}\rho_{11} \\
&= (\rho_{11} + \rho_{22})^2 = 1,
\end{aligned} \tag{4.41}$$

where we used the condition $\rho_{11} + \rho_{22} = 1$ and the relation $\rho_{12}\rho_{21} = \rho_{22}\rho_{11}$, which follows directly from Eq. (4.40).

4.3.4 Rotating-wave approximation

The motion of the Feynman–Bloch vector $\boldsymbol{\beta}$ is governed by the Rabi vector $\boldsymbol{\Omega}(t)$ through the relation (4.36). Owing to the presence of the rapidly time-varying term $2\Omega\cos(\omega t)$, where $\omega \approx \omega_0$ because of the assumed resonance situation, the Rabi vector oscillates at a high frequency. However, the amplitude of such oscillations is relatively small in most experimental situations, where the Rabi frequency Ω is a small fraction of ω_0 (typically $<0.1\%$). This enables us to decompose the vector $\boldsymbol{\Omega}$ in the form

$$\boldsymbol{\Omega}(t) = \boldsymbol{\Omega}_0 + \boldsymbol{\Omega}_+(t) + \boldsymbol{\Omega}_-(t), \tag{4.42}$$

where the three vectors are introduced as

$$\boldsymbol{\Omega}_0 = [0, 0, \omega_0], \tag{4.43a}$$

$$\boldsymbol{\Omega}_+(t) = [\Omega\cos(\omega t), \Omega\sin(\omega t), 0], \tag{4.43b}$$

$$\boldsymbol{\Omega}_-(t) = [\Omega\cos(\omega t), -\Omega\sin(\omega t), 0]. \tag{4.43c}$$

The constant vector $\boldsymbol{\Omega}_0$ points along the z axis of a Cartesian coordinate system and constitutes the dominant component. The other two much smaller vectors lie in the plane perpendicular to it. In particular, $\boldsymbol{\Omega}_+(t)$ provides counterclockwise rotation for increasing t around the vector $\boldsymbol{\Omega}_0$, while the vector $\boldsymbol{\Omega}_-(t)$ rotates in the opposite direction, as illustrated in Figure 4.4. Now, if we choose a reference frame co-rotating at a frequency ω with the vector $\boldsymbol{\Omega}_+(t)$, then $\boldsymbol{\Omega}_+(t)$ becomes a constant and $\boldsymbol{\Omega}_-(t)$ counter-rotates with a frequency 2ω. Owing to its high oscillation rate, the influence of $\boldsymbol{\Omega}_-(t)$ on the vector $\boldsymbol{\Omega}(t)$ is not significant in the rotating frame and can be safely neglected. This approximation is known as the *rotating-wave approximation* and has proven to be very useful in understanding the interaction of light with atomic media.

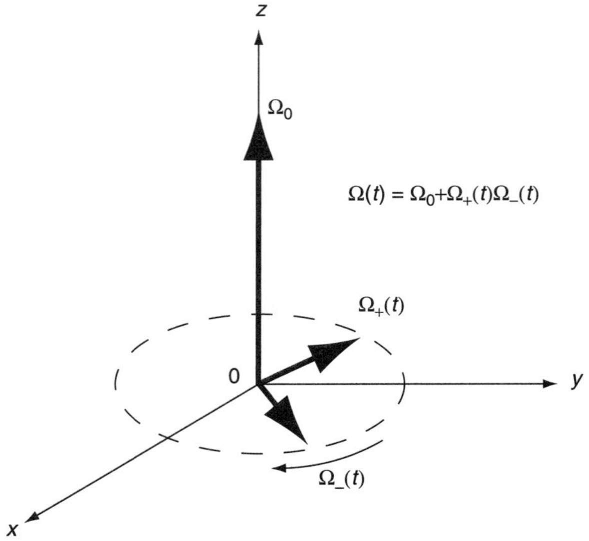

Figure 4.4 Schematic decomposition of the Bloch vector into three vectors. Vectors $\boldsymbol{\Omega}_+$ and $\boldsymbol{\Omega}_-$ lie in the x–y plane while $\boldsymbol{\Omega}_0$ points along the z axis.

The evolution of the Feynman–Bloch vector in Eq. (4.36) with $\boldsymbol{\Omega}(t) \approx \boldsymbol{\Omega}_0 + \boldsymbol{\Omega}_+(t)$ takes the form

$$\frac{d\boldsymbol{\beta}}{dt} = \boldsymbol{\Omega}_0 \times \boldsymbol{\beta} + \boldsymbol{\Omega}_+(t) \times \boldsymbol{\beta}. \qquad (4.44)$$

To map this equation to a frame rotating with frequency ω, we introduce the three unit vectors in this rotating frame as

$$\boldsymbol{e}_u = [\cos(\omega t), \sin(\omega t), 0], \qquad (4.45a)$$

$$\boldsymbol{e}_v = [-\sin(\omega t), \cos(\omega t), 0], \qquad (4.45b)$$

$$\boldsymbol{e}_w = [0, 0, 1], \qquad (4.45c)$$

and express the Feynman–Bloch vector in the new coordinate system:

$$\boldsymbol{\beta} = u\boldsymbol{e}_u + v\boldsymbol{e}_v + w\boldsymbol{e}_w. \qquad (4.46)$$

If the unitary rotation operator responsible for this rotation is given by $\mathbb{R}_\omega$, then we can write the transformation in the form $\boldsymbol{\beta}_\omega = \mathbb{R}_\omega\boldsymbol{\beta}$, where $\boldsymbol{\beta}_\omega$ is the Feynman–Bloch vector in the rotated frame and the matrix representation of $\mathbb{R}_\omega$ is given by

$$\mathbb{R}_\omega = \begin{bmatrix} \cos(\omega t) & \sin(\omega t) & 0 \\ -\sin(\omega t) & \cos(\omega t) & 0 \\ 0 & 0 & 1 \end{bmatrix}. \qquad (4.47)$$

Being a unitary matrix, its inverse can be written as

$$\mathbb{R}_\omega^{-1} = \mathbb{R}_\omega^\dagger = \begin{bmatrix} \cos(\omega t) & -\sin(\omega t) & 0 \\ \sin(\omega t) & \cos(\omega t) & 0 \\ 0 & 0 & 1 \end{bmatrix}. \tag{4.48}$$

The equation of motion for the rotated vector $\boldsymbol{\beta}_\omega$ is obtained by applying the preceding transformation to Eq. (4.44). After considerable algebra, the result is given by

$$\frac{d\boldsymbol{\beta}_\omega}{dt} = \frac{d}{dt}(\mathbb{R}_\omega \boldsymbol{\beta}) = \boldsymbol{\Omega}_\omega \times \boldsymbol{\beta}_\omega, \tag{4.49}$$

where $\boldsymbol{\Omega}_\omega = [\Omega, 0, \omega_0 - \omega]$ is a constant Rabi vector in the rotating frame and points along the z axis. As this vector is time-independent, we can use the precession picture shown in Figure 4.3 and apply the familiar interpretation from classical mechanics [4].

4.4 Optical Bloch equations

The treatment of a two-level system in the preceding section includes only transitions induced by the incoming radiation and ignores any other mechanism that can make an atom change its state. In practice, an atom can change its state through several mechanisms, such as spontaneous emission and collisions with the container walls or with other atoms. Moreover, collisions not only change atomic populations in the two levels of interest but can also affect the off-diagonal elements of the density matrix (decay of coherence). Their impact is included in practice through phenomenological damping terms [4].

To gain some insight into the way damping terms contribute to the density-matrix equations, consider the general state of a two-level atom given in Eq. (4.38) and assume that the coefficients c_1 and c_2 decay with time exponentially as

$$c_1(t) = c_{10} \exp(-\gamma_1 t), \tag{4.50a}$$

$$c_2(t) = c_{20} \exp(-\gamma_2 t), \tag{4.50b}$$

where c_{10} and c_{20} are proportionality constants and γ_1 and γ_2 are the effective decay rates for the two atomic states. The state $|\psi\rangle$ in Eq. (4.38) can be written as

$$|\psi\rangle = c_{10} \exp(-\gamma_1 t) |1\rangle + c_{20} \exp(-\gamma_2 t) |2\rangle. \tag{4.51}$$

The density operator of this state $|\psi\rangle$ has the following form:

$$\rho \equiv |\psi\rangle\langle\psi|$$

$$= |c_{10}(t)|^2 \exp(-2\gamma_1 t)\,|1\rangle\langle 1| + c_{10}(t)c_{20}^*(t)\exp(-\gamma_1 t - \gamma_2 t)\,|1\rangle\langle 2|$$

$$+ c_{20}(t)c_{10}^*(t)\exp(-\gamma_1 t - \gamma_2 t)\,|2\rangle\langle 1| + |c_{20}(t)|^2 \exp(-2\gamma_2 t)\,|2\rangle\langle 2|.$$

$$(4.52)$$

Noting that the elements of the density-matrix operator are defined as $\rho_{mn} \equiv \langle m|\rho|n\rangle$, the following expressions can be written from Eq. (4.52):

$$\rho_{11} = |c_{10}(t)|^2 \exp(-2\gamma_1 t), \tag{4.53a}$$

$$\rho_{12} = c_{10}(t)c_{20}^*(t)\exp(-\gamma_1 t - \gamma_2 t), \tag{4.53b}$$

$$\rho_{21} = c_{20}(t)c_{10}^*(t)\exp(-\gamma_1 t - \gamma_2 t), \tag{4.53c}$$

$$\rho_{22} = |c_{20}(t)|^2 \exp(-2\gamma_2 t). \tag{4.53d}$$

These expressions show that the off-diagonal terms of the density operator (ρ_{12} and ρ_{21}) decay at a rate $(\gamma_1 + \gamma_2)$, while the two diagonal terms decay with rates of $2\gamma_1$ and $2\gamma_2$. The important point to note is that ρ_{12} and ρ_{21} have the same decay rate. This conclusion can also be reached by recalling that ρ is an Hermitian operator, so that $\rho_{12} = \rho_{21}^*$ must hold. These observations guide us to formulating/introducing phenomenological decay rates in Eq. (4.36).

Figure 4.5 shows schematically how phenomenological decay rates Γ_1 and Γ_2 can be introduced into the optical Bloch equations in Eq. (4.36). For simplicity (and based on our simple analysis of exponential decay rates in Eq. (4.53)), it is common to assume that the diagonal elements (ρ_{11} and ρ_{22}) have the same decay rate Γ_1, different from the decay rate Γ_2 of the off-diagonal elements. Using these rates, the density-matrix equations, Eq. (4.31a) through Eq. (4.31d), are modified

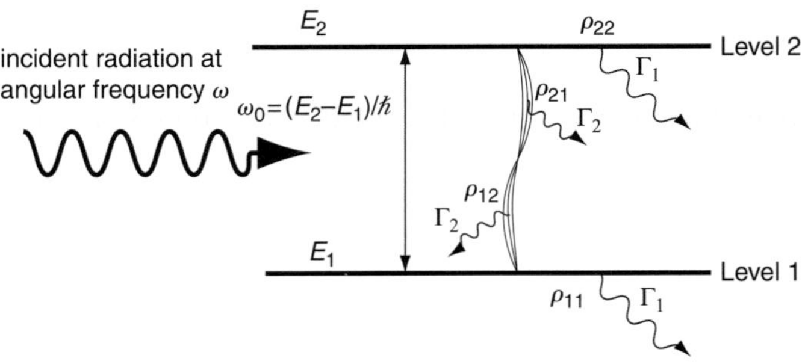

Figure 4.5 Schematic illustration of relaxation rates Γ_1 and Γ_2 associated with a two-level system.

to become

$$\frac{\partial \rho_{11}}{\partial t} = j\Omega \cos(\omega t)\,(\rho_{12} - \rho_{21}) - \Gamma_1 \rho_{11}, \tag{4.54a}$$

$$\frac{\partial \rho_{12}}{\partial t} = j\omega_0 \rho_{12} + j\Omega \cos(\omega t)\,(\rho_{11} - \rho_{22}) - \Gamma_2 \rho_{12}, \tag{4.54b}$$

$$\frac{\partial \rho_{21}}{\partial t} = -j\omega_0 \rho_{21} + j\Omega \cos(\omega t)\,(\rho_{22} - \rho_{11}) - \Gamma_2 \rho_{21}, \tag{4.54c}$$

$$\frac{\partial \rho_{22}}{\partial t} = j\Omega \cos(\omega t)\,(\rho_{21} - \rho_{12}) - \Gamma_1 \rho_{22}. \tag{4.54d}$$

When these equations are used for the Feynman–Bloch vector components defined in Eq. (4.34), we obtain the final optical Bloch equations in the form

$$\frac{d\beta_1}{dt} = -\Gamma_2 \beta_1 + \omega_0 \beta_2, \tag{4.55a}$$

$$\frac{d\beta_2}{dt} = \omega_0 \beta_1 - \Gamma_2 \beta_2 - 2\Omega \cos(\omega t)\beta_3, \tag{4.55b}$$

$$\frac{d\beta_3}{dt} = 2\Omega \cos(\omega t)\beta_2 - \Gamma_1 \left(\beta_3 - w_{eq}\right), \tag{4.55c}$$

where w_{eq} denotes the equilibrium value of β_3 in the absence of the incident optical field: its value is -1 if all atoms are in the lower energy state but can be between -1 and 1 if some atoms are in the excited state initially. Further details can be found in Refs. [12, 13].

It is important to stress that this phenomenological model is inadequate for describing certain solids and liquids undergoing relaxation under external electromagnetic stimulae [14, 15]. Even though such a failure can be attributed to many different scenarios, physical considerations alone lead to two general limits [16]. Inherent in the optical Bloch equations is the Markovian approximation. If this approximation is not valid, the associated differential equations cannot describe the atomic dynamics properly. One such scenario is attributed to the failure of the impact approximation in Ref. [16]. The reason for failure is related to changes in the transition frequency ω_0 attributed to relaxation processes. Such changes can take place in vapors because of Doppler shifts and in liquids because of fluctuations in local magnetic fields. Optical Bloch equations can fail even in the absence of an external field if relaxation processes include velocity-changing collisions [17].

4.5 Maxwell–Bloch equations

As seen in Chapters 1 to 3, light propagation in active or passive media is governed by Maxwell's equations. Therefore, if we are to analyze a medium made up of two-level atoms, such as a doped-fiber amplifier [18, 19], we need to couple Maxwell's equations with the optical Bloch equations. This is done using the Maxwell–Bloch formalism.

Although it is possible to provide a vectorial formulation of the Maxwell–Bloch equations, we resort to a scalar description of the optical field in this section to present underlying concepts as simply as possible. This is not a severe limitation in practice because the polarization effects are relatively weak and only provide a second-order correction to the results of this section. We also ignore the spatial transverse effects and start with the one-dimensional, scalar wave equation

$$\frac{\partial^2 E(z,t)}{\partial z^2} - \frac{1}{c^2}\frac{\partial^2 E(z,t)}{\partial t^2} = \frac{1}{\varepsilon_0 c^2}\frac{\partial^2 P(z,t)}{\partial t^2}, \tag{4.56}$$

where $P(z,t)$ is the material polarization induced by the electric field $E(z,t)$ in the atomic medium and is responsible for its dielectric susceptibility.

Similarly to the rotating-wave approximation used for reducing the Bloch equations to a slowly varying form, we now introduce the slowly-varying-envelope approximation (SVEA) for both the electric and the material polarization fields:

$$E(z,t) = \text{Re}[\mathscr{E}(z,t)\exp(jkz - j\omega_i t)], \tag{4.57a}$$

$$P(z,t) = \text{Re}[\mathscr{P}(z,t)\exp(jkz - j\omega_i t)], \tag{4.57b}$$

where the slowly varying variables, $\mathscr{E}(z,t)$ and $\mathscr{P}(z,t)$, vary at a rate much slower than the carrier frequency ω_i, and $k = c\omega_i$ is the propagation constant at this frequency. These quantities vary slowly in both time and space such that the following relations are satisfied:

$$k^2|\mathscr{E}(z,t)| \gg k\left|\frac{\partial \mathscr{E}(z,t)}{\partial z}\right| \gg \left|\frac{\partial^2 \mathscr{E}(z,t)}{\partial z^2}\right|, \tag{4.58a}$$

$$k^2|\mathscr{P}(z,t)| \gg k\left|\frac{\partial \mathscr{P}(z,t)}{\partial z}\right| \gg \left|\frac{\partial^2 \mathscr{P}(z,t)}{\partial z^2}\right|, \tag{4.58b}$$

$$\omega_i^2|\mathscr{E}(z,t)| \gg \omega_i\left|\frac{\partial \mathscr{E}(z,t)}{\partial t}\right| \gg \left|\frac{\partial^2 \mathscr{E}(z,t)}{\partial t^2}\right|, \tag{4.58c}$$

$$\omega_i^2|\mathscr{P}(z,t)| \gg \omega_i\left|\frac{\partial \mathscr{P}(z,t)}{\partial t}\right| \gg \left|\frac{\partial^2 \mathscr{P}(z,t)}{\partial t^2}\right|. \tag{4.58d}$$

Using the preceding relations, one can reduce Eq. (4.56) from a second-order to a first-order differential equation. More specifically, the second derivatives of the electric field can be approximated as

$$\frac{\partial^2 E(z,t)}{\partial z^2} = \frac{1}{2}\left(\frac{\partial^2 \mathscr{E}(z,t)}{\partial z^2} - k^2 \mathscr{E}(z,t) + 2jk\frac{\partial \mathscr{E}(z,t)}{\partial z}\right)\exp(jkz - j\omega_i t) + \text{c.c.}$$

$$\approx \frac{1}{2}\left(-k^2 \mathscr{E}(z,t) + 2jk\frac{\partial \mathscr{E}(z,t)}{\partial z}\right)\exp(jkz - j\omega_i t) + \text{c.c.} \qquad (4.59)$$

$$\frac{\partial^2 E(z,t)}{\partial t^2} = \frac{1}{2}\left(\frac{\partial^2 \mathscr{E}(z,t)}{\partial t^2} - \omega_i^2 \mathscr{E}(z,t) - 2j\omega_i\frac{\partial \mathscr{E}(z,t)}{\partial t}\right)\exp(jkz - j\omega_i t) + \text{c.c.}$$

$$\approx \left(-\omega_i^2 \mathscr{E}(z,t) - 2j\omega_i\frac{\partial \mathscr{E}(z,t)}{\partial t}\right)\exp(jkz - j\omega_i t) + \text{c.c.} \qquad (4.60)$$

Here, we have kept the first-order derivatives because the zeroth-order terms get cancelled owing to the dispersion relation $\omega_i^2 = k^2 c^2$. Because such a cancelation does not happen in the case of the polarization field, we can safely discard both the first- and the second-order time derivatives and keep only the zeroth-order term:

$$\frac{\partial^2 P(z,t)}{\partial t^2} = \frac{1}{2}\left(\frac{\partial^2 \mathscr{P}(z,t)}{\partial t^2} - \omega_i^2 \mathscr{P}(z,t) + 2jk\frac{\partial \mathscr{P}(z,t)}{\partial t}\right)\exp(jkz - j\omega_i t) + \text{c.c.}$$

$$\approx -\frac{1}{2}\omega_i^2 \mathscr{P}(z,t)\exp(jkz - j\omega_i t) + \text{c.c.} \qquad (4.61)$$

Substituting Eqs. (4.59), (4.60), and (4.61) into the wave equation (4.56), we obtain the first-order partial differential equation valid in the SVEA,

$$\frac{\partial \mathscr{E}(z,t)}{\partial z} + \frac{1}{v_g}\frac{\partial \mathscr{E}(z,t)}{\partial t} = \frac{jk}{2\varepsilon_0}\mathscr{P}(z,t), \qquad (4.62)$$

where we have replaced c with v_g in the time-derivative term to account for the group velocity associated with the host medium in which the two-level atoms are embedded. This is the case for fiber amplifiers in which silica glass acts as a host medium.

The optical Bloch equations obtained in the preceding section used a real-valued field E_0 appearing in the Rabi frequency Ω. We can extend them to include phase variations by adding the phase $\varphi(z,t) = \arg[\mathscr{E}(z,t)]$ to the exponential term associated with the electric field $E(z,t)$. The instantaneous frequency $\omega(t)$ of the incident field is then given by [20]

$$\omega(t) = \omega_i - \frac{\partial \varphi(z,t)}{\partial t}. \qquad (4.63)$$

When we use this frequency as the frequency of the rotating frame, the optical Bloch equations take the form

$$\frac{\partial u(z,t)}{\partial t} = -\Gamma_2 u(z,t) + \left(\omega_i - \omega_0 - \frac{\partial \varphi(z,t)}{\partial t}\right) v(z,t), \tag{4.64a}$$

$$\frac{\partial v(z,t)}{\partial t} = -\Gamma_2 v(z,t) - \left(\omega_i - \omega_0 - \frac{\partial \varphi(z,t)}{\partial t}\right) u(z,t) + \kappa \mathcal{E}(z,t)w(z,t), \tag{4.64b}$$

$$\frac{\partial w(z,t)}{\partial t} = -\Gamma_1[w(z,t) - w_{eq}] + \kappa \mathcal{E}(z,t)v(z,t), \tag{4.64c}$$

where $\kappa = -(\boldsymbol{d} \cdot \boldsymbol{e}_p)/\hbar$ is related to the dipole moment and the polarization direction $\boldsymbol{e}_p$ of the incident optical field.

The only remaining step now is to relate the Bloch vector components u and v to the induced polarization $\mathcal{P}(z,t)$ appearing in the wave equation (4.62). To do this, recall first that $P(\boldsymbol{r},t)$ appearing in Maxwell's equations represents the dipole moment per unit volume and can be written in the form $P = -\eta_d \langle \psi | \boldsymbol{d} | \psi \rangle$, where η_d is the density of atoms. Using $|\psi\rangle$ from Eq. (4.38) and recalling that the diagonal matrix elements of $\boldsymbol{d}$ vanish, we obtain

$$P(z,t) = \eta_d \hbar \kappa [\rho_{21}(z,t) + \rho_{12}(z,t)], \tag{4.65}$$

where we have used Eq. (4.40) for the off-diagonal density-matrix elements. By using Eqs. (4.34) and (4.46), we obtain

$$\begin{bmatrix} \beta_1 \\ \beta_2 \end{bmatrix} \equiv \begin{bmatrix} \rho_{21}(z,t) + \rho_{12}(z,t) \\ j\rho_{21}(z,t) - j\rho_{12}(z,t) \end{bmatrix} = \begin{bmatrix} \cos(\omega t) & -\sin(\omega t) \\ \sin(\omega t) & \cos(\omega t) \end{bmatrix} \begin{bmatrix} u(z,t) \\ v(z,t) \end{bmatrix}. \tag{4.66}$$

It follows from this equation that

$$\rho_{21}(z,t) = \frac{1}{2}[u(z,t) - jv(z,t)] \exp(-j\omega t), \tag{4.67a}$$

$$\rho_{12}(z,t) = \frac{1}{2}[u(z,t) + jv(z,t)] \exp(+j\omega t). \tag{4.67b}$$

Using the preceding relations in Eq. (4.65), the slowly varying part of the induced polarization is given by

$$\mathcal{P}(z,t) = \frac{1}{2}\eta_d \hbar \kappa [u(z,t) - jv(z,t)]. \tag{4.68}$$

If inhomogeneous broadening is present in the medium, the atomic transition frequency ω_0 becomes slightly different for different atoms, and the preceding expression needs to be modified [4, 21]. If we assume that this frequency shifts by

Δ, with the distribution function $g(\Delta)$, the contribution of $(u - jv)$ is modulated by this function, resulting in the expression

$$\mathscr{P}(z, t) = \frac{1}{2}\eta_d\hbar\kappa \int_{-\infty}^{+\infty} g(\Delta)[u(z, t) - jv(z, t)]d\Delta. \tag{4.69}$$

Since the important atomic quantity of interest is $u - jv$, and not the individual variables u and v, it is useful to introduce a new variable, $s = u - jv$. Using u and v from (4.64a) and (4.64b), we obtain the Maxwell–Bloch equations in their final form:

$$\frac{\partial s(z, t)}{\partial t} = -\left(\omega_i - \omega_0 - \frac{\partial\varphi(z, t)}{\partial t}\right) s(z, t) - \Gamma_2 s(z, t) - j\kappa\mathscr{E}(z, t)w(z, t), \tag{4.70a}$$

$$\frac{\partial w(z, t)}{\partial t} = j[\kappa\mathscr{E}(z, t)s^*(z, t) - \kappa^*\mathscr{E}^*(z, t)s(z, t)] - \Gamma_1[w(z, t) - w_{eq}], \tag{4.70b}$$

$$\frac{\partial\mathscr{E}(z, t)}{\partial z} + \frac{1}{v_g}\frac{\partial\mathscr{E}(z, t)}{\partial t} = \frac{jk}{4\varepsilon_0}\eta\hbar\kappa\mathscr{E}(z, t)s(z, t). \tag{4.70c}$$

These equations can be used to describe the operation of fiber amplifiers [19, 22], quantum-dot amplifiers [23], and semiconductor lasers [24]. We use them in later chapters for different types of amplifiers.

4.6 Numerical integration of Maxwell–Bloch equations

The preceding section has dealt with the formulation of the Maxwell–Bloch equations under the slowly-varying-envelope [25] and rotating-wave approximations [4]. These equations are adequate for describing a variety of linear and nonlinear optical phenomena, including self-induced transparency, pulse compression, and soliton-like propagation in atomic systems [4, 26]. However, there are situations in which the approximate form of the Maxwell–Bloch formulation fails. For example, if we consider the propagation of ultrashort few-cycle optical pulses through a two-level medium [27], it becomes necessary to solve Maxwell's equations numerically by the finite-difference time-domain (FDTD) method [28]. We refer the reader to Section 2.5 and describe in this section how the FDTD approach can be adopted for two-level systems.

As before, we consider a one-dimensional model of wave propagation that ignores the transverse variations of the optical fields. Three different sets of equations are needed for such an analysis.

- **Optical Bloch equations with relaxation terms:** We consider a plane electromagnetic wave propagating through an ensemble of two-level atoms. We map the density-matrix description of the medium to its equivalent Feynman–Bloch vector form given in Section 4.3.3. The resulting optical Bloch equations, supplemented with the appropriate relaxation terms, take the form

$$\frac{d\beta_1}{dt} = -\Gamma_2\beta_1 + \omega_0\beta_2, \tag{4.71a}$$

$$\frac{d\beta_2}{dt} = \omega_0\beta_1 - \Gamma_2\beta_2 - 2\Omega\cos(\omega t)\beta_3, \tag{4.71b}$$

$$\frac{d\beta_3}{dt} = 2\Omega\cos(\omega t)\beta_2 - \Gamma_1\left(\beta_3 - w_{eq}\right), \tag{4.71c}$$

where we have employed the notation used in Sections 4.3.3 and 4.4. This set of equations is preferred for implementing the FDTD algorithm because all variables and parameters are real-valued.

- **One-dimensional Maxwell's equations:** Assuming that the plane electromagnetic wave propagates through the nonmagnetic medium along the $+z$ direction and its electric field is polarized along the x axis, Maxwell's equations take the form

$$\frac{\partial E_x}{\partial z} = -\mu_0\frac{\partial H_y}{\partial t}, \tag{4.72a}$$

$$\frac{\partial H_y}{\partial z} = -\varepsilon_0\frac{\partial E_x}{\partial t} - \frac{\partial P_x}{\partial t}, \tag{4.72b}$$

where P_x is the x component of the material polarization induced by the electromagnetic field.

- **Induced material polarization:** As discussed in Section 4.5, when an electric field interacts with the two-level atoms, the motion of electrons creates dipoles. If d_x is the x component of the dipole moment, the material polarization is simply given by

$$P_x = \eta_d\hbar\kappa\beta_1, \tag{4.73}$$

where $\kappa = (d_x/\hbar)E_x$ and η_d is the density of two-level atoms within the medium.

If we substitute Eq. (4.73) into Eq. (4.72b) and use expression (4.71a), we obtain

$$\frac{\partial H_y}{\partial z} = -\varepsilon_0\frac{\partial E_x}{\partial t} - \eta_d\hbar\kappa\left(-\Gamma_2\beta_1 + \omega_0\beta_2\right). \tag{4.74}$$

This equation, together with Eqs. (4.71), and (4.72a), provides a self-consistent set of five equations that needs to be solved using the FDTD method.

Following the discussion in Section 2.5, we use Yee's staggered grid and adopt the following notation for the discrete sampled values in space and time:

$$\Upsilon(z, t) = \Upsilon(k\Delta z, n\Delta t) = \Upsilon|_k^n, \quad \text{where} \quad \Upsilon = E_x, H_y, \beta_1, \beta_2, \beta_3. \tag{4.75}$$

However, we treat the magnetic field variable H_x separately from the remaining four variables (E_x, β_1, β_2, and β_3) and assume that it is spatially separated by $\Delta z/2$ and temporarily separated by $\Delta t/2$ [28]. More specifically, the electric field and the Bloch vector components are located at the center of a Yee cell, while the magnetic field lies at the edges of a Yee cell. With this scheme, we obtain the following discrete form of the Maxwell–Bloch equations:

$$\left.\frac{d\beta_1}{dt}\right|_{(k+1/2)\Delta z}^{n\Delta t} = \left[-\Gamma_2\beta_1 + \omega_0\beta_2\right]\Big|_{(k+1/2)\Delta z}^{n\Delta t}, \tag{4.76a}$$

$$\left.\frac{d\beta_2}{dt}\right|_{(k+1/2)\Delta z}^{n\Delta t} = \left[\omega_0\beta_1 - \Gamma_2\beta_2 - 2\Omega\cos(\omega t)\beta_3\right]\Big|_{(k+1/2)\Delta z}^{n\Delta t}, \tag{4.76b}$$

$$\left.\frac{d\beta_3}{dt}\right|_{(k+1/2)\Delta z}^{n\Delta t} = \left[2\Omega\cos(\omega t)\beta_2 - \Gamma_1\left(\beta_3 - w_{eq}\right)\right]\Big|_{(k+1/2)\Delta z}^{n\Delta t}, \tag{4.76c}$$

$$\left.\frac{\partial E_x}{\partial z}\right|_{(k+1/2)\Delta z}^{n\Delta t} = -\mu_0\left.\frac{\partial H_y}{\partial t}\right|_{(k+1/2)\Delta z}^{n\Delta t}, \tag{4.76d}$$

$$\left.\frac{\partial H_y}{\partial z}\right|_{k\Delta z}^{(n+1/2)\Delta t} = -\varepsilon_0\left.\frac{\partial E_x}{\partial t}\right|_{k\Delta z}^{(n+1/2)\Delta t} + \eta_d\hbar\kappa\left[\Gamma_2\beta_1 - \omega_0\beta_2\right]\Big|_{k\Delta z}^{(n+1/2)\Delta t}. \tag{4.76e}$$

Employing central differences to approximate partial derivatives on both sides of these equations, we can easily derive the discrete versions of the equations suitable for numerical integration.

As a specific example, we consider at the propagation of a single-cycle light pulse through a two-level inverted medium [29]. It has been shown by McCall and Hahn [26] that the pulse area $\theta(z, t)$, defined as

$$\theta(z, t) = \int_{-\infty}^{t} \left(\frac{2d_x}{\hbar}\right) E_x(z, \tau)\, d\tau, \tag{4.77}$$

decays according to the simple equation

$$\frac{d\theta}{dz} = -\frac{\alpha}{2}\sin(\theta), \tag{4.78}$$

where α is the absorption coefficient of the medium.

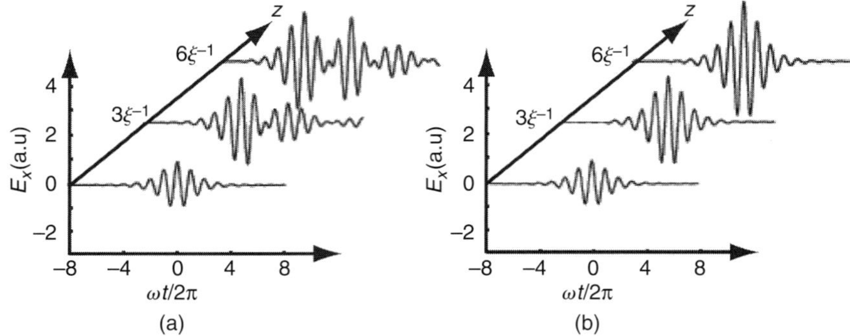

Figure 4.6 The evolution of single-cycle pulse waveform through a two-level amplifying medium with (a) uniform distribution of dipoles and (b) nonuniform distribution of dipoles (see Eq. (4.79)). The snapshots of the three different waveforms were taken at distances $z = 0$, $z = 3\xi^{-1}$, and $z = 6\xi^{-1}$ along the amplifying medium. (After Ref. [29]; © Elsevier 2001)

The most important consequence of the pulse-area theorem is that pulses with an initial area that is a multiple of π maintain the same area during propagation through the two-level medium, in spite of losses. Moreover, if a pulse does not have such a special value initially, it propagates through the medium obeying Eq. (4.78) until such a special value is reached. Figure 4.6 shows the propagation of a single-cycle pulse with an initial area of π through a two-level medium [29]. The dipole moment of the medium is uniformly distributed in the case of Figure 4.6(a) but varies in the case of Figure 4.6(b) as

$$d_x(z) = \frac{d_0}{\sqrt{1 + \xi z}}, \tag{4.79}$$

where $\xi = 4\pi \omega \eta_d / (nc\mathcal{E}_0)$ and $\mathcal{E}_0$ is the energy of the input pulse before entering the medium. The initial electric field of the pulse is $E_x(0, 0) = 0.159(\hbar\omega/2d_0)$ and the initial inversion level of the medium is given by $\beta_3(z, 0) = 0.12E_x(0, 0)/(4\pi d_0\eta_d)$. A Yee grid with 200 cells per wavelength was required for sufficiently accurate results. The time was measured from the moment when the center of the pulse passed through the input facet of the medium.

Figure 4.6(a) shows that a medium with uniform dipole distribution cannot efficiently amplify a single-cycle π pulse. In contrast to this, as shown in Figure 4.6(b), a nonuniform dipole distribution enables efficient amplification. Such a conclusion could not have been reached for a pulse of this short duration using the Maxwell–Bloch equations alone. Generally speaking, the FDTD technique is required for ultrashort pulses whose envelope contains only a few optical cycles.

References

[1] R. Loudon, *The Quantum Theory of Light*, 2nd ed. Oxford Science Publications, 1997.

[2] D. Suter, *The Physics of Laser-Atom Interactions*. Cambridge University Press, 1997.

[3] S.-H. Chen and M. Kotlarchyk, *Interactions of Photons and Neutrons with Matter*. World Scientific, 2007.

[4] L. Allen and J. H. Eberly, *Optical Resonance and Two-Level Atoms*. Wiley Inter-Science, 1975.

[5] J. J. Sakurai, *Modern Quantum Mechanics*, 2nd ed. Addison-Wesley, 2010.

[6] T. F. Jordan, "Assumptions implying the Schrödinger equation," *Am. J. Phys.*, vol. 59, pp. 606–608, 1991.

[7] C. R. Smith, "Operator techniques in three dimensions," *Am. J. Phys.*, vol. 44, pp. 989–993, 1976.

[8] E. Prugovecki, "The bra and ket formulation in extended Hilbert space," *J. Math. Phys.*, vol. 14, pp. 1410–1422, 1973.

[9] H.-A. Bachor and T. C. Ralph, *A Guide to Experiments in Quantum Optics*. Wiley, 2004.

[10] J. E. Roberts, "The Dirac bra and ket formalism," *J. Math. Phys.*, vol. 7, pp. 1097–1104, 1966.

[11] H. Goldstein, C. P. Poole, and J. L. Safko, *Classical Mechanics*. Addison-Wesley, 2001.

[12] R. W. Boyd, *Nonlinear Optics*, 3rd ed. Academic Press, 2008.

[13] B. Bidegaray, A. Bourgeade, and D. Reignier, "Introducing physical relaxation terms in Bloch equations," *J. Comput. Phys.*, vol. 170, pp. 603–613, 2001.

[14] A. G. Redfield, "Nuclear magnetic resonance saturation in solids," *Phys. Rev.*, vol. 98, pp. 1787–1809, 1955.

[15] R. G. DeVoe and R. G. Brewer, "Experimental test of the optical Bloch equations for solids," *Phys. Rev. Lett.*, vol. 50, pp. 1269–1272, 1983.

[16] P. R. Berman, "Validity conditions for the optical Bloch equations," *J. Opt. Soc. Am. B.*, vol. 3, pp. 564–571, 1986.

[17] A. P. Ghosh, C. D. Nabors, M. A. Attili, and J. E. Thomas, "$^3P_1$-orientation velocity-changing collision kernels studied by isolated multipole echoes," *Phys. Rev. Lett.*, vol. 54, pp. 1794–1797, 1985.

[18] J. T. Manassah and B. Gross, "Propagation of femtosecond pulses in a fiber amplifier," *Opt. Commun.*, vol. 122, pp. 71–82, 1995.

[19] B. Gross and J. T. Manassah, "Numerical solutions of the Maxwell–Bloch equations for a fiber amplifier," *Opt. Lett.*, vol. 17, pp. 340–342, 1992.

[20] G. P. Agrawal, *Nonlinear Fiber Optics*, 4th ed. Academic Press, 2007.

[21] T. Nakajima, "Pulse propagation through a coherently prepared two-level system," *Opt. Commun.*, vol. 136, pp. 273–276, 1997.

[22] L. W. Liou and G. P. Agrawal, "Solitons in fiber amplifiers beyond the parabolic-gain and rate-equation approximations," *Opt. Commun.*, vol. 124, pp. 500–504, 1996.

[23] M. van der Poel, E. Gehrig, O. Hess, D. Birkedal, and J. M. Hvam, "Ultrafast gain dynamics in quantum-dot amplifiers : Theoretical analysis and experimental investigations," *IEEE J. Quantum Electron.*, vol. 41, pp. 1115–1123, 2005.

[24] C. M. Bowden and G. P. Agrawal, "Maxwell–Bloch formulation for semiconductors: Effects of coherent Coulomb exchange," *Phys. Rev. A*, vol. 51, pp. 4132–4239, 1995.

[25] M. Sargent III, M. O. Scully, and W. E. Lamb Jr., *Laser Physics*. Addison-Wesley, 1974.

[26] S. L. McCall and E. L. Hahn, "Self-induced transparency by pulsed coherent light," *Phys. Rev. Lett.*, vol. 18, pp. 908–911, 1967.

[27] P. Kinser and G. H. C. New, "Few-cycle pulse propagation," *Phys. Rev. A*, vol. 67, p. 023813 (8 pages), 2003.

[28] R. W. Ziolkowski, J. M. Arnold, and D. M. Gogny, "Ultrafast pulse interaction with two-level atoms," *Phys. Rev. A*, vol. 52, pp. 3082–3094, 1995.

[29] A. V. Tarasishin, S. A. Magnitskii, and A. M. Zheltikov, "Propagation and amplification of ultrashort light pulses in a resonant two-level medium: Finite difference time-domain analysis," *Opt. Commun.*, vol. 193, pp. 187–196, 2001.

5

Fiber amplifiers

Modern optical fibers exhibit very low losses (≈ 0.2 dB/km) in the 1.55 µm wavelength region that is of interest for telecommunications applications. Even though light at wavelengths in this region can be transmitted over more than 100 km before its power degrades considerably, an optical amplifier is eventually needed for any telecommunications system to restore the signal power to its original level. Since a fiber-based amplifier is preferred for practical reasons, such amplifiers were developed during the 1980s by doping standard optical fibers with rare-earth elements (known as lanthanides), a group of 14 elements with atomic numbers in the range from 58 to 71. The term *rare* appears to be a historical misnomer because rare-earth elements are relatively abundant in nature. When these elements are doped into silica or other glass fibers, they become triply ionized. Many different rare-earth elements, such as erbium, holmium, neodymium, samarium, thulium, and ytterbium, can be used to make fiber amplifiers that operate at wavelengths covering a wide range from visible to infrared. Amplifier characteristics, such as the operating wavelength and the gain bandwidth, are determined by the dopants rather than by the fiber, which plays the role of a host medium. However, because of the tight confinement of light provided by guided modes, fiber amplifiers can provide high optical gains at moderate pump power levels over relatively large spectral bandwidths, making them suitable for many telecommunications and signal-processing applications [1–3].

After reviewing the generic properties of erbium-doped fiber amplifiers (EDFAs) in Section 5.1, we review in Section 5.2 the main features governing the performance of such amplifiers by considering the amplifier gain and its effective spectral bandwidth. We describe the rate equation model of EDFAs in Section 5.3, and use this model in Section 5.4 to analyze their continuous-wave (CW) performance. The propagation equation required for describing the amplification of picosecond pulses in EDFAs is discussed in Section 5.5 together with its numerical solution. The split-step Fourier method presented there can be applied to both the nonlinear

113

Schrödinger equation and the Ginzburg–Landau equation. We focus in Section 5.6 on special analytical solutions in the form of autosolitons and similaritons that preserve pulse shape during propagation through fiber amplifiers. Finally, we consider in Section 5.7 how this formalism needs to be modified when femtosecond pulses are passed through fiber amplifiers.

5.1 Erbium-doped fiber amplifiers

EDFAs have attracted the most attention for telecommunications applications because they operate in the spectral region near 1.55 μm [1–3]. To understand their properties, it is vital to understand details of erbium–glass spectroscopy. Relevant energy levels and the associated transitions are shown in Figure 5.1 for erbium ions doped into silica glass. Details about the spectroscopic notation used in this figure can be found in Ref. [4]. In general, electronic energy levels are denoted by the symbol $^{2S+1}L_J$, where S is the total spin of all electrons, L is the total orbital angular momentum, and J is the sum of two momenta resulting from spin–orbit coupling. The quantity $2S + 1$ denotes the multiplicity and provides the degeneracy that exists for a particular energy state. In the case of erbium ions with $L = 6$ and $S = 3/2$, the four lowest states correspond to J values ranging from $L + S$ to $L - S$.

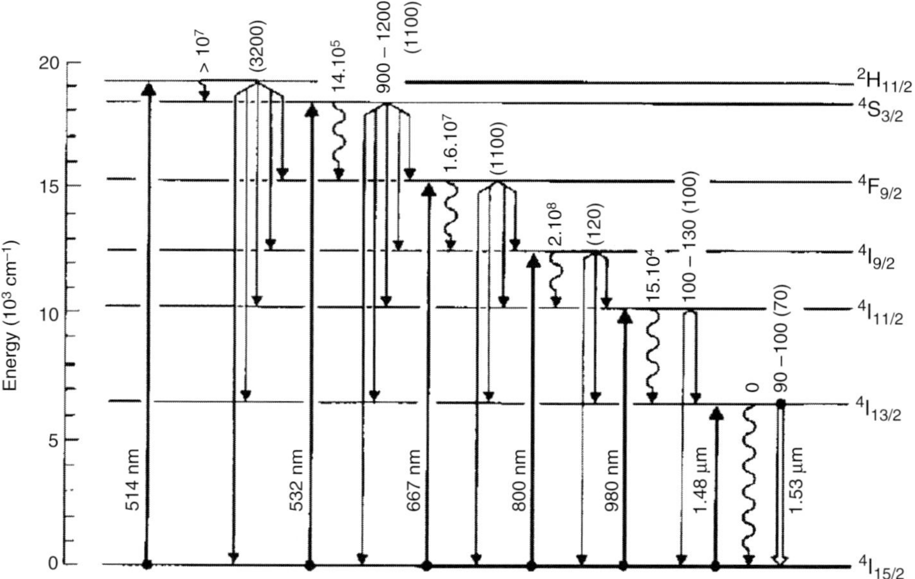

Figure 5.1 Energy-level diagram of erbium ions in a glass host and the associated optical transitions. (After Ref. [3]; © CRC Press 2001)

Note that the amorphous nature of glass broadens each energy level into an energy band.

It is evident from Figure 5.1 that pump light at several wavelengths can be used to achieve optical gain through stimulated emission between the two lowest energy levels, $^4I_{13/2}$ and $^4I_{15/2}$. The most common pump wavelength for EDFAs is 980 nm because semiconductor lasers operating at this wavelength are available commercially. The performance of any amplifier also depends on the rates of radiative and nonradiative decay caused by several mechanisms related to lattice vibrations, ion–ion interactions, and cooperative upconversion to higher levels [3]. Figure 5.1 shows the range of such decay rates (in units of s^{-1}) corresponding to various glasses, including silicate, fluorophosphate [5], and fluorozirconate glasses [6]. The important point to note is that the decay from $^4I_{13/2}$ and $^4I_{15/2}$ is completely radiative with a relatively long fluorescence lifetime of around 10 ms [3]. In view of this, when an EDFA is pumped at 980 nm, one can replace the actual energy-level diagram in Figure 4.1 with a three-level system, shown schematically in Figure 5.2, where we now depict each level as an energy band [7]. Moreover, if the EDFA is pumped at 1480 nm, the amplifier can be described using a two-level model [8,9].

Efficient pumping is possible at both 980 and 1480 nm using semiconductor lasers [10–13]. Amplification factors in the range of 30 to 35 dB can be obtained with pump powers of $\sim$10 mW when 980-nm pumping is employed. The transition $^4I_{15/2} \to\, ^4I_{9/2}$ allows the use of GaAs pump lasers operating near 800 nm, but the pumping efficiency is relatively poor [14]. It can be improved by co-doping the fiber with aluminum and phosphorus [15]. In all cases, a two-level model is a good approximation if the pump power is less than 1 W since the number of atoms in the pump level reamins relatively small owing to its short lifetime [8, 16]. In practice, total pump power is kept below 1 W and a two-level model is adequate to explain most of the observed phenomena in EDFAs.

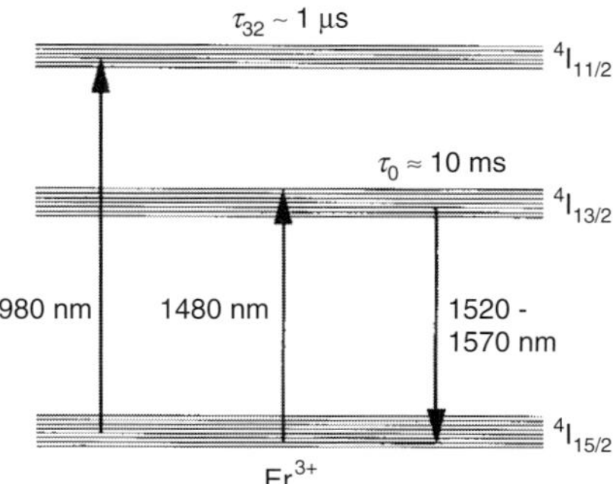

Figure 5.2 Simplified energy-level diagram of erbium ions for an EDFA pumped near 980 nm. (After Ref. [16]; © IEEE 1997)

5.2 Amplifier gain and its bandwidth

All doped-fiber amplifiers amplify incident light through stimulated emission. Energy for amplifying the incident beam comes from the pump light, whose absorption raises rare-earth ions to an excited state, thereby storing the energy that is eventually transferred to the signal beam being amplified. The emission and absorption between the two energy levels are characterized by the emission and absorption cross-sections, $\sigma_{21}(\omega)$ and $\sigma_{12}(\omega)$, which satisfy the relation [17, 18]

$$\sigma_{21}(\omega) = \sigma_{12}(\omega) \exp\left(\frac{\varepsilon(T) - \hbar\omega}{k_B T}\right), \tag{5.1}$$

where k_B is the Boltzmann constant, T is the absolute temperature in Kelvin, and $\varepsilon(T)$ is the excitation energy required to excite one ion to its excited state at temperature T. As an example, Figure 5.3 shows the spectra of $\sigma_{21}(\omega)$ and $\sigma_{12}(\omega)$ at $T = 295$ K (typical room temperature) for an EDFA whose fiber core is also doped with aluminium [19].

The emission and absorption cross-sections can be used to introduce an important amplifier parameter, called the gain coefficient and defined as

$$g(\omega) = \sigma_{21}(\omega)N_2 - \sigma_{12}(\omega)N_1, \tag{5.2}$$

where N_1 and N_2 denote the atomic densities of the lower and upper states, respectively. This expression can be simplified by making some simple assumptions, often satisfied in practice. For example, if we assume that the emission and absorption cross-sections have similar shapes governed by $\sigma(\omega)$ but different peak values,

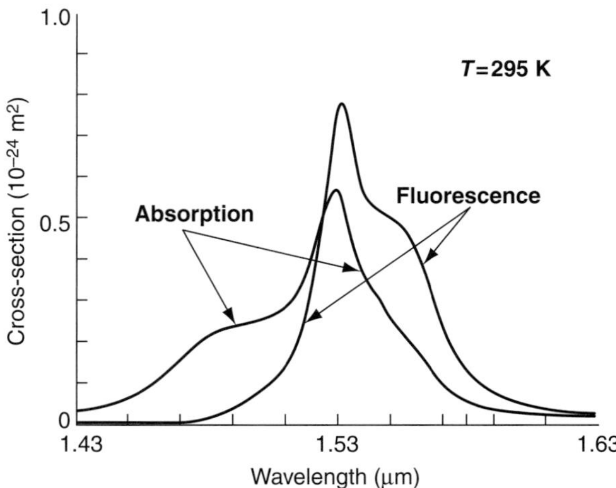

Figure 5.3 Absorption and emission-cross section spectra of an erbium-doped aluminosilcate fiber at $T = 295$ K. (After Ref. [19]; © OSA 1990)

then it is possible to write the gain coefficient as [3]

$$g(\omega) = \sigma(\omega)\left(N_2 - \frac{g_2}{g_1}N_1\right),\tag{5.3}$$

where g_1 and g_2 are the degeneracies of the lower and upper atomic states, respectively. For erbium ions, it is known that $g_2/g_1 = 7/8$ [18, 20]. Therefore, it is possible to approximate the gain coefficient by

$$g(\omega) = \sigma(\omega)(N_2 - N_1),\tag{5.4}$$

where $\sigma(\omega)$ is referred to as the transition cross-section. This formula holds regardless of whether the gain medium medium is modeled as a two- or three-level system.

The gain coefficient $g(\omega)$ in a fiber amplifier also depends on the distance z from its input end, and on the elapsed time time t if optical pulses are amplified. To calculate it, one needs to solve a coupled set of rate equations for the atomic densities; its final expression depends on the energy-level model employed. In the approximation of a two-level atomic model, the dynamic response of EDFAs is governed by the Maxwell–Bloch equations of Section 4.5.

Although a numerical solution of the Maxwell–Bloch equations becomes essential when ultrashort optical pulses are amplified in EDFAs, considerable simplification occurs in the CW or quasi-CW regime. In this case, the steady-state solution of optical Bloch equations can be obtained readily to find $s(z)$ and $w(z)$ by neglecting time derivatives. Noting that the inversion density $N_2 - N_1$ is related to $-w$, we can obtain an approximate analytic expression for $g(\omega)$ in the CW regime (no time dependence). This expression provides a very good approximation for many practical applications and has the form

$$g(z, \omega) = \frac{g_0(z)}{1 + (\omega - \omega_a)^2 T_2^2 + P(z)/P_s},\tag{5.5}$$

where $g_0(z)$ is the maximum value of the gain at a distance z, ω_a is the atomic transition frequency, and $P(z)$ is the optical power of the CW signal being amplified. The saturation intensity $I_s = (\hbar/\mu)^2 \Gamma_1 \Gamma_2$ depends on several dopant parameters, such as the dipole moment d and the two relaxation rates, Γ_1 and Γ_2, and can be used to obtain the saturation power P_s. It is common to introduce the population relaxation time (also called the fluorescence time) as $T_1 = 1/\Gamma_1$ and the dipole relaxation time as $T_2 = 1/\Gamma_2$. The population relaxation time T_1 varies in the range from 0.1 μs to 10 ms, depending on the dopant, and is about 10 ms in the case of EDFAs. The dipole relaxation time T_2 is quite small (~ 0.1 ps) for all fiber amplifiers.

The role of the term P/P_s in Eq. (5.5) is to reduce the gain as signal power increases in the fiber. This phenomenon is common to all amplifiers and is referred to as *gain saturation*. If the amplifier operates at power levels such that $P(z)/P_s \ll 1$ for all z, the amplifier is said to operate in the unsaturated regime. The unsaturated gain coefficient has a Lorentzian shape that is characteristic of homogenously broadened systems [21, 22]:

$$g(z, \omega) = \frac{g_0(z)}{1 + (\omega - \omega_a)^2 T_2^2}.$$
(5.6)

Clearly, $g(z, \omega)$ is at a maximum when the signal frequency ω coincides with the atomic transition frequency ω_a. The gain gets smaller as the signal frequency moves away from ω_a. The effective 3 dB bandwidth of the gain is identified using the frequency range over which gain exceeds $g_0/2$. This gain bandwidth, $\Delta\omega_g$, can be used to find the full width at half maximum (FWHM) of the gain spectrum, and is given by

$$\Delta\nu_g = \frac{\Delta\omega_g}{2\pi} = \frac{1}{\pi T_2}.$$
(5.7)

As an example, $\Delta\nu_g \approx 3\,\text{THz}$ when $T_2 = 0.1$ ps. As we saw in Figure 5.3, the actual gain spectrum of fiber amplifiers can deviate considerably from a Lorentzian profile. However, the insight and flexibility provided by this model justify its adoption in practice.

It is very important that we make a clear distinction between the bandwidth of the gain spectrum (see Eq. (5.7)) and that of the amplifier itself. The difference becomes clear when one considers the overall gain of the amplifier, defined as

$$G = P_{\text{out}}/P_{\text{in}},$$
(5.8)

where P_{in} is the input power fed into the amplifier and P_{out} is the signal power out of the amplifier. Noting that the gain coefficients represent the local gain seen by the signal, signal-power evolution along the amplifier in the CW case is governed by the simple differential equation

$$\frac{dP}{dz} = g(z, \omega)P(z),$$
(5.9)

where $P(z)$ is the optical power of a signal with mean frequency ω at a distance z from the input end of the amplifier. A straightforward integration with the conditions $P(0) = P_{\text{in}}$ and $P(L) = P_{\text{out}}$ shows that the CW gain of the amplifier is given by

$$G(\omega) = \exp\left(\int_0^L g(z, \omega)\, dz\right),$$
(5.10)

where L is the length of the amplifier. If we ignore the z dependence of the gain coefficient, we obtain the simple relation $G(\omega) = \exp[g(\omega)L]$.

Both $G(\omega)$ and $g(\omega)$ are maximum at $\omega = \omega_a$ and decrease when $\omega \neq \omega_a$. However, $G(\omega)$ decreases much faster than $g(\omega)$ because of its exponential dependence on the latter, and the bandwidths of $G(\omega)$ and $g(\omega)$ differ from each other. The amplifier bandwidth $\Delta \nu_A$ is defined as the FWHM of $G(\omega)$ and is related to the gain bandwidth $\Delta \nu_g$ by

$$\Delta \nu_A = \Delta \nu_g \left(\frac{\ln 2}{g(\omega_a)L - \ln 2} \right)^{1/2}. \tag{5.11}$$

We briefly discuss the issue of gain saturation. The origin of gain saturation lies in the power dependence of the gain coefficient in Eq. (5.5). Since g is reduced when P becomes comparable to P_s, the amplification factor G is also expected to decrease. To simplify the discussion, let us consider the case in which the signal frequency is exactly tuned to the atomic transition frequency ($\omega = \omega_a$). Using Eqs. (5.9) and (5.5) we obtain

$$\frac{dP}{dz} = \frac{g_0(z)P}{1 + P/P_s}. \tag{5.12}$$

This equation can be easily integrated over the amplifier length. By using the initial condition $P(0) = P_{\text{in}}$, the amplifier gain is given by the implicit relation

$$G = \exp\left[\int_0^L g_0(z)\,dz - (G - 1)P_{\text{in}}/P_s \right]. \tag{5.13}$$

This transcendental equation does not have an analytical solution and must be solved numerically to calculate the saturated gain of the amplifier.

5.3 Rate equations for EDFAs

As mentioned earlier and as shown in Figure 5.2, EDFAs can be pumped at a wavelength of 980 or 1480 nm. Since excited-state absorption does not occur for these two pumping bands [8], we can restrict ourselves to the three energy bands shown in Figure 5.2. The sublevels within each energy band result from the Stark effect, and their occupancy in thermal equilibrium is given by the Boltzmann distribution. In the following simplified analysis, we focus on the case of 980 nm pumping and ignore the sublevels by considering the total atomic densities, N_1, N_2, and N_3, associated with each energy band. When an EDFA is pumped by injecting light from one end, a continuous loss in the pump power because of fiber losses and absorption by dopants makes the atomic densities nonuniform along the EDFA length. It is important to include such axial variations of the pump power and atomic densities. We thus allow N_1, N_2, and N_3 to vary with both z and t.

In the three-level rate-equation model, we can write the atomic rate equations by considering all processes through which erbium ions appear or disappear in a given energy state [1, 23]:

$$\frac{\partial N_1}{\partial t} = -R_{13}N_1 - W_{12}N_1 + W_{21}N_2 + \frac{N_2}{T_1} + R_{31}N_3 + \frac{N_3}{T_1'}, \tag{5.14a}$$

$$\frac{\partial N_2}{\partial t} = W_{12}N_1 - W_{21}N_2 + R_{32}N_3 - \frac{N_2}{T_1}, \tag{5.14b}$$

$$\frac{\partial N_3}{\partial t} = R_{13}N_1 - R_{32}N_3 - R_{31}N_3 - \frac{N_3}{T_1'}, \tag{5.14c}$$

where T_1 and T_1' are the population relaxation times of levels 2 and 3, respectively, W_{ab} is the transition rate associated with the transition from state a to state b for the signal, and R_{ab} denotes this same rate for the pump.

In the case of erbium ions, $R_{32} \gg R_{31}$ because transitions from level 3 to level 2 are much more likely than transitions from level 3 to level 1 [1]. Moreover, the rate of spontaneous emission from state 3 to state 1 is also much weaker than R_{32} (i.e., $R_{32} \gg 1/T_1'$). With these simplifications, Eq. (5.14c) is reduced to

$$\frac{\partial N_3}{\partial t} \approx R_{13}N_1 - R_{32}N_3. \tag{5.15}$$

This equation shows that under the steady-state condition in which the time derivative vanishes, the population density of level 3 is given by

$$R_{32}N_3 \approx (R_{13}/R_{32})N_1 \approx 0, \tag{5.16}$$

where we have used the condition $R_{32} \gg R_{13}$.

If we use this result in Eqs. (5.14a), we obtain

$$\frac{\partial N_1}{\partial t} = -R_{13}N_1 - W_{12}N_1 + W_{21}N_2 + \frac{N_2}{T_1}. \tag{5.17}$$

Under the same conditions, Eqs. (5.14b) becomes

$$\frac{\partial N_2}{\partial t} = R_{13}N_1 + W_{12}N_1 - W_{21}N_2 - \frac{N_2}{T_1} = -\frac{\partial N_1}{\partial t}. \tag{5.18}$$

It follows that N_1 and N_2 change in such a way that $N_1 + N_2 = \rho_0$, where ρ_0 is the total dopant density. Physically, this relation shows the conservation of overall atomic populations during the pumping and stimulated-emission processes.

The parameters appearing in Eqs. (5.17) and (5.18) depend on the absorption and emission cross-sections as well as on the signal and pump powers, P_s and P_p, respectively. Their explicit expressions are [1, 23]

$$W_{12} = \frac{\Gamma_s \sigma_{12} P_s}{A_c h \nu_s},$$ (5.19a)

$$W_{21} = \frac{\Gamma_s \sigma_{21} P_s}{A_c h \nu_{-s}},$$ (5.19b)

$$R_{13} = \frac{\Gamma_p \sigma_{13} P_p}{A_c h \nu_p},$$ (5.19c)

where A_c is the cross-sectional area of the fiber core, Γ_s and Γ_p are the mode confinement factors, and ν_s and ν_p are the frequencies of the signal and pump waves, respectively.

To proceed further, we need to consider the evolution of pump and signal powers along the fiber. The corresponding equations can be written by considering all processes through which pump and signal fields lose or gain photons, resulting in [24, 25]

$$\frac{\partial P_p}{\partial z} = -\Gamma_p \sigma_{13} N_1 P_p - \alpha_p P_p,$$ (5.20a)

$$\frac{\partial P_s}{\partial z} = -\Gamma_s \sigma_{12} N_1 P_s + \Gamma_s \sigma_{21} N_2 P_s - \alpha_s P_s,$$ (5.20b)

where α_p and α_s represent fiber losses at the pump and signal wavelengths, respectively. If we ignore these losses, an assumption justified in practice for typical amplifier lengths (<0.1 km), it is possible to solve this coupled set of equations analytically [24, 25]. We first rewrite them, using definitions given in Eqs. (5.19a) through (5.20), in the form

$$\frac{\partial P_p}{\partial z} = -R_{13} A_c h \nu_p N_1,$$ (5.21a)

$$\frac{\partial P_s}{\partial z} = (W_{21} N_2 - W_{12} N_1) A_c h \nu_s.$$ (5.21b)

Noting that $N_1 = \rho_0 - N_2$, these equations can be easily integrated over the entire amplifier length L. The result is given by [25]

$$P_p = P_p^{\text{in}} \exp\left(B_p \int_0^L N_2 A_c dz - C_p\right),$$ (5.22a)

$$P_s = P_s^{\text{in}} \exp\left(B_s \int_0^L N_2 A_c dz - C_s\right),$$ (5.22b)

where P_p^{in} and P_s^{in} are pump and signal powers at the input end of the EDFA, respectively. The other coefficients are defined as

$$B_p = \Gamma_p \sigma_{13}/(A_c h \nu_p), \tag{5.23a}$$

$$B_s = \Gamma_s (\sigma_{12} + \sigma_{12})/(A_c h \nu_s), \tag{5.23b}$$

$$C_p = \Gamma_p \sigma_{13} \rho L, \tag{5.23c}$$

$$C_s = \Gamma_s \sigma_{12} \rho L. \tag{5.23d}$$

Substituting Eq. (5.21) into Eq. (5.18) and integrating along the fiber length, we obtain

$$\frac{\partial}{\partial t} \int_0^L N_2 A_c dz = -\int_0^L \frac{\partial P_s}{\partial z} dz - \int_0^L \frac{\partial P_p}{\partial z} dz - \frac{1}{T_1} \int_0^L N_2 A_c dz$$
$$= P_s^{\text{in}}(t) \left[1 - \exp\left(B_s \int_0^L N_2 A_c dz - C_s \right) \right]$$
$$+ P_p^{\text{in}}(t) \left[1 - \exp\left(B_p \int_0^L N_2 A_c dz - C_p \right) \right] - \frac{1}{T_1} \int_0^L N_2 A_c dz. \tag{5.24}$$

This equation can be written in compact form by introducing the variable ρ, representing the total excited erbium population at some time t and defined as $\rho = \int_0^L N_2 A_c \, dz$. The resulting equation is

$$\frac{\partial \rho}{\partial t} = P_s^{\text{in}}(t) \left[1 - \exp\left(B_s \rho - C_s \right) \right] + P_p^{\text{in}}(t) \left[1 - \exp\left(B_p \rho - C_p \right) \right] - \frac{\rho}{T_1}. \tag{5.25}$$

Being an ordinary differential equation for a single variable ρ, this equation can be easily solved numerically to characterize the time evolution of the EDFA gain [24, 25].

5.4 Amplification under CW conditions

It is instructive to consider the operation of an EDFA under steady-state conditions. In this case, we can set the time derivative to zero in Eq. (5.18). Using $N_1 = \rho_0 - N_2$, the excited-state population is given by

$$N_2 = \left(\frac{R_{13} + W_{12}}{R_{13} + W_{12} + W_{21} + \frac{1}{T_1}} \right) \rho_0. \tag{5.26}$$

Using the definitions in Eq. (5.19a), this equation can be written in the form

$$N_2 = \left(\frac{P_p/P_p^{\text{sat}} + P_s/P_s^{\text{sat}}}{1 + P_s/P_s^{\text{sat}} + P_s/\bar{P}_s^{\text{sat}} + P_p/P_p^{\text{sat}}} \right) \rho_0, \tag{5.27}$$

where we have defined the three saturation powers as

$$P_p^{\text{sat}} = \frac{Ahc}{\Gamma_p \sigma_{13} \lambda_p T_1}, \tag{5.28a}$$

$$P_s^{\text{sat}} = \frac{Ahc}{\Gamma_s \sigma_{12} \lambda_s T_1}, \tag{5.28b}$$

$$\bar{P}_s^{\text{sat}} = \frac{Ahc}{\Gamma_s \sigma_{21} \lambda_s T_1}. \tag{5.28c}$$

The ground-state population is obtained from the relation $N_1 = \rho_0 - N_2$.

The pump and signal powers vary along the amplifier length because of absorption, stimulated emission, and spontaneous emission. Their variations also depend on whether the signal and pump fields propagate in the same or opposite directions. If the contribution of spontaneous emission is neglected and forward pumping is assumed, P_p and P_s satisfy Eqs. (5.20). Substitution of N_1 and N_2 then leads to a set of coupled equations which can be readily solved numerically. Their predictions are in good agreement with experiment as long as the amplified spontaneous emission (ASE) remains negligible [26].

More insight can be gained by making further simplifications to this model. First, we neglect the difference between the emission and absorption cross-sections and set $\sigma_{12}(\omega) = \sigma_{21}(\omega)$. Since $\bar{P}_s^{\text{sat}} = P_s^{\text{sat}}$ in this case, N_1 and N_2 can be written in the form

$$N_1 = \left(\frac{1 + P_s/P_s^{\text{sat}}}{1 + 2P_s/P_s^{\text{sat}} + P_p/P_p^{\text{sat}}} \right) \rho_0, \tag{5.29a}$$

$$N_2 = \left(\frac{P_p/P_p^{\text{sat}} + P_s/P_s^{\text{sat}}}{1 + 2P_s/P_s^{\text{sat}} + P_p/P_p^{\text{sat}}} \right) \rho_0. \tag{5.29b}$$

Second, for lumped amplifiers with fiber lengths under 1 km, losses α_p and α_s can be set to zero. The results shown in Figure 5.4 are obtained with these simplifications using parameter values from Ref. [8]. This figure shows the amplification factor of an EDFA at 1.55 μm, under small-signal conditions, as a function of (a) pump power and (b) amplifier length.

It is evident from Figure 5.4 that, for a given amplifier length L, the amplification factor increases exponentially with pump power initially, but grows at a much reduced rate when pump power exceeds a certain value (corresponding to the "knee"

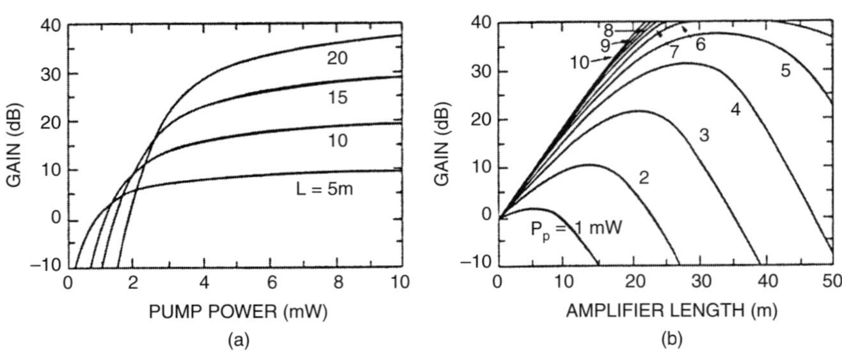

Figure 5.4 Small-signal gain at 1.55 μm as a function of (a) pump power and (b) amplifier length for an EDFA pumped at 1.48 μm. (After Ref. [8]; © IEEE 1991)

in Figure 5.4(a)). For a given pump power, amplifier gain becomes maximum at an optimum value of L and drops sharply when L exceeds this optimum value. The reason for this behavior is that the end portion of the amplifier remains unpumped and absorbs the amplified signal. Since the optimum value of L depends on the pump power P_p, it is necessary to choose both L and P_p appropriately. Figure 5.4(b) shows that, for 1.48 μm pumping, a 35 dB gain can be achieved at a pump power of 5 mW for $L = 30$ m. It is possible to design high-gain EDFAs using fiber lengths as short as a few meters when pumping is done near 980 nm.

The preceding analysis is based on the CW operation of an EDFA, and caution needs to be exercised when extending it to the amplification of pulse trains. In telecommunications networks, EDFAs amplify pulses corresponding to binary modulation. In this case, it is required that all pulses experience the same gain. Fortunately, this occurs naturally in EDFAs for pulses shorter than a few microseconds. The reason is related to the relatively large value of the fluorescence time associated with erbium ions ($T_1 \approx 10$ ms). When the time scale of signal-power variations is much shorter than T_1, erbium ions are unable to follow such fast variations. Since single-pulse energies are typically much below the saturation energy (~ 10 μJ), EDFAs respond to the average power. As a result, gain saturation is governed by the average signal power, and amplifier gain does not vary from pulse to pulse.

5.5 Amplification of picosecond pulses

Propagation of optical pulses in undoped (i.e., standard) optical fibers has been studied extensively using an envelope equation [27]. Here, we generalize that equation so that it can be applied for EDFAs. In this approach, one derives an equation called the nonlinear Schrödinger (NLS) equation using the slowly-varying-envelope and rotating-wave approximations. If the slowly varying amplitude A of the optical pulse is normalized in such a way that $|A|^2$ represents optical power, the NLS equation for a standard optical fiber has the form [27]

$$\frac{\partial A}{\partial z} + \beta_1 \frac{\partial A}{\partial t} + j\frac{\beta_2}{2}\frac{\partial^2 A}{\partial t^2} + \frac{\alpha}{2}A = j\gamma|A|^2 A, \tag{5.30}$$

where α accounts for fiber losses. The nonlinear coefficient γ is given by

$$\gamma = \frac{n_2\omega_0}{cA_{\text{eff}}}, \tag{5.31}$$

where ω_0 is the reference frequency relative to which the slowly varying envelop approximation was made [27], n_2 is the nonlinear index coefficient, and A_{eff} the effective mode area.

5.5.1 Inclusion of dopant's susceptibility

The dispersive effects taking place in the fiber are included in Eq. (5.30) by the parameter β_2 governing group-velocity dispersion (GVD). Since the exact functional form of the propagation constant $\beta(\omega)$ is rarely known, one employs its Taylor series expansion around the carrier frequency ω_0 in the form

$$\beta(\omega) = \beta_0 + \beta_1(\omega - \omega_0) + \frac{1}{2}\beta_2(\omega - \omega_0)^2 + \cdots. \tag{5.32}$$

The NLS equation (5.30) is obtained when this expansion is truncated after the quadratic term. This equation is responsible for a variety of nonlinear effects that often limit the transmission capacity of optical fibers. Nonlinear effects detrimental to transmission include self-phase modulation (SPM), cross-phase modulation, four-wave mixing (FWM), and stimulated Raman scattering (SRS). The last two nonlinear effects (FWM and SRS) can also be used to make the parametric and Raman fiber amplifiers discussed in later chapters. When dopants such as erbium are introduced into a fiber, all these effects of the host medium are inherited by the doped fibers and can affect the performance of EDFAs. In addition, modifications must be introduced to the standard NLS equation to account for the presence of dopants.

We saw in Section 5.2 that an erbium-doped fiber can be modeled using a two-level system. This permits us to employ the optical Bloch equations of Section 4.4 by modeling the erbium ions as an ensemble of two-level atoms. Such a model requires two relaxation times introduced earlier in Section 5.2: the population relaxation time T_1 and the dipole relaxation time T_2. The two-level system can be assumed to respond instantaneously if T_2 is much shorter than the width of the optical pulse being amplified; this holds true for pulses with widths in the picosecond range because T_2 is <100 fs whereas T_1 is close to 10 ms for EDFAs [28]. For pulses having widths in the picosecond range, the susceptibility of the dopants can be written as [21]

$$\chi_a(\omega) = \frac{\Gamma_s \sigma_{21}(N_2 - N_1)}{j + (\omega - \omega_a)T_2} \frac{c}{\omega} n(\omega), \tag{5.33}$$

where $n(\omega)$ is the refractive index of the fiber core in the absence of doping. A major assumption made in obtaining Eq. (5.33) is that the gain medium is homogeneously broadened. Because of the amorphous nature of silica glass, inhomogenous broadening cannot be totally avoided. However, the inhomogenous contribution is relatively small for aluminosilicate glasses, and T_2 can be related to the homogenous linewidth of the atomic transition as $\Delta\omega_a = 1/T_2$. The inhomogenous component appears to be larger for germanosilicate glasses. In this chapter we assume that T_2 is a fitting parameter determined by taking the gain characteristics of different dopants into account.

The relative permittivity or dielectric constant of a doped fiber is obtained by adding the contribution of the dopants to that of the undoped fiber, and is given by

$$\varepsilon(\omega) = n^2(\omega) + 2jn(\omega)c\alpha/\omega + \chi_a(\omega). \tag{5.34}$$

The absorptive and dispersive properties of the doped fiber result from the frequency dependence of $\varepsilon(\omega)$. The propagation constant of the signal within the doped-fiber medium can be written as $\beta(\omega) = \sqrt{\varepsilon(\omega)}\omega/c$. Assuming that the imaginary part is much smaller than the real part, we obtain

$$\beta(\omega) = [n(\omega) + n_2|A|^2]\omega/c + j\alpha + \chi_a(\omega)\omega/c, \tag{5.35}$$

where we have added the nonlinear contribution through the n_2 term.

This expression can be further simplified by expanding around the reference frequency ω_0 used for introducing the slowly varying envelope of the optical pulse. However, care must be exercised because both n_2 and α also vary with frequency. In practice, their variation with frequency is slow enough that they can be treated as constants over the entire bandwidth of picosecond pulses (< 1 THz). However, it is not possible to treat $\chi_a(\omega)$ as constant within the bandwidth of the pulse. Noting that α, $n_2|A|^2$, and $|\chi_a(\omega)|$ are much smaller than $n(\omega)$, and using Eq. (5.32), we can approximate Eq. (5.35) as [28]

$$\beta(\omega) = \beta_0 + \beta_1(\omega - \omega_0) + \frac{1}{2}\beta_2(\omega - \omega_0)^2 + \frac{\omega_0}{c}n_2|A|^2 + j\frac{\alpha}{2} + \frac{\omega\chi_a(\omega)}{2cn(\omega)}. \tag{5.36}$$

5.5.2 Derivation of the Ginzburg–Landau equation

Owing to the finite gain bandwidth associated with $\chi_a(\omega)$, all spectral components of the pulse do not experience the same gain, a phenomenon referred to as *gain dispersion.* Moreover, owing to the Kramers–Kronig relations, any gain is accompanied by a corresponding index change, which provides an additional contribution to the dispersion of the medium. Expanding $\chi_a(\omega)$ in a Taylor series around ω_0 and keeping only terms up to second order, we then obtain

$$\frac{\omega \chi_a(\omega)}{cn(\omega)} = g_0 \left(\frac{\delta - j}{1 + \delta^2} \right) + g_0 \left(\frac{1 - \delta^2 + 2j\delta}{1 + 2\delta^2 + \delta^4} \right) T_2(\omega - \omega_0)$$
$$+ g_0 \left(\frac{j - 3\delta - 3j\delta^2 + \delta^3}{1 + 3\delta + 3\delta^2 + \delta^3} \right) T_2^2(\omega - \omega_0)^2, \tag{5.37}$$

where $g_0 = \Gamma_s \sigma (N_2 - N_1)$ is the gain coefficient and $\delta = (\omega_0 - \omega_a)T_2$ represents a detuning of the carrier frequency ω_0 of the pulse from the atomic transition frequency ω_a. Substituting Eq. (5.37) into Eq. (5.32), we finally obtain

$$\beta(\omega) = \beta_0 + \beta_1^{\text{eff}}(\omega - \omega_0) + \frac{1}{2}\beta_2^{\text{eff}}(\omega - \omega_0)^2 + \frac{\omega_0}{c}n_2|A|^2 + j\frac{\alpha^{\text{eff}}}{2}, \tag{5.38}$$

where

$$\beta_1^{\text{eff}} = \beta_1 + \frac{g_0 T_2}{2} \left(\frac{1 - \delta^2 + 2j\delta}{1 + 2\delta^2 + \delta^4} \right), \tag{5.39a}$$

$$\beta_2^{\text{eff}} = \beta_2 + g_0 T_2^2 \left(\frac{j - 3\delta - 3j\delta^2 + \delta^3}{1 + 3\delta + 3\delta^2 + \delta^3} \right), \tag{5.39b}$$

$$\alpha^{\text{eff}} = \alpha - jg_0 \left(\frac{\delta - j}{1 + \delta^2} \right). \tag{5.39c}$$

With the introduction of the preceding effective parameters, we can write the propagation equation for an EDFA by replacing the corresponding parameters in the standard NLS equation (5.30) with the effective ones. The resulting equation is given by

$$\frac{\partial A}{\partial z} + \beta_1^{\text{eff}}\frac{\partial A}{\partial t} + j\frac{\beta_2^{\text{eff}}}{2}\frac{\partial^2 A}{\partial t^2} + \frac{\alpha}{2}A = j\gamma|A|^2 A + \frac{g_0}{2}\left(\frac{1 + j\delta}{1 + \delta^2} \right)A. \tag{5.40}$$

This equation does not include the simultaneous absorption of two photons. For silica fibers, two-photon absorption (TPA) is indeed negligible. However, it becomes important for fibers made using chalcogenide materials [29]. We can include the effects of two-photon absorption in such fibers by replacing α with $\alpha + \alpha_2|A|^2$, where α_2 is a material parameter responsible for two-photon absorption.

It is clear from Eq. (5.40) that the group velocity of the pulse, $v_g = 1/\beta_1^{\text{eff}}$, is affected by the dopants. However, the dopant-induced change in the group velocity is negligible in practice because the difference $(\beta_1^{\text{eff}}/\beta_1) - 1$ is close to 10^{-4} under typical operating conditions. In contrast, changes in β_2 are not negligible, especially near the zero-dispersion wavelength of the amplifier. It is instructive to look at the behavior of β_2^{eff} when the the carrier frequency ω_0 of the pulse coincides with the atomic transition frequency ω_0; this amounts to operating at the gain peak with $\delta = 0$. Even at this point, β_2^{eff} does not coincide with β_2 because of the finite gain bandwidth of the amplifier and the related gain-dispersion effects.

Much can be learned from Eq. (5.40) by looking at its behavior at the gain peak of the amplifier, so that $\delta = 0$. Replacing β_1^{eff} with β_1 and introducing a reduced time in a frame moving with the pulse as $T = t - \beta_1 z$, we obtain

$$\frac{\partial A}{\partial z} + j\frac{1}{2}(\beta_2 + jg_0 T_2^2)\frac{\partial^2 A}{\partial T^2} = j\gamma |A|^2 A + \frac{g_0}{2}A - \frac{1}{2}(\alpha + \alpha_2|A|^2)A, \quad (5.41)$$

where we have added the two-photon contribution to keep the following analysis general. As expected, if the gain and two-photon absorption are taken out, this equation reduces to the usual NLS equation.

In cases in which the mathematical structure of the preceding equation is paramount, it is better to recast it in normalized units. Suppose P_0 is the peak power and T_0 is the width of the input pulse. Adopting the notation used for solitons in undoped fibers, we introduce *soliton units* as

$$\xi = z/L_D, \quad \tau = T/T_0, \quad u = \sqrt{\gamma L_D}A, \quad (5.42)$$

where $L_D = T_0^2/|\beta_2|$ is the dispersion length. Equation (5.41) then takes the following normalized form:

$$j\frac{\partial u}{\partial \xi} - \frac{1}{2}(s + jd)\frac{\partial^2 u}{\partial \tau^2} + (1 + j\mu_2)|u|^2 u = \frac{j}{2}\mu u, \quad (5.43)$$

where $s = \text{sgn}(\beta_2) = \pm 1$ and the other parameters are defined as

$$d = g_0 L_D (T_2/T_0)^2, \quad (5.44a)$$

$$\mu = (g_0 - \alpha)L_D, \quad (5.44b)$$

$$\mu_2 = \alpha_2/2\gamma. \quad (5.44c)$$

The impact of doping in this equation is represented by the two parameters d and μ. Physically, d is related to the amplifier bandwidth (through the parameter T_2), μ is related to the amplifier gain, and μ_2 governs the effect of two-photon absorption. Numerical values of these parameters for most EDFAs are $\mu \sim 1$, $d \sim 10^{-3}$, and $\mu_2 \sim 10^{-4}$ when $T_0 \sim 1$ ps. Equation (5.43) is known as the Ginzburg–Landau equation and has its origins in superconductivity theory. In nonlinear optics, the Ginzburg–Landau equation is used for analyzing pulse evolution in amplifying nonlinear media in the form of bright and dark solitons.

5.5.3 Numerical solution of Ginzburg–Landau equation

It is possible to solve Eq. (5.43) using the well-known split-step Fourier method (SSFM) used for solving the NLS equation in nonlinear fiber optics [27]. Here, we show how to adapt this method to the more general equation (5.43) with complex coefficients. It is easy to see that this equation reduces to the NLS equation when $d = 0$ and $\mu_2 = 0$.

The underlying philosophy of the SSFM is to split the nonlinear differential equation into two parts where one part is linear and the other part is nonlinear. Let $\mathscr{L}$ be the linear part and $\mathscr{N}\{u\}$ the nonlinear part. Then Eq. (5.43) can be written as

$$\frac{\partial u}{\partial \xi} = [\mathscr{L} + \mathscr{N}\{u\}]\, u, \tag{5.45}$$

where

$$\mathscr{L}u = \frac{\mu}{2}u - \frac{j}{2}(s + jd)\frac{\partial^2 u}{\partial \tau^2}, \tag{5.46a}$$

$$\mathscr{N}\{u\}u = j(1 + j\mu_2)|u|^2 u. \tag{5.46b}$$

The functionals $\mathscr{L}$ and $\mathscr{N}\{u\}$ are known as the linear and nonlinear operators of the corresponding differential equation. It is important to realize that any differential equation that can be split this way can be solved using the SSFM method discussed here.

To solve Eq. (5.45) numerically, we create a uniform grid with spatial steps of $\Delta\xi$ and time steps of $\Delta\tau$. Within this grid, integration of Eq. (5.45) admits the following space-marching solution [27]:

$$u(\xi + \Delta\xi, \tau) \approx \exp\left(\mathscr{L}\Delta\xi + \mathscr{N}\left\{u\left(\xi + \frac{\Delta\xi}{2}, \tau\right)\right\}\Delta\xi\right) u(\xi, \tau). \tag{5.47}$$

Owing to the fact that $\mathscr{L}$ and $\mathscr{N}\{u\}$ are noncommuting operators, we cannot expand the exponential in the preceding equation in the conventional form. Instead, we must use the Zassenhaus product formula [30] for the noncommuting operators $\mathscr{L}$ and $\mathscr{N}$. When the commutator is a c number, the result is given by

$$\exp\left(\mathscr{L}\Delta\xi + \mathscr{N}\Delta\xi\right) = \exp\left(\mathscr{L}\Delta\xi\right)\exp\left(\mathscr{N}\Delta\xi\right)\exp\left(\frac{\Delta\xi^2}{2}[\mathscr{L},\mathscr{N}]\right). \tag{5.48}$$

To order $O(\Delta\xi^2)$, Eq. (5.47) can be approximated as

$$u(\xi + \Delta\xi, \tau) \approx \exp\left(\mathscr{L}\Delta\xi\right)\exp\left(\mathscr{N}\left\{u\left(\xi + \frac{\Delta\xi}{2}, \tau\right)\right\}\Delta\xi\right) u(\xi, \tau). \tag{5.49}$$

Further improvement in the accuracy of the numerical procedure can be made by symmetrizing this splitting as [31]

$$
u(\xi + \Delta\xi, \tau) \approx
$$

$$
\exp\left(\mathscr{L}\frac{\Delta\xi}{2}\right) \exp\left(\mathscr{N}\left\{u\left(\xi + \frac{\Delta\xi}{2}, \tau\right)\right\}\Delta\xi\right) \exp\left(\mathscr{L}\frac{\Delta\xi}{2}\right) u(\xi, \tau),
$$

$$(5.50)$$

which is accurate to order $\mathrm{O}(\Delta\xi^3)$. If we apply the operator $\exp\left(\mathscr{L}\frac{\Delta\xi}{2}\right)$ to both sides of the expression, we obtain

$$
\exp\left(\mathscr{L}\frac{\Delta\xi}{2}\right) u(\xi + \Delta\xi, \tau) \approx
$$

$$
\exp\left(\mathscr{L}\Delta\xi\right) \exp\left(\mathscr{N}\left\{u\left(\xi + \frac{\Delta\xi}{2}, \tau\right)\right\}\Delta\xi\right) \exp\left(\mathscr{L}\frac{\Delta\xi}{2}\right) u(\xi, \tau).
$$

$$(5.51)$$

Use of this equation reduces computational time.

We can reduce the computational burden further by noting that the following relations hold approximately:

$$
u(\xi + 3\Delta\xi/2, \tau) \approx \exp\left(\mathscr{L}\Delta\xi/2\right) u(\xi + \Delta\xi, \tau),
$$

$$
u(\xi + \Delta\xi/2, \tau) \approx \exp\left(\mathscr{L}\Delta/2\right) u(\xi, \tau).
$$

$$(5.52)$$

Using these relations, we obtain

$$
u\left(\xi + \frac{3\Delta\xi}{2}, \tau\right) \approx \exp\left(\mathscr{L}\Delta\xi\right) \exp\left(\mathscr{N}\left\{u\left(\xi + \frac{\Delta\xi}{2}, \tau\right)\right\}\Delta\xi\right) u\left(\xi + \frac{\Delta\xi}{2}, \tau\right),
$$

$$(5.53)$$

which is also accurate to order $\mathrm{O}(\Delta\xi^3)$.

The linear part of Eq. (5.53) can be numerically evaluated in a straightforward way by using the standard Fast Fourier Transform (FFT) and its inverse (IFFT) as [27]

$$
u\left(\xi + \frac{3\Delta\xi}{2}, \tau\right) = \mathrm{IFFT}\left\{\mathscr{F}_{\tau-}\left\{\exp\left(\mathscr{L}\Delta\xi\right)\right\}(\omega)\mathrm{FFT}\left\{u_{NL}(\xi, \Delta\xi, \tau)\right\}\right\}, \quad (5.54)
$$

where we have used the Fourier-transform operator, $\mathscr{F}_{\tau-}\{\ldots\}(\omega)$ (see Section 1.1.2), to include linear effects, with ω acting as the frequency variable. We have also introduced a new variable, $u_{NL}(\xi + \frac{\Delta\xi}{2}, \tau)$, as

$$
u_{NL}\left(\xi, \Delta\xi, \tau\right) = \exp\left(\mathscr{N}\left\{u(\xi + \frac{\Delta\xi}{2}, \tau)\right\}\Delta\xi\right) u\left(\xi + \frac{\Delta\xi}{2}, \tau\right). \quad (5.55)
$$

To evaluate u_{NL}, we write it as a partial differential equation,

$$
\frac{\partial u_{NL}}{\partial \xi} = j(1 + j\mu_2)|u_{NL}|^2 u_{NL}, \quad (5.56)
$$

and integrate this equation from ξ to $\xi + \Delta\xi$. The integration can be carried out easily if we write this equation in terms of two real variables using the polar form

$$u_{NL} \equiv A_{NL}\exp(j\phi_{NL}). \qquad (5.57)$$

Substituting Eq. (5.57) into Eq. (5.56) and collecting the real and imaginary parts, we obtain

$$\frac{\partial A_{NL}}{\partial \xi} = -\mu_2 A_{NL}^3, \qquad (5.58a)$$

$$\frac{\partial \phi_{NL}}{\partial \xi} = A_{NL}^2. \qquad (5.58b)$$

Since Eq. (5.58a) does not depend on the variable ϕ_{NL}, it can be solved using standard analytical methods. The resulting solution is then substituted into Eq. (5.58b) to finally construct the complex envelope u_{NL} using Eq. (5.57).

5.6 Autosolitons and similaritons

Solitons are wave packets that can propagate in a nonlinear dispersive medium without changing their shape or size. Owing to this property, they appear to have particle-like properties. Like elementary particles, solitons not only survive collisions but undergo elastic collisions. All solitons associated with the NLS equation share this property. However, it is important to note that the Ginzburg–Landau equation is not integrable by the inverse scattering method, and therefore does not support solitons in a strict mathematical sense. Nevertheless, its solitary-wave solutions represent optical pulses whose shape does not change on propagation [32]. Such solitary-wave solutions of the Ginzburg–Landau equation, often called dissipative solitons, have been studied extensively in recent years [33, 34]. In the context of optical amplifiers, they are also called *autosolitons* because all input pulses evolve toward a specific pulse whose width and other properties are set by the amplifying medium [35].

The existence of solitons in a fiber amplifier is somewhat surprising because energy is constantly fed to a propagating pulse during its amplification. Clearly, for a pulse to maintain its width and shape, this energy has to be dissipated through some mechanism. Two-photon absorption provides such a mechanism. However, it is not sufficient to create a perfect energy balance along an amplifier. A second loss mechanism originates from the finite bandwidth of the gain spectrum. If the spectral width of a pulse becomes larger than the gain bandwidth, its spectral components that lie outside the gain bandwidth may experience a net loss. Even if the spectrum of the input pulse is narrower than the gain bandwidth, SPM-induced spectral broadening of the pulse during its amplification in an amplifier may broaden it enough for the pulse to experience no net gain in the presence of linear and two-photon absorption. Therefore, a fiber amplifier could support autosolitons even in the normal-dispersion region, a phenomenon not possible in a standard undoped fiber.

5.6.1 Autosolitons

Since there is no systematic way to construct soliton solutions for the Ginzburg–Landau equation, we need to use a technique in which a solution with several parameters is first postulated and these parameters are then chosen to ensure that the Ginzburg–Landau equation is indeed satisfied. Based on the known soliton solutions of the NLS equation in a standard undoped fiber, the postulated pulse-like solution (corresponding to a *bright soliton*, which behaves similarly to a standard soliton but has a characteristic localized intensity peak above a CW background) of the Ginzburg–Landau equation (5.43) has the following form [36]:

$$
\begin{aligned}
u(\xi, \tau) &= N_s[\mathrm{sech}(p\tau)]^{1+jq} \exp(jK_s\xi) \\
&\equiv N_s \mathrm{sech}(p\tau) \exp[jK_s\xi - jq \ln(\cosh p\tau)].
\end{aligned}
\tag{5.59}
$$

The parameters N_s, p, q, and K_s are found by substituting this solution back into Eq. (5.43) and are given by [27]

$$
N_s^2 = \tfrac{1}{2}p^2[s(q^2 - 2) + 3qd], \tag{5.60a}
$$

$$
p^2 = \mu[d(q^2 - 1) - 2sq]^{-1}, \tag{5.60b}
$$

$$
K_s = \tfrac{1}{2}p^2[s(q^2 - 1) + 2qd], \tag{5.60c}
$$

where q is a solution of the following quadratic equation:

$$
(d - \mu_2 s)q^2 - 3(s + \mu_2 d)q - 2(d - \mu_2 s) = 0. \tag{5.61}
$$

It is clear from Eq. (5.61) that $q \neq 0$ only when either d or μ_2 is nonzero. For silica fiber amplifiers, μ_2 is small enough that it can be set to zero. The parameter q is then given by

$$
q = [3s \pm (9 + 8d^2)^{1/2}]/2d, \tag{5.62}
$$

where the sign is chosen so that both p and N_s are real.

The general solution (5.59) of the Ginzburg–Landau equation (5.43) exists for both positive and negative values of β_2. It is easy to verify that, when $s = -1$ (anomalous GVD) and d, μ, and μ_2 are set to zero, this solution reduces to the standard soliton of the NLS equation [27]. The parameter p remains undetermined in that limit because the NLS equation supports a whole family of fundamental solitons such that $N_s = p$. By contrast, both p and N_s are fixed for the Ginzburg–Landau equation by the amplifier parameters μ and d. This is a fundamental difference introduced by the dopants: fiber amplifiers select a single soliton from the entire family of solitons supported by the undoped fiber. The width and the peak power of this soliton are uniquely determined by the amplifier parameters (such as its gain and bandwidth), justifying the name autosoliton given to such pulses [35].

The new feature of the autosolitons forming in fiber amplifiers is that their phase $\phi(\tau)$ is not constant but varies with time along the pulse. The parameter q may be called a chirp parameter because the phase varies only with time when $q \neq 0$. Having a quantitative understanding of this phase variation gives us a better understanding of soliton-like pulses forming within a fiber amplifier. From Eq. (5.59), we find the following expression for the autosoliton phase:

$$\phi(\tau) = K_s \xi - q \ln(\cosh p\tau). \tag{5.63}$$

The associated chirp $\delta\omega(\tau)$, or the frequency variation experienced by this pulse around its central reference frequency, is found using $\delta\omega = -\partial\phi/\partial\tau$. This frequency chirp is given by

$$\delta\omega(\tau) = qp \tanh(p\tau). \tag{5.64}$$

Figure 5.5 compares the intensity and chirp profiles of an autosoliton in the cases of normal (dashed curve) and anomalous (solid curve) GVD, using $d = 0.5$, $\mu = 0.5$, and $\mu_2 = 0$. In both cases, the chirp is nearly linear over most of the intensity profile, but the soliton is considerably broader in the case of normal GVD. The dependence of soliton parameters on the gain-dispersion parameter d is shown

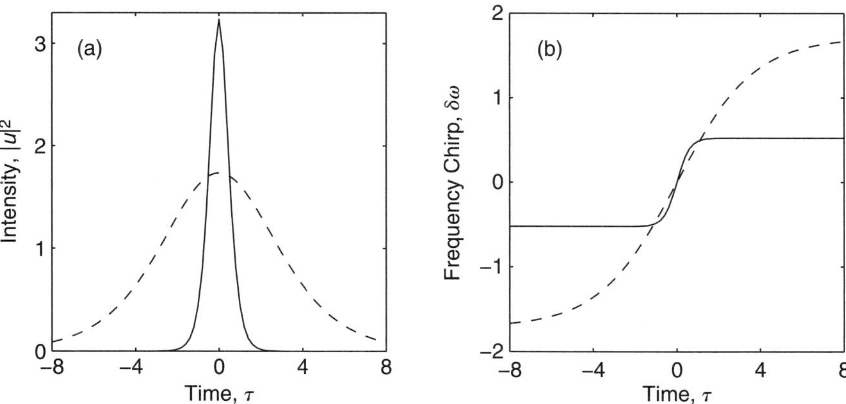

Figure 5.5 (a) Intensity and (b) chirp profiles of an autosoliton when $d = 0.5$. Solid and dashed curves correspond to the cases of normal and anomalous GVD, respectively.

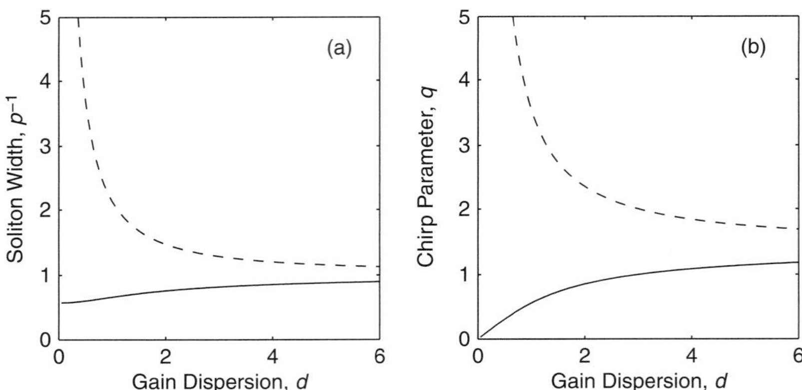

Figure 5.6 (a) Soliton width p^{-1} and (b) chirp parameter q plotted as functions of d. Solid and dashed curves correspond to the cases of normal and anomalous GVD, respectively.

in Figure 5.6, where the width parameter p^{-1} and the chirp parameter q are plotted as functions of d using $\mu = d$ and $\mu_2 = 0$. Solid and dashed curves correspond to the cases of normal ($s = 1$) and anomalous ($s = -1$) GVD, respectively. For large values of d, the difference between normal and anomalous GVD disappears since the soliton behavior is determined by gain dispersion (rather than the frequency dependence of the refractive index of the silica host). In contrast, both the width and the chirp parameters are much larger in the case of normal GVD when $d < 1$. Indeed, both of these parameters tend to infinity as $d \rightarrow 0$ since undoped fibers do not support bright solitons in the case of normal GVD. In the presence of two-photon absorption, the soliton amplitude decreases and its width increases. For most fiber amplifiers μ_2 is so small that its effects can be ignored.

Since gain dispersion and two-photon absorption permit the existence of bright solitons in the normal-GVD region, one may ask whether the Ginzburg–Landau equation has solutions in the form of dark solitary waves that exist in both the normal- and anomalous-GVD regions. This turns out to be the case. Recalling that $\text{sech}(\tau)$ is replaced by $\tanh(\tau)$ for dark solitons in undoped fibers [27], we can postulate the following solution for the Ginzburg–Landau equation:

$$u(\xi, \tau) = N_s[\tanh(p\tau)]^{1+jq} \exp(jK_s\xi)$$
$$\equiv N_s \tanh(p\tau) \exp[jK_s\xi - jq \ln(\cosh p\tau)]. \tag{5.65}$$

As before, the parameters N_s, p, q, and K_s are determined by substituting this solution back into the Ginzburg–Landau equation and are given by a set of equations similar to Eqs. (5.60a) through (5.61). The qualitative behavior of dark autosolitons is also similar to that of bright autosolitons. In particular, gain dispersion determines the frequency chirp imposed on the dark soliton, and the values of p and N_s are fixed by μ and d.

An important issue related to stability needs to be stressed in the case of autosolitons. Because $\mu > 0$ for an amplifier, the background of an autosoliton is not always stable because any small fluctuation in the wings of the pulse (or dip in the case of dark autosolitons) can be amplified by the fiber gain [37]. This instability, arising from the amplification of background noise, has important implications for fiber amplifiers and lasers.

5.6.2 Similaritons

As mentioned earlier, solitons have much resemblance to elementary particles, as they preserve their shape and size during propagation in a dispersive nonlinear medium. However, if we relax the size part and insist only that the shape be maintained during propagation, we land on a more general class of waves known as self-similar pulses, whose amplitude and width vary continuously during their propagation [38]. It was discovered during the 1990s that parabolic pulses could behave as self-similar pulses in EDFAs operating in the normal dispersion regime [39–41]. Interestingly, self-similar pulses tend to behave like an attracting manifold because input pulses with differing shapes tend to become nearly parabolic asymptotically as they are amplified, while maintaining a linear frequency chirp across them. Pulses that evolve in a self-similar fashion are often called *similaritons*.

It is easy to find similariton solutions for the Ginzburg–Landau equation when two-photon absorption and dipole relaxation times are ignored, by setting $T_2 = 0$ and $\alpha_2 = 0$. Equation (5.41) then takes the form

$$\frac{\partial A}{\partial z} + \frac{j\beta_2}{2}\frac{\partial^2 A}{\partial T^2} = j\gamma|A|^2 A + \frac{g}{2}A, \tag{5.66}$$

where net gain g is defined as $g = g_0 - \alpha$. We now use the following ansatz to seek an asymptotic solution in the limit $z \to \infty$ [40]:

$$A(z, T) = a_p(z)F(\tau)\exp[jc_p(z)T^2 + \phi_p(z)], \tag{5.67}$$

where the three pulse parameters, $a_p(z)$, $c_p(z)$, and $\phi_p(z)$, are allowed to evolve with z. The function $F(\tau)$, with $\tau = T(a_p^2 e^{-gz})$ acting as the self-similarity variable, governs the pulse shape. The phase term quadratic in T ensures that the chirp remains linear even though the chirp parameter c_p itself changes with z.

When the preceding form of the solution is substituted back into Eq. (5.66), it leads to the following equation:

$$\left(\frac{dc_p}{dz} - 2\beta_2 c_p^2\right)\frac{\tau^2}{a_p^6}e^{2gz} + \frac{1}{a_p^2}\frac{d\phi_p}{dz} = \gamma F^2(\tau) - \frac{\beta_2}{2}\frac{a_p^2}{F}\frac{d^2 F}{d\tau^2}e^{-2gz}, \qquad (5.68)$$

together with an ordinary differential equation for $a_p(z)$,

$$\frac{da_p}{dz} = \beta_2 c_p a_p + \frac{g}{2}a_p. \qquad (5.69)$$

In the asymptotic limit $z \to \infty$, the last term in Eq. (5.68) tends to zero and thus can be safely discarded. The resulting equation is

$$\left(\frac{dc_p}{dz} - 2\beta_2 c_p^2\right)\frac{\tau^2}{a_p^6}e^{2gz} + \frac{1}{a_p^2}\frac{d\phi_p}{dz} = \gamma F^2(\tau). \qquad (5.70)$$

The right side of this equation is only a function of τ, whereas the left side depends on both z and τ. The only way to ensure that this equality is satisfied for all τ values is to assume that $F^2(\tau) = (1 - \tau^2/\tau_0^2)$, where τ_0 is an arbitrary constant, and equate the two terms on both sides. The resulting two equations are [42]

$$\frac{dc_p}{dz} = 2\beta_2 c_p^2 - \frac{\gamma}{\tau_0^2}a_p^6 e^{-2gz}, \qquad (5.71a)$$

$$\frac{d\phi_p}{dz} = \gamma a_p^2. \qquad (5.71b)$$

The asymptotic solution exists only for a specific pulse shape, $F(\tau) = (1 - \tau^2/\tau_0^2)^{1/2}$ for $|\tau| \le \tau_0$; outside of this range, $F(\tau)$ must vanish for physical reasons.

We have thus found a pulse whose intensity profile $F^2(\tau)$ becomes a perfect parabola for large z and whose parameters evolve with z in a self-similar fashion (i.e., a similariton). When dispersion is normal ($\beta_2 > 0$), it is possible to integrate Eqs. (5.69) and (5.71) and obtain [40]

$$a_p(z) = \tfrac{1}{2}(gE_0)^{1/3}(\gamma\beta_2/2)^{-1/6}\exp(gz/3), \qquad (5.72a)$$

$$T_p(z) = 6g^{-1}(\gamma\beta_2/2)^{1/2}a_p(z), \qquad (5.72b)$$

$$c_p(z) = g/(6\beta_2), \qquad (5.72c)$$

$$\phi_p(z) = \phi_0 + (3\gamma/2g)a_p^2(z), \qquad (5.72d)$$

where ϕ_0 is an arbitrary constant phase and we have used the fact that $\gamma > 0$ for doped fiber amplifiers. The parameter $\tau = T/T_p(z)$ appearing in the pulse shape $F(\tau)$ changes with z, and $T_p(z)$ can be interpreted as a pulse width that scales

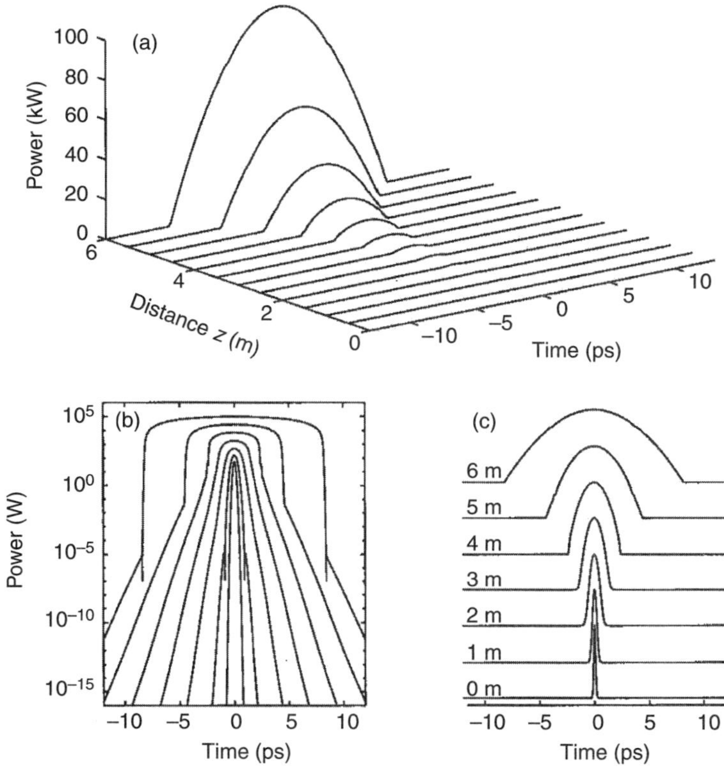

Figure 5.7 (a) Evolution of a 200-fs Gaussian input pulse over a 6-m-long Yb-doped fiber amplifier providing a 50-dB gain. Intensity profiles of the amplified pulse are plotted on (b) logarithmic and (c) linear scales. (After Ref. [42]; © OSA 2002)

linearly with the amplitude $a_p(z)$. The preceding expressions show that both the width and the amplitude of the parabolic pulse increase exponentially with z in the asymptotic regime. Neither the initial pulse shape nor its width has any effect on the final parabolic pulse. Only the energy E_0 of input pulse appears in the final pulse parameters.

The evolution toward a parabolic shape can be verified numerically by solving Eq. (5.66) using the split-step Fourier method [42]. Figure 5.7 shows the simulated evolution of a 200-fs Gaussian input pulse with 12 pJ of energy, launched in a 6-m-long fiber amplifier doped with ytterbium and pumped to provide a net amplification of 50 dB at its output ($g = 1.92$ m^{-1}). Realistic values of $\beta_2 = 25$ ps^2 km^{-1} and $\gamma = 5.8$ W^{-1} km^{-1} were used in numerical simulations. It is clear from Figure 5.7 that the pulse shape indeed becomes parabolic near the output end of the amplifier. Also, the pulse width increases with amplification, as predicted theoretically. Indeed, the shape, the chirp profile, and the spectrum of the output pulse agree well with the predictions of the analytic self-similar solution.

Parabolic pulses have been observed in several experiments in which picosecond or femtosecond pulses were amplified in the normal-dispersion regime of a fiber amplifier. In a 2000 experiment, a 200-fs pulse with 12 pJ of energy was launched into a 3.6-m-long Yb-doped fiber amplifier pumped to provide a 30 dB gain [40]. The intensity and phase profiles of the amplified output pulses were deduced using the technique of frequency-resolved optical gating (FROG). The experimental results agreed well with both the numerical results obtained by solving the NLS equation and the asymptotic parabolic-pulse solution.

5.7 Amplification of femtosecond pulses

We saw in Section 5.5 that one has to consider the nonlinear interaction of dopants with the optical field to accurately characterize the pulse propagation through fiber amplifiers. In the case of picosecond pulses, we could simplify the problem because pulse width is much larger than atomic dipole dephasing time T_2. In the parabolic gain approximation in which the gain spectrum is approximated by a parabola, propagation of picosecond pulses is governed by the Ginzburg–Landau equation. However, if input pulses are much shorter than 1 ps, it may become necessary to include the higher-order nonlinear and dispersive effects, or solve Maxwell's equations and optical Bloch equations simultaneously using a numerical approach (see Section 4.6 for details).

The spectral bandwidth of femtosecond pulses may exceed the gain bandwidth if pulses are shorter than the time scale T_2 associated with the dopants. Moreover, spectral bandwidth is likely to increase in the amplifier because of SPM and other nonlinear effects [43]. Because spectral tails of a pulse are then amplified much less than its central part, femtosecond pulses will inevitably experience temporal and spectral distortion. It has been found numerically that the Ginzburg–Landau equation introduces a systematic error in the energy extracted by the pulse from the gain medium [44] because it approximates the actual spectrum by a parabola. One may ask how the autosoliton associated with the Ginzburg–Landau equation, obtained in the parabolic-gain approximation, changes when the Lorentzian shape of the gain spectrum is taken into account.

An approximate approach for dealing with femtoseond pulses relaxes the parabolic-gain approximation made in deriving the Ginzburg–Landau equation. The resulting generalized Ginzburg–Landau equation can be written, in terms of soliton units, as [27, 44]

$$
j\frac{\partial u}{\partial \xi} \pm \frac{1}{2}\frac{\partial^2 u}{\partial \tau^2} + |u|^2 u - j\delta_3 \frac{\partial^3 u}{\partial \tau^3} + js_0 \frac{\partial |u|^2 u}{\partial \tau} - \tau_R u \frac{\partial |u|^2}{\partial \tau}
$$
$$
= \frac{j}{2} g_0 L_d \int_{-\infty}^{\infty} \frac{\tilde{u}(\xi, f)\exp(-jf\tau)df}{1 - j(f - \omega_0 T_0)(T_2/T_0)} - \frac{j}{2}\alpha L_D u,
$$

(5.73)

where $\widetilde{u}(\xi, f)$ is the Fourier transform of $u(\xi, \tau)$. The simple Ginzburg–Landau equation of Section 5.5 can be recovered from this equation by expanding the integrand around $\omega_0 T_0$ up to second order in f. The parameters δ_3, s_0, and τ_R, responsible for third-order dispersion, self-steepening, and intrapulse Raman scattering, respectively, are defined as [27]

$$\delta_3 = \frac{\beta_3}{6|\beta_2|T_0}, \quad s_0 = \frac{1}{\omega_0 T_0}, \quad \tau_R = \frac{T_R}{T_0}, \tag{5.74a}$$

where T_0 is the pulse width and T_R is the Raman parameter (about 3 fs for a silica fiber). The self-steepening parameter s_0 is negligible except for extremely short pulses (<10 fs wide). Third-order dispersive effects are also negligible unless a fiber amplifier operates very close to the zero-dispersion wavelength of the fiber. In contrast, the parameter τ_R governs the frequency shift induced by intrapulse Raman scattering, and its effects should be included for pulse widths below 5 ps.

As an example, Figure 5.8 shows the pulse evolution over 2.5 dispersion lengths when a 50-fs-wide fundamental soliton is amplified by an EDFA providing a 10 dB gain over each dispersion length ($g_0 L_D = 2.3$). The parameter values used for this figure were $\tau_R = 0.1$, $\delta_3 = 0.01$, $s_0 = 0$, and $T_2/T_0 = 0.2$. The pulse becomes narrower as it is amplified and is compressed by a factor of about 5 after one dispersion length ($\xi = 1$). It splits into two subpulses soon after $\xi = 1$ and evolves toward four pulses at a distance of $\xi = 2.5$. Each pulse shifts toward the right side as it slows down because of Raman-induced spectral shift. The widths and amplitudes of these subpulses are still governed approximately by the autosoliton solution given earlier. The formation of subpulses can be understood by noting that each autosoliton has a fixed energy because its width is set by the time scale T_2. If the amplifier provides so much gain that the pulse energy begins to exceed this value, the pulse splits into multiple subpulses, each of which evolves toward an autosoliton.

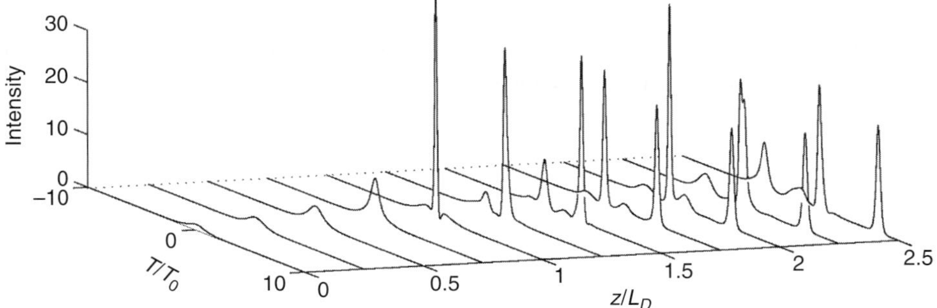

Figure 5.8 Evolution of a 50-fs-wide fundamental soliton over 2.5 dispersion lengths, showing pulse splitting and the Raman-induced temporal delay of subpulses.

For optical pulses much shorter than T_2, in principle one should use the complete set of Maxwell–Bloch equations given in Section 4.7. If the induced polarization of the dopants is given by P_d, then optical Bloch equations can be written in the form

$$\frac{\partial P_d}{\partial t} = -\frac{P_d}{T_2} - i(\omega_a - \omega_0)P_d - \frac{id^2}{\hbar}AW, \tag{5.75a}$$

$$\frac{\partial W}{\partial t} = \frac{W_0 - W}{T_1} + \frac{1}{\hbar}\mathrm{Im}(A^*P_d), \tag{5.75b}$$

where d is the dipole moment, ω_a is the atomic transition frequency, $W = N_2 - N_1$ is the density of population inversion with its initial value W_0, T_1 and T_2 are the population and dipole-relaxation times, and $A(z, t)$ is the slowly varying amplitude associated with the optical field. In terms of induced polarization, Eq. (5.73) takes the form

$$\begin{aligned}
j\frac{\partial u}{\partial \xi} &\pm \frac{1}{2}\frac{\partial^2 u}{\partial \tau^2} + |u|^2 u - j\delta_3\frac{\partial^3 u}{\partial \tau^3} + js_0\frac{\partial |u|^2 u}{\partial \tau} - \tau_R u\frac{\partial |u|^2}{\partial \tau} \\
&= \frac{j\omega_0 L_d}{\varepsilon_0 c}\langle P_d \exp(-i\beta_0 z)\rangle,
\end{aligned} \tag{5.76}$$

where angle brackets denote spatial averaging over the mode profile of the optical fiber. An average over the atomic transition frequencies should also be performed if one wants to include the effects of inhomogeneous broadening. The complete set of Maxwell–Bloch equations can be solved using the split-step Fourier method [27]. The predictions of such a model agree well with experiments dealing with the amplification of femtosecond pulses in fiber amplifiers [45, 46].

References

[1] E. Desurvire, *Erbium-Doped Fiber Amplifiers: Principles and Applications*. Wiley, 1994.

[2] P. C. Becker, N. A. Olsson, and J. R. Simpson, *Erbium-Doped Fiber Amplifiers: Fundamentals and Technology*. Academic Press, 1999.

[3] M. J. Digonnet, ed., *Rare-Earth-Doped Fiber Lasers and Amplifiers*, 2nd ed. CRC Press, 2001.

[4] K. Patek, *Glass Lasers*. CRC Press, 1970.

[5] C. B. Layne, W. H. Lowdermilk, and M. J. Weber., "Multiphon relaxation of rare earth ions in oxide glasses," *Phys. Rev. B.*, vol. 16, pp. 10–20, 1977.

[6] M. D. Shin, W. A. Sibley, M. G. Drexhage, and R. N. Brown, "Optical transitions of Er^3 ions in fluorozirconate glass," *Phys. Rev. B*, vol. 27, pp. 6635–6648, 1983.

[7] C. R. Giles, E. Desurvire, and J. R. Simpson, "Transient gain and cross talk in erbium-doped fiber amplifiers," *Opt. Lett.*, vol. 14, pp. 880–882, 1989.

[8] C. R. Giles and E. Desurvire, "Modeling erbium-doped fiber amplifiers," *J. Lightw. Technol.*, vol. 9, pp. 271–283, 1991.

[9] A. A. M. Saleh, R. M. Jopson, J. D. Evankow, and J. Aspell, "Modeling of gain in erbium-doped fiber amplifiers," *IEEE Photon. Technol. Lett.*, vol. 2, pp. 714–717, 1990.

[10] M. Nakazawa, Y. Kimura, and K. Suzuki, "Efficient Er^{3+}-doped optical fiber amplifier pumped by a 1.48 μm InGaAsP laser diode," *Appl. Phys. Lett.*, vol. 54, pp. 295–297, 1989.

[11] P. C. Becker, J. R. Simpson, N. A. Olsson, and N. K. Dutta, "High-gain and high-efficiency diode laser pumped fiber amplifier at 1.56 μm," *IEEE Photon. Technol. Lett.*, vol. 1, pp. 267–269, 1989.

[12] M. Yamada, M. Shimizu, T. Takeshita, *et al.*, "Er^{3+}-doped fiber amplifier pumped by 0.98 μm laser diodes," *IEEE Photon. Technol. Lett.*, vol. 1, pp. 422–424, 1989.

[13] M. Shimizu, M. Yamada, M. Horiguchi, T. Takeshita, and M. Okayasu, "Erbium-doped fibre amplifiers with an extremely high gain coefficient of 11.0 dB/mW," *Electron. Lett.*, vol. 26, pp. 1641–1643, 1990.

[14] M. Nakazawa, Y. Kimura, E. Yoshida, and K. Suzuki, "Efficient erbium-doped fibre amplifier pumped at 820 nm," *Electron. Lett.*, vol. 26, pp. 1936–1938, 1990.

[15] B. Pedersen, A. Bjarklev, H. Vendeltorp-Pommer, and J. H. Povlsen, "Erbium doped fibre amplifier: Efficient pumping at 807 nm," *Opt. Commun.*, vol. 81, pp. 23–26, 1991.

[16] Y. Sun, J. L. Zyskind, and A. K. Srivastava, "Average inversion level, modeling, and physics of erbium-doped fiber amplifiers," *IEEE J. Sel. Topics Quantum Electron.*, vol. 3, pp. 991–1007, 1997.

[17] D. E. McCumber, "Theory of phonon-terminated optical masers," *Phys. Rev.*, vol. 134, pp. A299–A306, 1964.

[18] W. J. Miniscalco and R. S. Quimby, "General procedure for the analysis of Er^{3+} cross-section," *Opt. Lett.*, vol. 16, pp. 258–260, 1991.

[19] E. Desurvire and J. R. Simpson, "Evaluation of $^4I_{15/2}$ and $^4I_{13/2}$ Stark level energies in erbium-doped aluminosilicate fibers," *Opt. Lett.*, vol. 15, pp. 547–549, 1990.

[20] W. L. Barnes, R. I. Laming, P. R. Morkel, and E. J. Tarbox, "Absorption-emission cross-section ratio for Er^{3+} doped fibers at 1.5 μm," in *Proc. Conference on Lasers and Electro-Optics*, paper JTUA3. Optical Society of America, 1990.

[21] A. E. Siegman, *Lasers*. University Science Books, 1986.

[22] P. W. Milonni and J. H. Eberly, *Lasers*, 2nd ed. Wiley, 2010.

[23] S. Novak and A. Moesle, "Analytic model for gain modulation in EDFAs," *J. Lightw. Technol.*, vol. 20, pp. 975–985, 2002.

[24] Y. Sun, G. Luo, J. L. Zyskind, A. A. M. Saleh, A. K. Sirivastava, and J. W. Sulhoff, "Model for gain dynamics in erbium-doped fiber amplifiers," *Electron. Lett.*, vol. 32, pp. 1490–1491, 1996.

[25] A. Bononi and L. A. Rusch, "Doped-fiber amplifier dynamics: A system perspective," *J. Lightw. Technol.*, vol. 16, pp. 945–956, 1998.

[26] K. Nakagawa, S. Nishi, K. Aida, and E. Yoneda, "Trunk and distribution network application of erbium-doped fiber amplifier," *J. Lightw. Technol.*, vol. 9, pp. 198–208, 1991.

[27] G. P. Agrawal, *Nonlinear Fiber Optics*, 4th ed. Academic Press, 2007.

[28] ——, "Optical pulse propagation in doped fiber amplifiers," *Phys. Rev. A*, vol. 44, pp. 7493–7501, 1991.

[29] M. Asobe, T. Kanamori, and K. Kubodera, "Applications of highly nonlinear chalcogenide glass fibers in ultrafast all-optical switches," *IEEE J. Quantum Electron.*, vol. 2, pp. 2325–2333, 1993.

[30] D. Scholz and M. Weyrauch, "A note on the Zassenhaus product formula," *J. Phys. A*, vol. 47, p. 033505 (7 pages), 2006.

[31] G. Strang, "On the construction and comparison of difference schemes," *SIAM J. Numer. Anal.*, vol. 5, pp. 506–517, 1968.

[32] P. A. Belanger, L. Gagnon, and C. Pare, "Solitary pulses in an amplified nonlinear dispersive medium," *Opt. Lett.*, vol. 14, pp. 943–945, 1989.

[33] W. van Saarloos and P. Hohenberg, "Fronts, pulses, sources and sinks in generalized complex Ginzburg–Landau equations," *Physica D*, vol. 56, pp. 303–367, 1992.

[34] N. N. Akhmediev and A. Ankiewicz, eds., *Dissipative Solitons*. Springer, 2005.

[35] V. S. Grigoryan and T. S. Muradyan, "Evolution of light pulses into autosolitons in nonlinear amplifying media," *J. Opt. Soc. Am. B*, vol. 8, pp. 1757–1765, 1991.

[36] N. Pereira and L. Stenflo, "Nonlinear Schrödinger equation including growth and damping," *Phys. Fluids*, vol. 20, pp. 1733–1734, 1977.

[37] N. N. Akhmediev and A. Ankiewicz, *Solitons: Nonlinear Pulses and Beams*. Chapman and Hall, 1997.

[38] G. I. Barenblatt, *Scaling, Self-Similarity, and Intermediate Asymptotics*. Cambridge University Press, 1996.

[39] K. Tamura and M. Nakazawa, "Pulse compression by nonlinear pulse evolution with reduced optical wave breaking in erbium-doped fiber amplifiers," *Opt. Lett.*, vol. 21, pp. 68–70, 1996.

[40] M. E. Fermann, V. I. Kruglov, B. C. Thomsen, J. M. Dudley, and J. D. Harvey, "Self-similar propagation and amplification of parabolic pulses in optical fibers," *Phys. Rev. Lett.*, vol. 84, pp. 6010–6013, 2000.

[41] S. Wabnitz, "Analytical dynamics of parabolic pulses in nonlinear optical fiber amplifiers," *IEEE Photon. Technol. Lett.*, vol. 19, pp. 507–509, 2007.

[42] V. I. Kruglov, A. C. Peacock, J. D. Harvey, and J. M. Dudley, "Self-similar propagation of parabolic pulses in normal-dispersion fiber amplifiers," *J. Opt. Soc. Am. B*, vol. 19, pp. 461–469, 2002.

[43] V. V. Afanasjev, V. N. Serkin, and V. A. Vysloukh, "Amplification and compression of femtosecond optical solitons in active fibers," *Sov. Lightwave Commun.*, vol. 2, pp. 35–58, 1992.

[44] C. Paré and P. A. Bélanger, "Optical solitary waves in the presence of a Lorentzian gain line: Limitations of the Ginzburg–Landau model," *Opt. Commun.*, vol. 145, pp. 385–392, 1998.

[45] M. Nakazawa, K. Kurokawa, H. Kubota, K. Suzuki, and Y. Kimura, "Femtosecond erbium-doped optical fiber amplifier," *Appl. Phys. Lett.*, vol. 57, pp. 653–655, 1990.

[46] K. Kurokawa and M. Nakazawa, "Wavelength-dependent amplification characteristics of femtosecond erbium-doped optical fiber amplifiers," *Appl. Phys. Lett.*, vol. 58, pp. 2871–2873, 1991.

6

Semiconductor optical amplifiers

Semiconductor optical amplifiers (SOAs) are increasingly used for optical signal processing applications in all-optical integrated circuitry [1, 2]. Research on SOAs started just after the invention of semiconductor lasers in 1962 [3]. However, it was only after the 1980s that SOAs found widespread applications [4, 5]. The effectiveness of SOAs in photonic integrated circuits results from their high gain coefficient and a relatively low saturation power [6, 7]. In addition, SOAs are often used for constructing functional devices such as nonlinear optical loop mirrors [8, 9], clock-recovery circuits [10, 11], pulse-delay discriminators [12–14], and logic elements [15, 16].

A semiconductor, as its name implies, has a conductivity in between that of a conductor and an insulator. Some examples of elemental semiconductors include silicon, germanium, selenium, and tellurium. Such group-IV semiconductors have a crystal structure similar to that of diamond (a unit cell with tetrahedral geometry) and the same average number of valence electrons per atom as the atoms in diamond. Compound semiconductors can be made by combining elements from groups III and V or groups II and VI in the periodic table. Two group III–V semiconductors commonly used for making SOAs are gallium arsenide (GaAs) and indium phosphide (InP). These semiconductors enable one to manipulate properties such as conductivity by doping them with impurities, and allow the formation of the p–n junctions required for the electrical pumping of SOAs.

After discussing in Section 6.1 the material and design aspects of SOAs, we provide a rate-equation description in Section 6.2 that governs how the density of injected electrons and holes changes with time in the presence of light. We solve the rate equations approximately in Section 6.3 for the amplification of picosecond pulses in SOAs, using an analytical approach. We use a versatile technique based on multiple scales to construct the analytical solution. The method used here can be easily generalized to other types of amplifiers discussed in this book. We adopt in Section 6.4 the Maxwell–Bloch formulation of Section 4.5 to study amplification of femtosecond pulses in SOAs and provide an intuitive derivation of the density-matrix formulation.

6.1 Material aspects of SOAs

The most important property of a semiconductor from the standpoint of optics applications is the bandgap of the semiconductor. Figure 6.1 shows the two types of energy-band diagrams for naturally occurring semiconductors, plotting the electron's energy as a function of its momentum p (or wave vector, since $p = \hbar k$). All semiconductors exhibit a bandgap that separates their valence band from the conduction band. In the indirect-bandgap semiconductor shown in Figure 6.1(a), the minimum energy point of the conduction band is not aligned with the maximum energy point of the valence band. Physically, this means that an electron in the conduction band cannot drop down to the valence band while emitting some if its energy in the form of a photon, because total momentum cannot be conserved during this process. For this reason, light emission in indirect-bandgap materials (such as silicon) is an inefficient process because it requires the participation of lattice vibrations (or phonons) to conserve momentum.

In contrast, in a direct-bandgap semiconductor, with an energy-band diagram like that shown in Figure 6.1(b), the conduction-band minimum is perfectly aligned with the valence-band maximum. As a result, radiative transitions can take place without the assistance of phonons. Clearly, semiconductors with a direct bandgap are preferred for all applications that require optical amplification, and are used routinely for making semiconductor lasers and amplifiers. An exception occurs in the case

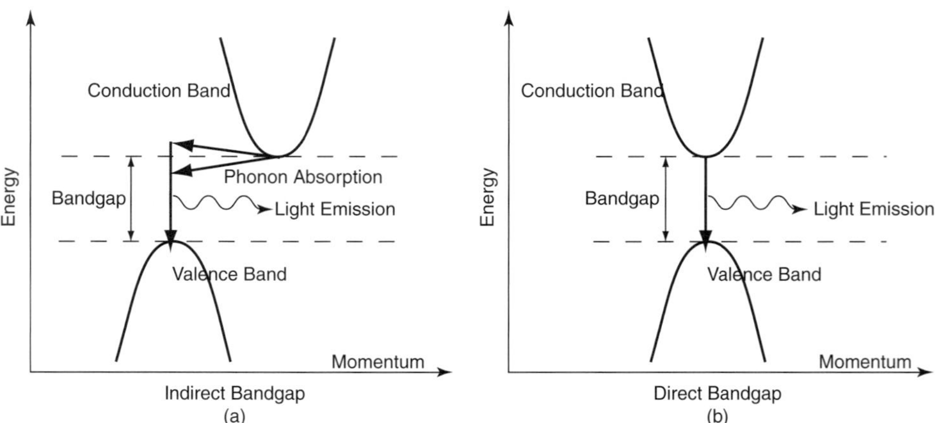

Figure 6.1 Schematic illustration of (a) indirect and (b) direct bandgap semiconductors. The direct transition is preferred because it does not need the assistance of phonons.

of silicon optical amplifiers, but they make use of Raman gain and require optical pumping. SOAs provide optical gain through electrical pumping with relatively low currents and must make use of direct-bandgap materials. Electrical pumping injects electrons into the conduction band and creates a population inversion when the density of electron–hole pairs exceeds a certain value. The maximum gain under normal conditions occurs close to the bandgap E_g, and the SOA provides amplification at wavelengths near $\lambda = hc/E_g$. Clearly, the semiconductor material must be chosen to match the operating wavelength of the application. Compound semiconductors used to make SOAs are ternary (containing three elements) or quaternary (containing four elements) crystalline alloys grown on lattice-matched binary substrates (such as GaAs or InP). By varying the composition of elements making up the ternary or quaternary compound, it is possible to vary the bandgap of the material and the wavelength at which most amplification occurs.

Once the operating wavelength is determined, the other most important criterion in selecting semiconductor materials and their composition is related to lattice matching. SOAs make use of a planar waveguide to confine light during its amplification and thus consist of multiple semiconductor layers with different bandgaps and refractive indices. However, the various semiconducting materials making up an SOA device must have lattice constants that match to better than 0.1%. If such matching is not achieved, the quality of the semiconductor heterostructure making up the device tends to deteriorate because it develops lattice defects at interfaces, which serve as nonradiative recombination centers.

Figure 6.2 shows the relationship between the bandgap and the lattice constant for several ternary and quaternary compounds [17]. Solid dots represent binary semiconductors, and the lines connecting them correspond to ternary compounds. The dashed portion of each line indicates that the relevant ternary compound has an indirect bandgap. The area of a closed polygon corresponds to quaternary compound. The bandgap is not necessarily direct for such semiconductors. The shaded area in Figure 6.2 represents the ternary and quaternary compounds with a direct bandgap formed from the elements indium (In), gallium (Ga), arsenic (As), and phosphorus (P).

SOAs operating near 800 nm make use of ternary compounds $Al_x Ga_{1-x} As$, made by replacing a fraction x of the Ga atoms by Al atoms. The resulting semiconductors have nearly the same lattice constant as GaAs but greater bandgaps. The horizontal line in Figure 6.2 connecting GaAs and AlAs corresponds to such ternary compounds, whose bandgaps are direct for values of x up to about 0.45. The bandgap depends on the fraction x and can be approximated by a simple linear relation

$$E_g(x) = 1.424 + 1.247x, \qquad 0 < x < 0.45, \tag{6.1}$$

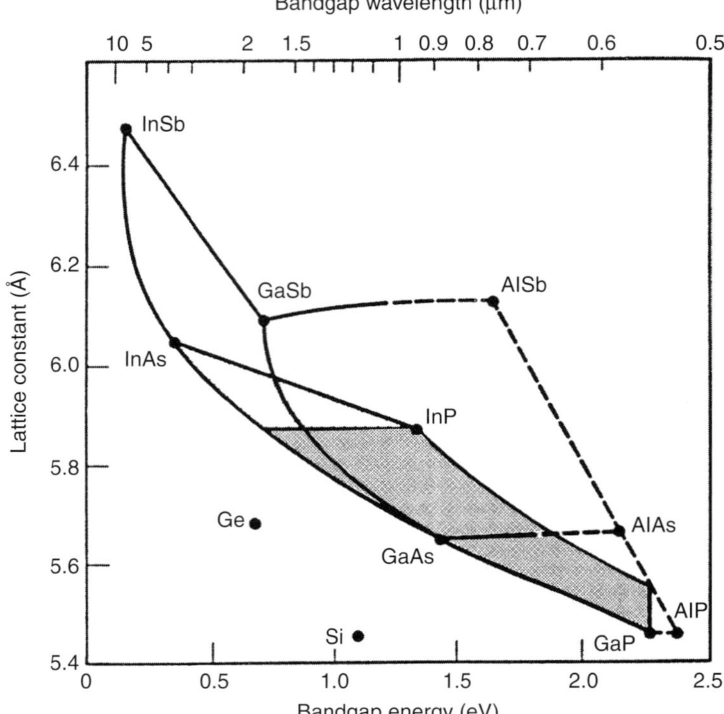

Figure 6.2 Lattice constants and bandgap energies for ternary and quaternary compounds formed from nine group III–V semiconductors. The shaded area corresponds to possible In GaAsP and AlGaAs structures. The horizontal lines passing through InP and GaAs show lattice-matched designs.

where E_g is expressed in units of electron-volts (eV). The core and cladding layers of the SOA waveguide are formed so that x is larger for the cladding layers than for the core layer. Using $E_g \approx h\nu = hc/\lambda$, one finds that $\lambda \approx 0.87$ μm for a core layer made of GaAs ($E_g = 1.424$ eV). The wavelength can be reduced to near 0.8 μm by using an active layer with $x = 0.1$.

The wavelength range of 1.3–1.6 μm is important for telecommunications applications because both dispersion and losses in silica fibers are considerably reduced compared with the values in the 0.85 μm region. The InP semiconductor is the base material for SOAs operating in this wavelength region. As seen in Figure 6.2 from the horizontal line passing through InP, the bandgap of InP can be reduced considerably using a quaternary compound $In_{1-x}Ga_xAs_yP_{1-y}$ whose lattice constant remains matched to InP. The fractions x and y cannot be chosen arbitrarily but are related by $x/y = 0.45$ to ensure matching of the lattice constant. The bandgap of the quaternary compound can be expressed in terms of y only and is well approximated by

$$E_g(y) = 1.35 - 0.72y + 0.12y^2, \tag{6.2}$$

where $0 \leq y \leq 1$. By a suitable choice of the mixing fractions x and y, $In_{1-x}Ga_xAs_yP_{1-y}$ SOAs can be designed to work in a wide wavelength range that includes the 1.3–1.6 µm region that is important for telecommunications applications.

The fabrication of SOAs requires epitaxial growth of multiple layers on a base substrate (GaAs or InP). The thickness and composition of each layer need to be controlled precisely. Several epitaxial growth techniques can be used for this purpose. The three primary techniques are known as liquid-phase epitaxy (LPE), vapor-phase epitaxy (VPE), and molecular-beam epitaxy (MBE), depending on whether the constituents of various layers are in liquid form, vapor form, or in the form of a molecular beam. The VPE technique is also called chemical vapor deposition. A variant of this technique is metal-organic chemical vapor deposition (MOCVD), in which metal alkalis are used as the mixing compounds. An SOA is made by sandwiching a direct bandgap material between two cladding layers with slightly higher bandgaps. The cladding layers are doped to form a p–i–n junction (i stands for intrinsic). The central layer is where amplification occurs, through stimulated emission when current is injected into such a device [5].

6.2 Carrier density and optical gain

When current is injected into an SOA, charge carriers (electrons and holes) enter the active region (the middle i-layer of the p–i–n junction), increasing in density N as the applied current increases. These electron–hole pairs can recombine through several different radiative and nonradiative processes and, in the absence of an optical signal, they typically last for a duration called the carrier lifetime, denoted by τ_e. Without an optical signal that needs to be amplified, the only radiative recombination process that can create photons is spontaneous emission. The incoherent nature of this process creates noise that is amplified by the SOA and results in a relatively weak broadband output called amplified spontaneous emission (ASE). However, when a coherent optical signal is injected at one end of the SOA, it is amplified exponentially in the SOA by stimulated emission, resulting in an amplified output beam corrupted somewhat by the ASE.

Since the rate of stimulated emission depends on the number of signal photons present, it can be made fast enough that most electron–hole pairs recombine though this coherent process. However, it is clear that the distribution of carrier density $N(\boldsymbol{r}, t)$ in the active region will then depend on the intensity of light, which increases with z because of signal amplification. Thus, one must solve a coupled set of equations along the SOA length, similar to the Maxwell–Bloch equations needed for fiber amplifiers. In practice, a rate equation for the carrier density replaces the optical Bloch equations associated with the two-level atoms. In this section we provide details of this rate equation.

6.2.1 Rate equation for carrier density

Similarly to the case of fiber amplifiers, the transverse distribution of light does not change in an SOA because light is transported in the form of a waveguide mode. This feature reduces the problem to one spatial dimension, and we only need to consider how N changes with distance z and time t. The rate equation is written by considering all mechanisms by which carriers are generated locally at z and all processes by which carriers can recombine and thus disappear. This approach leads to the following carrier-density rate equation [18]:

$$\frac{\partial}{\partial t}N(z,t) = \wp(z) - [AN(z,t) + BN^2(z,t) + CN^3(z,t)] - \frac{g(z,t)}{h\nu}|F(z,t)|^2,$$

(6.3)

where $F(z,t)$ is the optical field, $\wp(z)$ is the carrier-injection rate, and $h\nu$ is the photon energy. The parameters A, B, and C account for the rates of three recombination processes that produce a finite carrier lifetime in the absence of an optical signal. More specifically, A is the nonradiative recombination coefficient, B is the spontaneous recombination coefficient, and C is the Auger recombination coefficient. The rate of stimulated emission is governed by the gain coefficient, defined as $g(z,t) = \Gamma a[N(z,t) - N_0]$, where Γ is the mode confinement factor, a is the differential gain coefficient, and N_0 is the carrier density required to achieve transparency. The optical field is normalized such that a pulse launched with energy E_p at the input of the SOA satisfies the relation

$$E_p = A_m \int_{-\infty}^{+\infty} |F(0,t)|^2 \, dt,$$

(6.4)

where A_m is the effective mode area of the SOA active region.

A major assumption behind Eq. (6.3) is that the electron and hole densities are equal to one another (i.e., charge neutrality is maintained) and given by $N(z,t)$. Another assumption is that diffusion of minority carriers can be neglected. Moreover, we have neglected the frequency dependence of the optical gain over the pulse bandwidth and approximated the gain dependency on minority carrier density using a simple linear expression. In practice, Eq. (6.3) is rewritten as [18]

$$\frac{\partial}{\partial t}N(z,t) = \wp(z) - \frac{N(z,t)}{\tau_e} - \frac{\Gamma a}{h\nu}[N(z,t) - N_0]|F(z,t)|^2,$$

(6.5)

where the carrier lifetime τ_e is defined as $1/\tau_e = A + BN + CN^2$.

6.2.2 Gain characteristics of SOAs

We saw earlier in this section that optical gain is approximately a linear function of the carrier density. However, optical amplifier gain is also dependent on the frequency. This functional dependence on frequency is usually referred to as the gain spectrum. The gain spectrum of an SOA is nonzero only for a limited range of frequencies, and is commonly characterized using a finite bandwidth value. Interestingly, the bandwidth of an SOAs gain spectrum is not constant for a given material composition of the active medium, but changes with the amount of current injected into it. Under normal operating conditions this bandwidth is greater than 4 THz, making it a relatively broadband amplifier. The shape and width of the gain spectrum depends also on the material composition of the SOA and thus can be manipulated by varying the fractions of constituting elements. However, the optical gain must be calculated numerically as a function of frequency, injection signal, and signal power because it also depends on details of the band structure. At moderate power levels, the optical gain of a semiconductor can be approximated in the same form as that of a homogeneously broadened two-level system [19, 20]:

$$g(\omega) = \frac{g_0}{1 + (\omega - \omega_0)^2 T_2^2 + P/P_s},$$ (6.6)

where g_0 is the peak value of the gain determined by the pumping level of the amplifier, ω is the frequency of the amplified signal, ω_0 is the frequency at which gain peaks, and P is the optical power of the signal being amplified. The saturation power P_s of the gain medium depends on the material parameters and is typically below 10 mW for SOAs. The parameter T_2, known as the dipole relaxation time in the atomic case, now depends on the intraband scattering time of the carriers. Typical values are below 100 fs for most semiconductors.

At low signal powers satisfying the condition $P/P_s \ll 1$, the gain coefficient in Eq. (6.6) can be approximated as

$$g(\omega) = \frac{g_0}{1 + (\omega - \omega_0)^2 T_2^2}.$$ (6.7)

This equation shows that the gain spectrum of SOAs can be approximated with a Lorentzian profile. The gain bandwidth, defined as the full width at half-maximum (FWHM) of the gain spectrum, is given by $\Delta v_g = 1/\pi T_2$, resulting in a value of about 5 THz if we use $T_2 = 60$ fs. This large bandwidth of SOAs suggests that they are capable of amplifying ultrashort optical pulses (as short as 1 ps) without significant pulse distortion. However, when the pulse width τ_p becomes shorter than the carrier lifetime τ_e, gain dynamics play an important role, since both the carrier density and the associated gain become time-dependent. This can lead to considerable spectral broadening, a phenomenon discussed in the next section.

It is useful to write an explicit rate equation governing the gain dynamics in SOAs. For this purpose we use Eq. (6.5) together with the definition $g(z, t) = \Gamma a[N(z, t) - N_0]$. Replacing N by $N_0 + g(z, t)/(\Gamma a)$ in Eq. (6.5), we obtain the following rate equation for the gain coefficient:

$$\frac{\partial g}{\partial t} = \frac{g_0 - g}{\tau_e} - \frac{g P}{\tau_e P_s},$$ (6.8)

where we have defined the two parameters appearing in the preceding equation as

$$g_0 = \Gamma a(\wp \tau_e - N_0), \qquad P_s = h\nu A_m/(a\tau_e).$$ (6.9)

As a simple application of the gain equation (6.8), we consider the amplification of a CW signal in an SOA of length L. Similarly to the case of fiber amplifiers in Chapter 5, the amplification process is described by the simple equation

$$\frac{dP}{dz} = g(z)P(z),$$ (6.10)

where $g(z)$ is obtained from Eq. (6.8) after setting the time derivative to zero:

$$g(z) = \frac{g_0}{1 + P(z)/P_s}.$$ (6.11)

The integration of Eq. (6.10) over the amplifier length is straightforward, and the amplification factor, defined as $G = P(L)/P(0)$, is obtained from the transcendental equation

$$\ln G = g_0 L - (G - 1)P(0)/P_s.$$ (6.12)

Notice that the last term in this equation results from gain saturation. If gain saturation is negligible, we recover the unsaturated amplifier gain $G_0 = \exp(g_0 L)$. Replacing G by G_0 in the saturation term, an approximate solution is given by

$$\ln G \approx \ln G_0 - (G_0 - 1)P(0)/P_s.$$ (6.13)

However, this solution should be used with care because it is valid only if the input power satisfies the condition $P(0)/P_s \ll \ln G_0/(G_0 - 1)$. We solve the gain equation (6.8) under quite general conditions in the next section, where we discuss the amplification of picosecond pulses in an SOA.

6.3 Picosecond pulse amplification

Propagation of picosecond pulses in SOAs has been studied extensively for applications in optical signal processing and optical communications areas. The problem is similar to that studied in Section 5.5 in the context of EDFAs except that the gain dynamics of SOAS are governed by Eq. (6.8), and we must solve this together with Maxwell's equations. Even though it is possible to solve these equations numerically, such an approach does not provide enough physical insight. In this section we present an analytic approach developed in 1989 by Agrawal and Olsson [6] and extended further in more recent work [18].

Pulse amplification in two-level atomic media was studied as early as 1963 [21], and this approach has attracted considerable attention since then [19, 22, 23]. In this technique, the optical Bloch equations, in order are reduced to a set of rate equations, in order to calculate the amplifier gain for a given input pulse energy without taking into account details of the pulse shape. It relies on the assumption that the stimulated-emission-induced gain depletion by a short pulse is so fast that spontaneous emission can be ignored. Siegman [22] showed how these results can be recast in terms of output pulse energies, and derived a transcendental equation relating the input and output pulse energies.

6.3.1 General theory of pulse amplification

In the context of SOAs, a similar approach can be applied to pulses whose full width at half maximum (FWHM) is smaller than the carrier lifetime ($\tau_e \sim 1$ ns). More specifically, the following analysis applies to picosecond pulses, but not to femtosecond pulses for which intraband relaxation processes such as electron–electron scattering and electron–phonon scattering may not be fast enough to thermalize the distribution of electrons within the conduction band. It was found in a 1989 study that amplification of picosecond pulses leads to both temporal distortion and spectral broadening [6]. Premaratne *et al.* [13, 18] have recently extended this technique to describe amplification of counterpropagating pulse trains in SOAs. Their results show that the spatial distribution of the carrier density can also be described accurately for the duration of pulse amplification and beyond.

Figure 6.3 shows a schematic of the amplification process considered in this section. Picosecond pulses are launched at $z = 0$ and amplify as they propagate within an SOA of length L. To simplify the following analysis, we do not consider any backward propagating waves, resulting from either amplified spontaneous emission (ASE) or partial reflections at the two facets. This amounts to neglecting noise added by ASE and to assuming that the SOA is a traveling-wave amplifier. Even though these approximations somewhat limit our analysis, past work has shown that relaxing them only introduces second-order effects of minor importance [14, 24].

As we saw in Section 5.5, the propagation of pulses in a dispersive nonlinear medium such as an optical fiber is governed by the NLS equation (5.30). We can start with this equation but it simplifies considerably for SOAs because neither the dispersive effects nor the nonlinear effects, governed by the parameters β_2 and γ, are important because of their extremely short lengths (1 mm or less). Setting these parameters to zero but adding the gain $g(z, t)$ provided by an SOA, we obtain

$$\frac{\partial}{\partial z} A(z, t) + \frac{1}{v_g} \frac{\partial}{\partial t} A(z, t) = \frac{1}{2}(1 - j\beta_c)g(z, t)A(z, t) - \alpha A(z, t), \qquad (6.14)$$

where the parameter β_c accounts for small changes in the refractive index which occur when the carrier density changes in response to the stimulated emission that amplifies the optical pulse. The inclusion of such carrier-induced index changes is essential because they lead to frequency chirping and spectral changes during pulse amplification [6]. This equation should be solved together with the rate equation (6.5) which describes carrier dynamics in SOAs.

Noting that the carrier rate equation (6.5) involves only the optical intensity $I(z, t) = |A(z, t)|^2$, we convert Eq. (6.14) into the following intensity equation [6]:

$$\frac{\partial}{\partial z} I(z, t) + \frac{1}{v_g} \frac{\partial}{\partial t} I(z, t) = g(z, t)I(z, t) - \alpha I(z, t). \qquad (6.15)$$

To make subsequent analysis easier, we make the coordinate transformations $\xi = z$ and $\tau = t - z/v_g$ so that we are in a reference plane that moves with the forward propagating pulse. The transformed equations take the form

$$\frac{\partial}{\partial \xi} I(\xi, \tau) = g(\xi, \tau)I(\xi, \tau) - \alpha I(\xi, \tau), \qquad (6.16)$$

$$\frac{\partial}{\partial \tau} N(\xi, \tau) = \wp(\xi) - \frac{N(\xi, \tau)}{\tau_e} - g(\xi, \tau)\frac{I(\xi, \tau)}{h\nu}. \qquad (6.17)$$

Equation (6.16) can easily be solved as an initial-value problem for which $I(0, \tau) = I_0(\tau)$ at $z = 0$, where $I_0(\tau)$ represents the intensity profile of the input

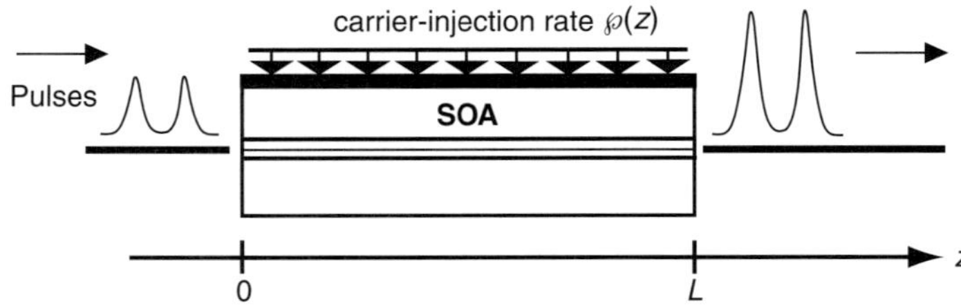

Figure 6.3 Schematic of amplification of picosecond pulses in an SOA of length L when pulses are incident on its left facet at $z = 0$.

pulse. The solution is given by

$$I(\xi, \tau) = I_0(\tau) \exp\left(\int_0^\xi [g(\xi, \tau) - \alpha]\, d\xi \right). \tag{6.18}$$

It is clear from the structure of this equation that subsequent calculations can be simplified by introducing a new variable $h(\xi, \tau)$ with the definition

$$h(\xi, \tau) = \int_0^\xi g(\xi, \tau)\, d\xi. \tag{6.19}$$

Substitution of Eq. (6.19) into Eq. (6.17) gives us the following integro-differential equation describing the gain dynamics of an SOA:

$$\begin{aligned}
\frac{\partial}{\partial \tau_n} h(\xi, \tau_n) = {} & \varepsilon \left[h_\wp - h(\xi, \tau_n) \right] - \beta(\tau_n) \left\{ \exp\left[h(\xi, \tau_n) - \alpha\xi \right] - 1 \right\} \\
& - \beta(\tau_n) \left(\alpha \int_0^\xi \exp\left[h(\xi, \tau_n) - \alpha\xi \right] d\xi \right),
\end{aligned} \tag{6.20}$$

where we have introduced a normalized time $\tau_n = \tau/T_s$, $T_s = L/v_g$ being the single-pass transit time through the SOA. The dimensionless parameter $\varepsilon = T_s/\tau_e$ is < 1 for SOAs in which transit time is typically below 5 ps. The parameters β and $h_\wp$ are defined as

$$\beta(\tau_n) = \frac{\Gamma a T_s}{h\nu} I_0(T_s \tau_n), \tag{6.21a}$$

$$h_\wp(\xi) = \int_0^\xi \left[\tau_e \Gamma a \wp(\xi) - \Gamma a N_0 \right] d\xi. \tag{6.21b}$$

The preceding integro-differential equation can be integrated numerically using well-known techniques [25]. However, such a numerical analysis does not provide physical insight because essential dynamical features are not readily evident. Although it is not possible to solve Eq. (6.20) exactly, in what follows we discuss an approximate solution that captures all essential features of the dynamics of gain recovery.

6.3.2 Method of multiple scales

For simplicity, we first assume that losses within the active region are negligible. This is a reasonable approximation for most SOAs with high internal gain. When losses are negligible compared with the gain, we can set $\alpha = 0$ in Eq. (6.20) and obtain

$$\frac{\partial}{\partial \tau_n} h(\xi, \tau_n) = \varepsilon \left[h_\wp - h(\xi, \tau_n) \right] - \beta(\tau_n) \left\{ \exp\left[h(\xi, \tau_n) \right] - 1 \right\}. \tag{6.22}$$

The most effective way of solving this equation approximately is to use the method of multiple scales. The impetus for this approach is the observation that stimulated emission and carrier recovery have two distinct time scales. More specifically, stimulated-emission-induced electron–hole recombinations occur on a fast time scale $T = \tau_n$, whereas the recovery of carrier density though current and nonradiative electron–hole recombinations occurs on a much slower time scale $U = \varepsilon\tau_n \equiv \tau_e$. The underlying idea behind the method of multiple scales is to formulate the original problem in terms of these two time scales from the outset and treat all physical quantities as a function of these two time variables. Even though T and U are interdependent, we treat them as two independent variables. The resulting solution is more general than the solution of the original problem but contains the original solution as a special case. This aspect can be understood by noting that because such a solution is valid in a two-dimensional region, it should also be valid along each and every path in this region.

Since $h(\xi, \tau_n) \equiv h(\xi, T, U, \varepsilon)$ in the multiple-scale description, the partial derivative in Eq. (6.22) is replaced with

$$\frac{\partial}{\partial \tau_n} = \frac{\partial}{\partial T} + \varepsilon \frac{\partial}{\partial U}. \tag{6.23}$$

Substituting this expression into Eq. (6.22), we obtain

$$\frac{\partial}{\partial T} h(\xi, T, U, \varepsilon) + \varepsilon \frac{\partial}{\partial U} h(\xi, T, U, \varepsilon) = $$
$$\varepsilon[h_\wp - h(\xi, T, U, \varepsilon)] - \beta(\tau_n)\{\exp[h(\xi, T, U, \varepsilon)] - 1\}. \tag{6.24}$$

Assuming that ε is a small parameter ($|\varepsilon| \ll 1$), we seek an approximate solution of the preceding equation as a power series expansion,

$$h(\xi, T, U, \varepsilon) = \sum_{n=0}^{\infty} h_n(\xi, T, U)\, \varepsilon^n. \tag{6.25}$$

Substituting Eq. (6.25) into Eq. (6.24) and noting that

$$\exp[h(\xi, T, U, \varepsilon)] \approx \exp[h_0(\xi, T, U)] \times [1 + \varepsilon h_1(\xi, T, U) + \cdots], \tag{6.26}$$

we obtain a partial differential equation as a power series in ε. Because each term in this expansion needs to be identically equal to zero, we obtain an infinite set of equations. The zeroth-order term in ε leads to an equation for h_0,

$$\frac{\partial}{\partial T} h_0(\xi, T, U) = -\beta(T)\{\exp[h_0(\xi, T, U)] - 1\}, \tag{6.27}$$

and the first-order term leads to

$$\frac{\partial}{\partial T} h_0(\xi, T, U) + \frac{\partial}{\partial U} h_1(\xi, T, U) = \left[h_\wp - h_0(\xi, T, U) \right]$$
$$- \beta(T) \exp\left[h_0(\xi, T, U) \right] h_1(\xi, T, U). \tag{6.28}$$

We find h_0 and h_1 in the (T, U) plane by seeking a solution that satisfies Eqs. (6.27) and (6.28) simultaneously.

To solve the preceding two equations, we assume that the initial condition for $h(\xi, T, U, \varepsilon)$ is independent of the small parameter ε. This is a reasonable assumption because the carrier recovery rate does not affect the initial state of the SOA. If $h_I(\xi)$ is the initial profile of $h(\xi, T, U, \varepsilon)$, the initial conditions become

$$h_n(\xi, 0, 0) = \begin{cases} h_I(\xi) & \text{if} \quad n = 0, \\ 0 & \text{otherwise.} \end{cases} \tag{6.29}$$

The differential equation (6.27) can be solved by multiplying it by the integrating factor $\exp\left[-h_0(\xi, T, U)\right]$. The resulting equation takes the form

$$\frac{\partial}{\partial T} \exp\left[-h_0(\xi, T, U\right] = \beta(T) \left\{ 1 - \exp\left[-h_0(\xi, T, U)\right] \right\}. \tag{6.30}$$

This equation can be easily integrated to obtain the solution

$$h_0(\xi, T, U) = -\ln\left(1 - \left\{ 1 - \exp\left[-h_0(\xi, 0, U)\right] \right\} \frac{\varphi(U)}{E_\beta(T)} \right), \tag{6.31}$$

where $\varphi(U)$ is an arbitrary function of the slow time scale U, and $E_\beta(T)$ is defined as

$$E_\beta(T) = \exp\left(\int_0^T \beta(T)\, dT \right). \tag{6.32}$$

A long time after an optical pulse has left the SOA, its internal gain will become independent of the small parameter ε. This observation leads to the condition that $h_n(\xi, \infty, \infty) = 0$ for $n = 1, 2, \cdots$. Using Eq. (6.29), it is possible to show that the general solution of Eq. (6.28) is given by

$$h_0(\xi, T, U) = \left[h_0(\xi, T, 0) + \vartheta(T) \right] \exp(-U) + h_\wp(\xi) \left[1 - \exp[(-U)] \right], \tag{6.33}$$

where $\vartheta(T)$ is an arbitrary function of the fast time scale T. However, this solution must satisfy Eq. (6.27). Enforcing this constraint allows us to obtain a specific functional form for $\vartheta(T)$.

The time-dependent gain G seen by the optical pulse is related to h_0 by the simple relation

$$G(\xi, T, U) = \exp[h_0(\xi, T, U)]. \tag{6.34}$$

Expressions (6.31) and (6.33) for $h_0(\xi, T, U)$ need to be identical in the entire (T, U) space. Considering this, we match the results at $T = U = 0$. This is conveniently done by calculating $G(\xi, 0, 0)$. Equation (6.31) leads to

$$G(\xi, 0, 0) = \frac{1}{1 - \{1 - \exp[-h_I(\xi)]\varphi(0)\}}, \tag{6.35}$$

while Eq. (6.33) gives us the relation

$$G(\xi, 0, 0) = \exp[h_I(\xi)] \exp(\vartheta(0)). \tag{6.36}$$

These two expressions for $G(\xi, 0, 0)$ match only if we choose

$$\varphi(0) = 1 \quad \text{and} \quad \vartheta(0) = 0. \tag{6.37}$$

Taking the exponential of both sides in Eq. (6.33), we obtain

$$G(\xi, T, U) = \exp\left[h_0(\xi, T, 0)\exp(-U)\right] \times \exp\left\{h_\wp(\xi)\left[1 - \exp(-U)\right]\right\} \times \exp\left[\vartheta(T)\exp(-U)\right]. \tag{6.38}$$

Using Eq. (6.37) in Eq. (6.31), we obtain the following expression for $h_0(\xi, T, 0)$:

$$h_0(\xi, T, 0) = -\ln\left(1 - \frac{\{1 - \exp[-h_I(\xi)]\}}{E_\beta(T)}\right). \tag{6.39}$$

Substitution of this result into Eq. (6.38) gives us, finally, the time-dependent gain along the SOA seen by a picosecond pulse:

$$G(\xi, T, U) = \exp[h_\wp(\xi)]\left(\frac{\exp[\vartheta(T) - h_\wp(\xi)]}{1 - \{1 - \exp[-h_I(\xi)]\}/E_\beta(T)}\right)^{\exp(-U)}. \tag{6.40}$$

To fully characterize the gain evolution, we still need to find the functional form of $\vartheta(T)$ in Eq. (6.40). For this purpose, we consider a path in the (T, U) plane along which $U = 0$, and evaluate the partial derivative of Eq. (6.40) along this path. The result is

$$\frac{\partial h_0(\xi, T, U)}{\partial T}\bigg|_{U=0} = \frac{d}{dT}\vartheta(T) - \beta(T)\frac{\{1 - \exp[-h_I(\xi)]\}/E_\beta(T)}{1 - \{1 - \exp[-h_I(\xi)]\}/E_\beta(T)}. \tag{6.41}$$

Substituting Eqs. (6.41) and (6.40) into Eq. (6.27), we obtain the following differential equation for the unknown variable $\vartheta(T)$:

$$\frac{d}{dT}\vartheta(T) = \frac{\beta(T)\{1 - \exp[\vartheta(T)]\}}{1 - \{1 - \exp[-h_I(\xi)]\}/E_\beta(T)}. \tag{6.42}$$

Multiplying Eq. (6.42) by $\exp[-\vartheta(T)]$ and noting that

$$\frac{\beta(T)E_\beta(T)}{E_\beta(T) - \{1 - \exp[-h_I(\xi)]\}} \equiv \frac{d}{dT}\ln\left(E_\beta(T) - \{1 - \exp[-h_I(\xi)]\}\right), \quad (6.43)$$

we obtain the following general solution of Eq. (6.42):

$$\exp(-\vartheta(T)) = 1 - C\left(E_\beta(T) - \{1 - \exp[h_I(\xi)]\}\right), \quad (6.44)$$

where C is a constant. The use of the initial condition $\vartheta(0) = 0$ results in $C = 0$. Hence, we obtain the simple result

$$\vartheta(T) = 0. \quad (6.45)$$

This leads to the following final expression for the signal gain:

$$G(\xi, T, U) = \frac{\exp\left\{h_\wp(\xi)\left[1 - \exp(-U)\right]\right\}}{\left(1 - \{1 - \exp[-h_I(\xi)]\}/E_\beta(T)\right)^{\exp(-U)}}. \quad (6.46)$$

We can now go back to the original single time scale τ by recalling that $\xi = z$, $T = \tau/T_s$, and $U = \tau/\tau_e$. Using these relations, the time-dependent gain seen by the optical pulse is given by

$$G(z, \tau) = \frac{\exp\left\{h_\wp(z)\left[1 - \exp(-\tau/\tau_e)\right]\right\}}{\left(1 - \{1 - \exp[-h_I(z)]\}/E_\beta(\tau)\right)^{\exp(-\tau/\tau_e)}}. \quad (6.47)$$

This equation shows how the amplification factor depends on the carrier lifetime τ_e. In the limit $\tau/\tau_e \ll 1$, it reduces to the one given in Ref. [6].

6.3.3 Pulse distortion and spectral broadening

In this section we focus on changes in the pulse shape and spectrum occurring when a picosecond pulse is amplified in an SOA. The gain expression in Eq. (6.47) can be used to calculate the output pulse shape using

$$I(L, \tau) = G(L, \tau)I(0, \tau). \quad (6.48)$$

In the limit $\tau/\tau_e \ll 1$, $G(L, \tau)$ can be written in the simple form [6]

$$G(L, \tau) = \frac{G_I}{G_I - (G_I - 1)\exp[-E_0(\tau)/E_{\text{sat}}]}, \quad (6.49)$$

where $G_I = \exp[h_I(L)]$ is the unsaturated SOA gain experienced by a low-power CW beam, E_{sat} the saturation energy $E_{\text{sat}} = h\nu(A_m/a)$, and $E_0(\tau) = \int_{-\infty}^{\tau} P_{\text{in}}(\tau)\,d\tau$ is the partial energy of the input pulse defined such that $E_0(\infty)$

equals the total pulse energy. The solution Eq. (6.48) shows that the amplifier gain is different for different parts of the pulse. The leading edge experiences the full gain G_I since the amplifier is not yet saturated. The trailing edge experiences the least gain since almost the entire pulse has saturated the amplifier gain by then.

To find the pulse spectrum, we need to take the Fourier transform of $A(L, t)$ using

$$\widetilde{A}(L, \omega) = \mathscr{F}_{\tau+}\left\{A(L, \tau)\right\}(\omega) \equiv \int_{-\infty}^{\infty} A(L, \tau)\exp(j\omega\tau)\,d\tau. \tag{6.50}$$

Before we can calculate this integral, we need to find the phase of the output pulse. As seen from the β_c term in Eq. (6.14), gain saturation will also produce a time-dependent phase shift across the pulse. If we use $A = \sqrt{I}\exp(j\phi)$ in this equation, we find that the phase $\phi(z, \tau)$ satisfies the simple equation

$$\frac{\partial\phi}{\partial z} = -\tfrac{1}{2}\beta_c g(z, \tau), \tag{6.51}$$

which can be integrated over the amplifier length to provide the following expression:

$$\phi(L, \tau) = -\tfrac{1}{2}\beta_c \int_0^L g(z, \tau)\,dz = -\tfrac{1}{2}\beta_c h(L, \tau) = -\tfrac{1}{2}\beta_c \ln[G(L, \tau)]. \tag{6.52}$$

Since the pulse modulates its own phase through gain saturation, this phenomenon is referred to as *saturation-induced* SPM [6].

We can now obtain the pulse spectrum $S(\omega) = |\widetilde{A}(L, \omega)|^2$ by taking the Fourier transform as indicated in Eq. (6.50). Because of SPM and the associated frequency chirping, we expect that the spectrum of a picosecond pulse will be modified by the SOA during its amplification. This behavior is similar to that occurring when an optical pulse propagates through an optical fiber, but major differences are expected because the SOA nonlinearity responds relatively slowly compared with an optical fiber, whose Kerr-type response is nearly instantaneous.

To illustrate the nature of pulse amplification in SOAs, we consider the amplification of a 10 ps unchirped (transform-limited) Gaussian input pulse. Figure 6.4 shows the shape and the spectrum of the amplified pulse for several values of the unsaturated amplifier gain G_I. The input pulse energy corresponds to $E_{\text{in}}/E_{\text{sat}} = 0.1$. The linewidth enhancement factor β_c may vary from one amplifier to another, as it depends on the relative position of the gain peak with respect to the operating wavelength. We use a value $\beta_c = 5$ in this section. Figure 6.4(a) shows clearly that the amplified pulse becomes asymmetric, with its leading edge sharper than its trailing edge. Sharpening of the leading edge is a common feature of all amplifiers and occurs because it experiences more gain than the trailing edge. The pulse spectra in Figure 6.4(b) show features which are unique to SOAs. In general, the spectrum

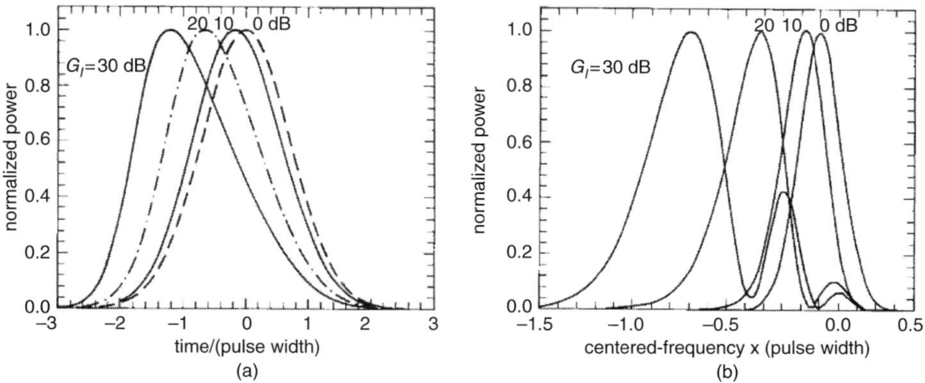

Figure 6.4 Interaction of a transform-limited Gaussian pulse with an SOA: (a) shows the characteristics of the amplified pulse under various SOA gain values; (b) shows the corresponding spectral changes in the amplified pulse (relative to the center frequency of the input pulse).

develops an asymmetric, multipeak structure. The dominant spectral peak shifts to the low-frequency side (red shift). The red shift increases with the amplifier gain G_I and can be as large as 3 to 5 times the spectral width of the input pulse. For 10 ps input pulses, the frequency shift can easily exceed 100 GHz. The temporal and spectral changes also depend on the level of amplifier gain. Experiments performed using picosecond pulses have confirmed the behavior seen in Figure 6.4.

It turns out that the frequency chirp imposed by the SOA is nearly linear over a considerable portion of the amplified pulse. For this reason, the amplified pulse can be compressed by propagating it in the anomalous-dispersion region of an optical fiber of suitable length. Such a compression has been observed in an experiment [6] in which 40 ps optical pulses were first amplified in a 1.52 μm SOA and then propagated through 18 km of single-mode fiber with $\beta_2 = -18$ ps^2 km^{-1}. This compression mechanism can be used to design fiber-optic communications systems in which SOAs are used to compensate simultaneously for both fiber loss and dispersion.

6.3.4 Impact of waveguide losses

The situation becomes much more complicated mathematically when waveguide losses are not negligible. However, the method of multiple scales can still provide an approximate expression for the time-dependent signal gain. As seen in Eq. (6.20), the α term contains an integral over the amplifier length. When this term is included in Eq. (6.24) and terms with various powers of ε are considered, the zeroth-order equation (6.27) is modified and becomes

$$\frac{\partial}{\partial T}h_0(\xi, T, U) = -\beta(T)\{\exp[h_0(\xi, T, U)] - 1\}$$
$$-\alpha\beta(T)\int_0^\xi \exp[h_0(\xi, T, U) - \alpha\xi]\,d\xi. \tag{6.53}$$

Similarly, the first-order equation (6.28) takes the form

$$\frac{\partial}{\partial T}h_0(\xi, T, U) + \frac{\partial}{\partial U}h_1(\xi, T, U) = \left[h_\wp - h_0(\xi, T, U)\right]$$
$$-\alpha\beta(T)\int_0^\xi h_1(\xi, T, U)\exp[h_0(\xi, T, U) - \alpha\xi]\,d\xi \tag{6.54}$$
$$-\beta(T)\exp[h_0(\xi, T, U)]h_1(\xi, T, U).$$

It is not possible to solve these two equations exactly, owing to the presence of the integrals. However, if we homogenize the signal gain by setting local gain to the averaged gain along the SOA, we can replace $h_0(\xi, T, U)$ with $\bar{g}(T, U)\xi$, where the average gain coefficient $\bar{g}(T, U)$ is independent of the spatial coordinate ξ. With this change, we can perform the integration in Eq. (6.53) as

$$\int_0^\xi \exp[h_0(\xi, T, U) - \alpha\xi]\,d\xi = \frac{\exp[\bar{g}(T, U)\xi - \alpha\xi] - 1}{\bar{g}(T, U) - \alpha}. \tag{6.55}$$

Noting this relation, we employ the following approximation for the preceding integral when the gain distribution is spatially nonuniform:

$$\int_0^\xi \exp[h_0[(\xi, T, U)] - \alpha\xi]\,d\xi \approx \frac{\exp[h_0(\xi, T, U) - \alpha\xi] - 1}{\tilde{g}(\xi, T, U) - \alpha}, \tag{6.56}$$

where $\tilde{g}(\xi, T, U)$ is calculated from Eq. (6.47) and is found to be

$$\tilde{g}(\xi, T, U) = \frac{h_\wp(\xi)}{\xi}\left[1 - e^{-U}\right] - \frac{\ln\left[\left(1 - \{1 - \exp[-h_I(\xi)]\}/E_\beta(T)\right)\right]}{\xi\exp(U)}. \tag{6.57}$$

As verified through numerical simulations, this turns out to be a good approximation for the integral in Eq. (6.56), with a relative error of $< 10\%$ if input pulse energy is below 30% of the saturation energy of the amplifier. Substitution of Eq. (6.56) into Eq. (6.53) then leads to

$$\frac{\partial}{\partial T}h_0(\xi, T, U) = -\frac{\beta(T)\tilde{g}(\xi, T, U)}{\tilde{g}(\xi, T, U) - \alpha}\left(\exp[h_0(\xi, T, U) - \alpha\xi] - 1\right). \tag{6.58}$$

In the method of multiple scales, lower-order coefficients in the expansion $h(\varepsilon) = \sum_{n=0}^\infty h_n\varepsilon^n$ affect the higher-order ones, but not the other way around. Using this

Table 6.1. *Parameters used for numerical simulations*

SOA length (L)	378 µm
Active region width (w)	2.5 µm
Active region thickness (d)	0.2 µm
Confinement factor (Γ)	0.3
Group index (n_g)	3.7
Loss coefficient (α)	30 cm^{-1}
Carrier lifetime (τ_e)	300 ps
Carrier injection rate ($\wp$)	1.177×10^{34} s^{-1}m^{-3}
Material differential gain (a)	2.5×10^{-20} m^2
Transparency carrier density (N_0)	1.5×10^{24} m^{-3}
Linewidth enhancement factor (β_c)	5.0
Operating wavelength (λ)	1552.5 nm

feature, we split Eq. (6.54) into two separate differential equations:

$$\frac{\partial}{\partial T} h_0(\xi, T, U) = h_\wp - h_0(\xi, T, U), \tag{6.59}$$

$$\frac{\partial}{\partial U} h_1(\xi, T, U) = -\beta(T) \exp[h_0(\xi, T, U) - \alpha\xi] h_1(\xi, T, U)$$

$$- \alpha\beta(T) \int_0^\xi h_1(\xi, T, U) \exp[h_0(\xi, T, U) - \alpha\xi] \, d\xi. \tag{6.60}$$

We solve these coupled equations using the procedure used earlier in this section and obtain the following expression for the time-dependent gain seen by the pulse:

$$G(\xi, T, U)e^{-\alpha\xi} = \exp\{[h_\wp(\xi) - \alpha\xi][(1 - \exp(-U)]\}$$

$$\times \left(1 - \{1 - \exp[-h_I(\xi) + \alpha\xi]\} \big/ E_\gamma(T)\right)^{-\exp(-U)}, \tag{6.61}$$

where $E_\gamma(T)$ is defined as

$$E_\gamma(T) = \exp\left(\int_0^T \frac{\beta(T)\widetilde{g}(\xi, T, U)}{\widetilde{g}(\xi, T, U) - \alpha} \, dT\right). \tag{6.62}$$

To check the accuracy of the preceding result, we compare its predictions with numerical simulations using the parameters given in Table 6.1. The input pulse is an unchirped Gaussian pulse of energy E_{in} with the intensity profile

$$I_0(t) = \frac{E_{\text{in}}}{A_m T_0 \sqrt{\pi}} \exp\left(-\frac{t^2}{T_0^2}\right). \tag{6.63}$$

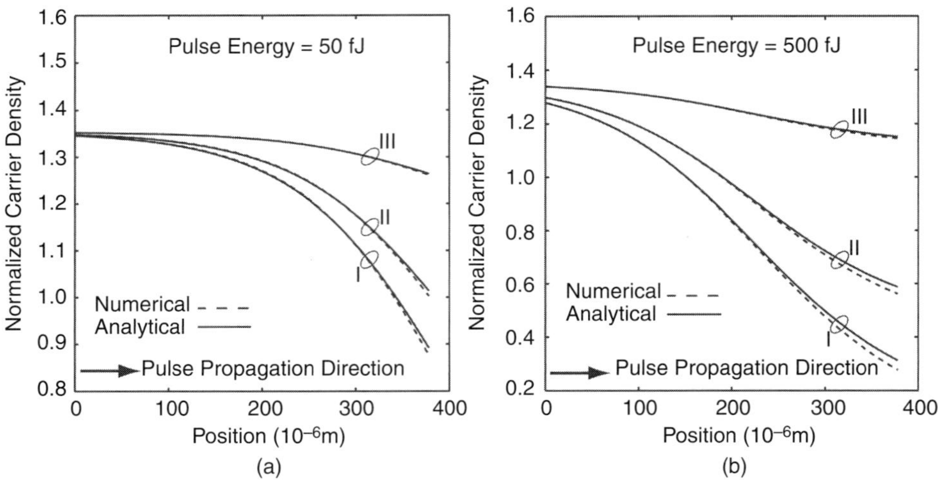

Figure 6.5 Variation of carrier density along the SOA length at time intervals of 0,100, and 500 ps (curves marked I to III) after a 2-ps-wide (FWHM) Gaussian pulse has completely passed through the SOA. Pulse energy is 50 fJ in part (a) and 500 fJ in part (b).

The numerical results were obtained by integrating Eqs. (6.15) and (6.5). The results provide the intensity profile $I(z, t)$ of the pulse, as it is being amplified, and the corresponding carrier-density profile $N(z, t)$, which is then used to calculate $h(z, t)$ by performing the integral indicated in (6.19). The amplification factor is then obtained using $G(z, t) = \exp[h(z, t]$. The saturation energy of the amplifier corresponding to the data in Table 6.1 is 5.5 pJ.

Before the pulse enters the SOA, the carrier density N has a constant value $(3.531 \times 10^{24} \text{ m}^{-3})$ set by the carrier-injection rate. After the pulse enters the SOA, gain saturation reduces the carrier density N all along the amplifier, making it a function of both z and t. After the pulse has passed through the SOA, the carrier density and the saturated gain g begin to recover as carriers are continuously injected into the active region. Figure 6.5 shows the gain-recovery dynamics by plotting $(N - N_0)/N_0$ as a function of z at intervals of 0 ps (I), 100 ps (II), and 500 ps (III) after a 2-ps-wide (FWHM) Gaussian pulse has completely passed through the SOA. Input pulse energy is 50 fJ in part (a) and 500 fJ in part (b). Dashed lines show the results of numerical simulations, while solid lines show the carrier density calculated using Eq. (6.61). Clearly, the analytical expression predicts the spatial and temporal gain dynamics of SOAs quite accurately.

Figure 6.6 shows (a) the pulse shape and (b) the pulse spectrum when a 20-ps-wide Gaussian pulse is amplified. Its energy is 50 fJ for trace I and 500 fJ for trace II. The dashed lines show numerical simulation results and the solid lines correspond to analytical results. Finally, Figure 6.7 shows the impact of waveguide

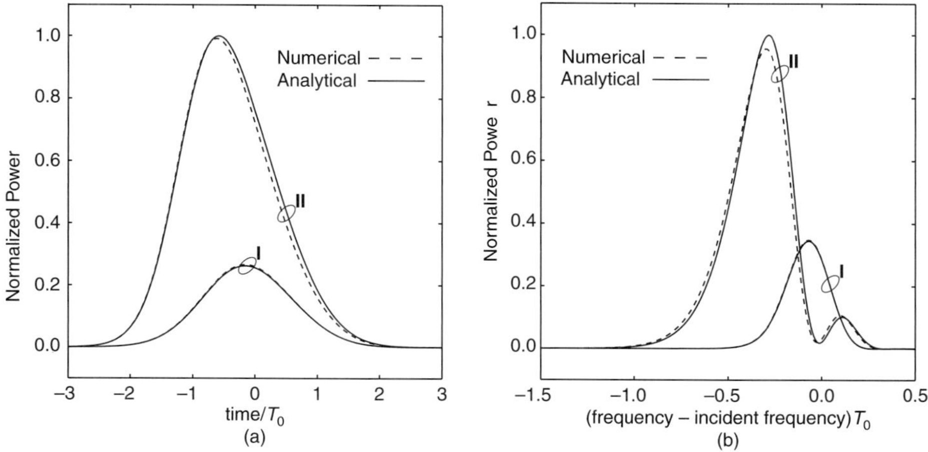

Figure 6.6 (a) Shape and (b) spectrum of output pulses when a 20-ps-wide input Gaussian pulse is amplified in the SOA. Pulse energy is 50 fJ for trace I and 500 fJ for trace II. (After Ref. [18]; © IEEE 2008)

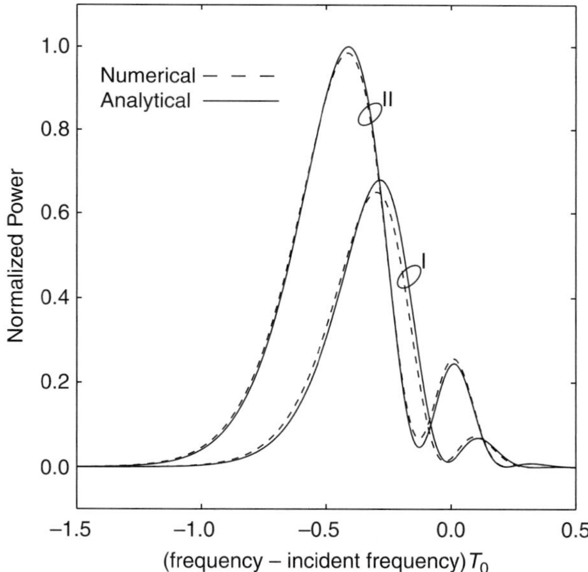

Figure 6.7 Output pulse spectra when a 20-ps-wide input Gaussian pulse with 500 fJ energy is amplified in the SOA. The waveguide loss is 30 cm^{-1} for trace I and zero for trace II.

losses on the spectrum of the amplified pulse when a 20 ps input Gaussian pulse with 500 fJ of energy is launched. The loss is 30 cm^{-1} for trace I but zero for trace II. As before, dashed lines show numerical results, while solid lines show the corresponding analytical results. Losses reduce the magnitude of red shift as well as peak heights. The good match obtained in all cases shows the accuracy of the approximate treatment based on the multiple-scales method.

6.4 Femtosecond pulse amplification

The theory presented in the preceding section is not adequate for femtosecond pulses or for pulses with high energies, >10 pJ. Under such conditions, the response of an SOA is significantly influenced by carrier heating, spectral hole burning, and two-photon absorption (TPA) [26, 27], phenomena that are not included in Section 6.3. If the input pulses have high peak powers, TPA-initiated hot carrier generation also plays a dominant role in the recovery dynamics of the SOA [28].

It is possible to analyze the operation of SOAs in the femtosecond and high-power regime using the density-matrix equations of Chapter 4 [29, 30]. In this approach, carrier–carrier and carrier–phonon scattering events are described using phenomenological relaxation rates, and carrier occupation probabilities are assumed not to deviate significantly from equilibrium Fermi distributions [31]. The main advantage of these approximations is that one can avoid solving a complicated many-body problem associated with the SOA dynamics, without a significant loss of accuracy [32]. Since such an approach provides valuable physical insight, we discuss it in this section.

Let ρ_{ck} be the probability that an electron occupies a state with momentum $\hbar k$ in the conduction band of the SOA material, and let ρ_{vk} represent this probability for holes in the valence band. The situation is similar to that of an inhomogeneously broadened two-level system except that the electron's momentum is used to distinguish various states, in place of atomic velocities. The total carrier densities in the conduction and valence bands can thus be written as

$$N_c(t) = \frac{1}{V}\sum_k \rho_{ck}(t), \qquad N_v(t) = \frac{1}{V}\sum_k \rho_{vk}(t). \qquad (6.64)$$

where V is the volume of the active region. Owing to charge neutrality in the semiconductor, these two densities must be equal to the single carrier density $N(t)$ used earlier, i.e., $N(t) \equiv N_c(t) = N_v(t)$.

If we assume that various carrier–carrier scattering processes take place with relaxation times τ_{1c} and τ_{1v} in the conduction and valence bands, respectively, we can include these processes through simple rate equations:

$$\left.\frac{\partial \rho_{ck}(t)}{\partial t}\right|_{\text{e-e scat.}} = -\frac{1}{\tau_{1c}}[\rho_{ck}(t) - f_{ck}(U_{ck}, F_c, T_c)], \qquad (6.65a)$$

$$\left.\frac{\partial \rho_{vk}(t)}{\partial t}\right|_{\text{e-e scat.}} = -\frac{1}{\tau_{1v}}[\rho_{vk}(t) - f_{vk}(U_{vk}, F_v, T_v)], \qquad (6.65b)$$

where f_{ck} and f_{vk} represent the corresponding quasi-equilibrium Fermi distributions given by [30]:

$$f_{ck} = \frac{1}{1 + \exp\left(\frac{U_{ck} - F_c}{k_B T_c}\right)}, \qquad f_{vk} = \frac{1}{1 + \exp\left(\frac{U_{vk} - F_v}{k_B T_v}\right)}. \qquad (6.66)$$

In these relations, k_B is the Boltzmann constant, T_s is the carrier temperature, U_{sk} is the carrier energy, and F_s is the quasi-Fermi level (with $s = c, v$).

The carrier temperatures T_c and T_v in the conduction and valence bands, respectively, relax toward a common lattice temperature, denoted by T_L, owing to carrier–phonon collisions. We denote the quasi-Fermi level at the lattice temperature by F_s^L, with $s = c, v$. Since intraband processes such as carrier–carrier scattering and carrier–phonon scattering do not change the total number of carriers within the conduction or valence band, we have the following constraints [30]:

$$\begin{aligned}
N(t) &= \frac{1}{V} \sum_k f_{ck}(U_{ck}, F_c, T_c) = \frac{1}{V} \sum_k f_{vk}(U_{vk}, F_v, T_v) \\
&= \frac{1}{V} \sum_k f_{ck}(U_{ck}, F_c^L, T_L) = \frac{1}{V} \sum_k f_{vk}(U_{vk}, F_v^L, T_L).
\end{aligned} \qquad (6.67)$$

These equations show that, once the carrier density $N(t)$ and temperatures T_c, T_v, and T_L are known, it is possible to find the quasi-Fermi levels and the associated quasi-Fermi distributions uniquely. If τ_{hc} and τ_{hv} denote the relaxation times for electron–phonon collisions in the conduction and valence bands, respectively, we can incorporate such collisions through

$$\left. \frac{\partial \rho_{ck}(t)}{\partial t} \right|_{\text{e-p scat.}} = -\frac{1}{\tau_{hc}} [\rho_{ck}(t) - f_{ck}(U_{ck}, F_c^L, T_L)], \qquad (6.68a)$$

$$\left. \frac{\partial \rho_{vk}(t)}{\partial t} \right|_{\text{e-p scat.}} = -\frac{1}{\tau_{hv}} [\rho_{vk}(t) - f_{vk}(U_{vk}, F_v^L, T_L)]. \qquad (6.68b)$$

Finally, if the external pumping is turned off, the carrier populations would relax toward their thermal equilibrium distribution. At thermal equilibrium, it is not possible to have any discontinuity in the Fermi levels of the two bands, F_c and F_v converge to a common value F_{eq}, and the carrier temperatures converge to a common value T_{eq}. We can calculate F_{eq} for a given equilibrium temperature by using the relation

$$\frac{1}{V} \sum_k f_{ck}(U_{ck}, F_{eq}, T_{eq}) \approx \frac{1}{V} \sum_k f_{vk}(U_{vk}, F^{eq}, T_{eq}) \approx 0. \qquad (6.69)$$

This relation assumes that there are no free carriers in the semiconductor when it has reached thermal equilibrium. If τ_s is the characteristic time for attaining thermal equilibrium, we have the relations

$$\left.\frac{\partial \rho_{ck}(t)}{\partial t}\right|_{\text{equ.}} = -\frac{1}{\tau_s}[\rho_{ck}(t) - f_{ck}(U_{ck}, F^{eq}, T_{eq})], \qquad (6.70a)$$

$$\left.\frac{\partial \rho_{vk}(t)}{\partial t}\right|_{\text{equ.}} = -\frac{1}{\tau_s}[\rho_{vk}(t) - f_{vk}(U_{vk}, F^{eq}, T_{eq})]. \qquad (6.70b)$$

This step takes care of all the processes through which carrier populations can relax spontaneously.

The amplification process in an SOA is dominated by carrier recombinations initiated by stimulated emission. Following the density-matrix approach of Chapter 4, it is possible to write the rate of such recombinations in the form[30]

$$\left.\frac{\partial \rho_{ck}(t)}{\partial t}\right|_{\text{stim-em}} = -\frac{j}{\hbar}[d_k^* \rho_{cv,k}(t) - d_k \rho_{vc,k}(t)]E(z,t), \qquad (6.71a)$$

$$\left.\frac{\partial \rho_{vk}(t)}{\partial t}\right|_{\text{stim-em}} = -\frac{j}{\hbar}[d_k^* \rho_{cv,k}(t) - d_k \rho_{vc,k}(t)]E(z,t), \qquad (6.71b)$$

where E is the electric field of the optical signal being amplified, and the cross-density matrix element $\rho_{cv,k}(t)$ satisfies

$$\frac{\partial \rho_{cv,k}(t)}{\partial t} = -\left(j\omega_k + \frac{1}{T_2}\right)\rho_{cv,k}(t) - j\frac{d_k}{\hbar}[\rho_{ck}(t) + \rho_{vk}(t) - 1]E(z,t). \qquad (6.72)$$

Here T_2 is the dipole relaxation time and ω_k is the transition frequency between the conduction and valence band states with wave number k. The dipole moment d_k for many semiconductors can be written in the form [33]

$$d_k^2 = \frac{e^2}{6m_0\omega_k^2}\left(\frac{m_0}{m_c} - 1\right)\frac{E_g(E_g + \Delta_0)}{E_g + 2\Delta_0/3}, \qquad (6.73)$$

where m_c is the effective mass of carriers in the conduction band, m_0 is the free-electron mass, Δ_0 is the spin–orbit splitting value, and E_g is the bandgap. The macroscopic polarization $P(z,t)$ induced by the electric field $E(z,t)$ can be calculated using

$$P(t) = \frac{1}{V}\sum_k d_k[\rho_{cv,k}(t) + \rho_{vc,k}(t)]. \qquad (6.74)$$

We can now combine all the recombination mechanisms to obtain the following rate equations for $\rho_{ck}(t)$ and $\rho_{vk}(t)$:

$$\frac{\partial \rho_{ck}}{\partial t} = \Lambda_{ck} - \frac{1}{\tau_{1c}}[\rho_{ck}(t) - f_{ck}(U_{ck}, F_c, T_c)] - \frac{1}{\tau_{hc}}[\rho_{ck}(t) - f_{ck}(U_{ck}, F_c^L, T_L)]$$

$$- \frac{1}{\tau_s}[\rho_{ck}(t) - f_{ck}(U_{ck}, F^{eq}, T_{eq})] - \frac{j}{\hbar}[d_k^* \rho_{cv,k}(t) - d_k \rho_{vc,k}(t)]E(z,t),$$

$$(6.75)$$

$$\frac{\partial \rho_{vk}}{\partial t} = \Lambda_{vk} - \frac{1}{\tau_{1v}}\rho_{vk}(t) - f_{vk}(U_{vk}, F_v, T_v)] - \frac{1}{\tau_{hv}}[\rho_{vk}(t) - f_{vk}(U_{vk}, F_v^L, T_L)]$$

$$- \frac{1}{\tau_s}[\rho_{vk}(t) - f_{vk}(U_{vk}, F^{eq}, T_{eq})] - \frac{j}{\hbar}[d_k^* \rho_{cv,k}(t) - d_k \rho_{vc,k}(t)]E(z,t),$$

$$(6.76)$$

where Λ_{ck} and Λ_{vk} denote the pumping rates that generate carriers in the conduction and valence bands, respectively.

Equations (6.72), (6.75), and (6.76) must be solved numerically for each value of k corresponding to the momentum $\hbar k$. Such solutions provide neither physical insight into the operation of an SOA nor a direct way to correlate the results with experimental observations. A simpler approach is to employ the carrier density $N(t)$, as defined in Eq. (6.67), and introduce the two variables representing the total energy densities in the conduction and valence bands:

$$U_c(t) = \frac{1}{V}\sum_k U_{ck}\rho_{ck}(t), \qquad U_v(t) = \frac{1}{V}\sum_k U_{vk}\rho_{vk}(t). \tag{6.77}$$

Noting that all intraband collisions of carriers are elastic and hence do not change the energy density of an energy band, we can write these energy densities using Fermi distributions as

$$U_c(t) = \frac{1}{V}\sum_k U_{ck} f_{ck}(U_{ck}, F_c, T_c), \tag{6.78a}$$

$$U_v(t) = \frac{1}{V}\sum_k U_{vk} f_{vk}(U_{vk}, F_v, T_v). \tag{6.78b}$$

Similarly, we define the lattice energy distributions $U_c^L(t)$ and $U_v^L(t)$ by using the corresponding Fermi distributions $f_{ck}(U_{ck}, F_c^L, T_c^L)$ and $f_{vk}(U_{vk}, F_v^L, T_v^L)$, and neglect any residual energy present at thermal equilibrium by using the approximation

$$\frac{1}{V}\sum_k U_{ck} f_{ck}(U_{ck}, F_{eq}, T_{eq}) \approx 0. \tag{6.79}$$

Now, summing Eq. (6.75) or Eq. (6.76) over all possible values of k, we obtain the carrier-density rate equation in the form

$$\frac{\partial N(t)}{\partial t} = \frac{I}{eV} - \frac{N(t)}{\tau_s} - \frac{j}{\hbar V} \sum_k \left(d_k^* \rho_{cv,k}(t) - d_k \rho_{vc,k}(t) \right) E(z,t). \qquad (6.80)$$

To simplify this equation further, we need an explicit expression for $\rho_{cv,k}(t)$. We can obtain such an expression using the steady-state solution of Eq. (6.72) when the widths of pulses being amplified by the SOA are sufficiently larger than the dipole relaxation time T_2. Under such conditions, we get from Eq. (6.72) the result

$$\rho_{cv,k}(t) \exp(j\omega_0 t) = -j \frac{d_k}{\hbar} \frac{[\rho_{ck}(t) + \rho_{vk}(t) - 1]}{j(\omega_k - \omega_0) + 1/T_2} \mathcal{E}(z,t), \qquad (6.81)$$

where we have assumed that the carrier frequency of the pulse is ω_0. The slowly varying amplitude $\mathcal{E}(z,t)$ of the electric field $E(z,t)$ is defined, as usual, by the relation

$$E(z,t) = \mathrm{Re}[\mathcal{E}(z,t) \exp(-j\omega_0 t)]. \qquad (6.82)$$

We are now in a position to calculate the gain coefficient at the carrier frequency ω_0. It is related to the imaginary part of the susceptibility χ by

$$g(\omega_0) = -\frac{\omega_0}{cn} \mathrm{Im}(\chi), \qquad (6.83)$$

where n is the modal refractive index. The susceptibility χ is related to the macroscopic polarization $P(z,t)$ given in (6.74) through the relation

$$\mathrm{Re}[\varepsilon_0 \chi \mathcal{E} \exp(-j\omega_0 t)] = \frac{1}{V} \sum_k d_k (\rho_{cv,k}(t) + \rho_{vc,k}(t)). \qquad (6.84)$$

We can write the last term of Eq. (6.80) in the form [34]

$$\frac{j}{\hbar V} \sum_k \left(d_k^* \rho_{cv,k}(t) - d_k \rho_{vc,k}(t) \right) E(z,t) =$$

$$v_g \left[g(\omega_0) - \frac{\Gamma_2 \varepsilon_0 n n_g}{\Gamma \hbar \omega_0} \beta_2 |\mathcal{E}(z,t)|^2 \right] \left[\frac{\varepsilon_0 n n_g}{\hbar \omega_0} |\mathcal{E}(z,t)|^2 \right], \qquad (6.85)$$

where $v_g = c/n_g$ is the group velocity, Γ is the confinement factor, β_2 is the TPA coefficient, and Γ_2 is the TPA confinement factor. Substitution of this expression into Eq. (6.80) leads to the final rate equation,

$$\frac{\partial N(t)}{\partial t} = \frac{I}{eV} - \frac{N(t)}{\tau_s} - v_g \left[g(\omega_0) - \frac{\Gamma_2 \varepsilon_0 n n_g}{\Gamma \hbar \omega_0} \beta_2 |\mathcal{E}(z,t)|^2 \right] \left[\frac{\varepsilon_0 n n_g}{\hbar \omega_0} |\mathcal{E}(z,t)|^2 \right].$$

$$(6.86)$$

Similarly, by multiplying Eqs. (6.75) and (6.76) by U_k and summing over all values of k, we obtain the following rate equations for the energy densities:

$$\frac{\partial U_c(t)}{\partial t} = -\frac{U_c(t)}{\tau_s} - \frac{1}{\tau_{hc}}[U_c(t) - U_c^L] + \sum_k U_{ck}\Lambda_{ck}$$
$$- v_g \left[g(\omega_0)E_{c,\omega_0} - \frac{\Gamma_2\varepsilon_0 n n_g}{\Gamma \hbar\omega_0}\beta_2 E_{c,2\omega_0}|\mathscr{E}(z,t)|^2 \right] \left[\frac{\varepsilon_0 n n_g}{\hbar\omega_0}|\mathscr{E}(z,t)|^2 \right], \tag{6.87}$$

$$\frac{\partial U_v(t)}{\partial t} = -\frac{U_v(t)}{\tau_s} - \frac{1}{\tau_{hv}}[U_v(t) - U_v^L] + \sum_k U_{vk}\Lambda_{vk}$$
$$- v_g \left[g(\omega_0)E_{v,\omega_0} - \frac{\Gamma_2\varepsilon_0 n n_g}{\Gamma \hbar\omega_0}\beta_2 E_{v,2\omega_0}|\mathscr{E}(z,t)|^2 \right] \left[\frac{\varepsilon_0 n n_g}{\hbar\omega_0}|\mathscr{E}(z,t)|^2 \right], \tag{6.88}$$

where E_{α,ω_0} and $E_{\alpha,2\omega_0}$ (with $\alpha = c, v$) correspond to carrier energies participating in optical transition at $\omega_k = \omega_0$ and $\omega_k = 2\omega_0$, respectively.

Experiments have shown that there are two types of carriers in an SOA [26]. The first category constitutes carriers that interact directly with the optical field, and the second category corresponds to those whose energies are so different that they cannot interact with the optical field directly. If the pulse width is significantly larger than the dipole relaxation time T_2, then one can isolate the carriers in the first category by grouping energies around the photon energy $\hbar\omega_0$ over a bandwidth $\approx \hbar/T_2$. If the effective carrier densities of this energy band are given by $n_c(t)$ and $n_v(t)$ for the conduction and valence bands, respectively, then using Eqs. (6.75) and (6.76) we obtain

$$\frac{\partial n_c(t)}{\partial t} = -\frac{1}{\tau_{1c}}[n_c(t) - \overline{n_c}] - v_g g(\omega_0) \left[\frac{\varepsilon_0 n n_g}{\hbar\omega_0}|\mathscr{E}(z,t)|^2 \right], \tag{6.89}$$

$$\frac{\partial n_v(t)}{\partial t} = -\frac{1}{\tau_{1v}}[(n_v(t) - \overline{n_v}] - v_g g(\omega_0) \left[\frac{\varepsilon_0 n n_g}{\hbar\omega_0}|\mathscr{E}(z,t)|^2 \right], \tag{6.90}$$

where $\overline{n_\alpha} = N_0 f_{\alpha,k}(U_{\alpha,k}, F_\alpha, T_\alpha)$ ($\alpha = c, v$) and N_0 is the carrier density of available states in the optically coupled region. These coupled carrier densities can then be used to write the following approximate expression for the gain [26]:

$$g(\omega_0) = \frac{1}{v_g}\frac{\omega_0 d_0^2 T_2}{\hbar\varepsilon_0 n n_g}(n_c + n_v - N_0), \tag{6.91}$$

where d_0 is obtained by evaluating the expression (6.73) at $\omega_k = \omega_0$.

The preceding analysis completes the derivation of equations governing the femtosecond (or subpecosecond) response of an SOA. Of course, it must be supplemented with the propagation equation (6.15),

$$\frac{\partial I(z,t)}{\partial z} + \frac{1}{v_g}\frac{\partial I(z,t)}{\partial t} = g(\omega_0)I(z,t) - \alpha I(z,t) - \beta_2 I(z,t), \qquad (6.92)$$

where the intensity $I(z,t)$ is related the electric field $\mathscr{E}(z,t)$ by

$$I(z,t) = \frac{\varepsilon_0 n n_g}{\hbar\omega_0}|\mathscr{E}(z,t)|^2. \qquad (6.93)$$

To illustrate the value of this type of a detailed model of SOAs, we consider the pump-probe configuration in which a femtosecond pump pulse is sent through an InGasAsP amplifier together with an orthogonal, polarized weak probe pulse [34]. In the following example, both pulses have widths of 120 fs. Figure 6.8 shows the experimentally observed probe intensities as a function of time, together with corresponding theoretical predictions for two different pump-pulse energies (input probe pulse energy is 1 fJ in both cases). The list of parameter values used in the simulations can be found in Ref. [34]. The agreement between theory and experiment is remarkable when the femtosecond model of this section is used. The picosecond model of Section 6.3 cannot explain the experimental observations. Physically, the narrow peak observed for 1.28 pJ pump energy is due to spectral hole burning, and the longer transient tail with a time constant of around 0.7 ps

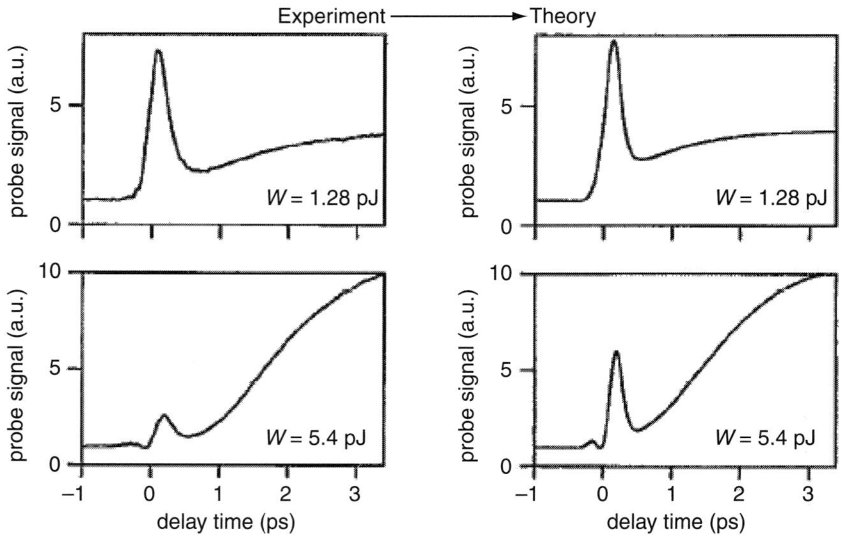

Figure 6.8 Experimental and theoretical probe signals for varying pump energy W. (After Ref. [34]; © AIP 1994)

reflects the relaxation of the carrier temperature towards the lattice temperature. Both of these features are observed in the case of 5.4 pJ pump pulses but with considerable quantitative differences. The peak due to spectral hole burning is less pronounced in the measured data then in the simulations. This may be due to the fact that intraband scattering times actually depend on carrier density but were considered constant in the theoretical model [35]. However, the model is powerful enough to capture most of the subtle features observed in the experiment. Further improvements to this model require self-consistent many-body calculations of the carrier dynamics [32].

References

[1] P. G. Eliseev and V. V. Luc, "Semiconductor optical amplifiers: Multifunctional possibilities, photoresponse and phase shift properties," *Pure Appl. Opt.*, vol. 4, pp. 295–313, 1995.

[2] B. Dagens, A. Labrousse, R. Brenot, B. Lavigne, and M. Renaud, "SOA-based devices for all optical signal processing," in *Proc. Opt. Fiber Commun. Conf. (OFC 2003)*, pp. 582–583. Optical Society of America, 2003.

[3] M. J. Coupland, K. G. Hambleton, and C. Hilsum, "Measurement of amplification in a GaAs injection laser," *Phys. Lett.*, vol. 7, pp. 231–232, 1963.

[4] Y. Yamamoto, "Characteristics of AlGaAs Fabry–Perot cavity type laser amplifiers," *IEEE J. Quantum Electron.*, vol. QE-16, pp. 1047–1052, 1980.

[5] G. P. Agrawal and N. K. Dutta, *Semiconductor Lasers*, 2nd ed. Van Nostrand Reinhold, 1993.

[6] G. P. Agrawal and N. A. Olsson, "Self-phase modulation and spectral broadening of optical pulses in semiconductor laser amplifiers," *IEEE J. Quantum Electron.*, vol. 25, pp. 2297–2306, 1989.

[7] A. Mecozzi and J. Mork, "Saturation induced by picosecond pulses in semiconductor optical amplifiers," *J. Opt. Soc. Am. B*, vol. 14, pp. 761–770, 1997.

[8] J. P. Sokoloff, P. R. Prucnal, I. Glesk, and M. Kane, "A terahertz optical asymmetric demultiplexer (TOAD)," *IEEE Photon. Technol. Lett.*, vol. 5, pp. 787–790, 1993.

[9] M. Eiselt, W. Pieper, and H. G. Weber, "SLALOM: Semiconductor laser amplifier in a loop mirror," *J. Lightw. Technol.*, vol. 13, pp. 2099–2112, 1995.

[10] T. Wang, Z. Li, C. Lou, Y. Wu, and Y. Gao, "Comb-like filter preprocessing to reduce the pattern effect in the clock recovery based on SOA," *IEEE Photon. Technol. Lett.*, vol. 14, pp. 855–857, 2002.

[11] M. Weiming, L. Yuhua, A. M. Mohammed, and G. Li, "All optical clock recovery for both RZ and NRZ data," *IEEE Photon. Technol. Lett.*, vol. 14, pp. 873–875, 2002.

[12] E. S. Awad, C. J. K. Richardson, P. S. Cho, N. Moulton, and J. Godhar, "Optical clock recovery using SOA for relative timing extraction between counterpropagating short picosecond pulses," *IEEE Photon. Technol. Lett.*, vol. 14, pp. 396–398, 2002.

[13] M. Premaratne and A. J. Lowery, "Analytical characterization of SOA based optical pulse delay discriminator," *J. Lightw. Technol.*, vol. 23, pp. 2778–2787, 2005.

[14] ——, "Semiclassical analysis of the impact of noise in SOA-based optical pulse delay discriminator," *IEEE J. Sel. Topics Quantum Electron.*, vol. 12, pp. 708–716, 2006.

[15] K. L. Hall and K. A. Rauschenbach, "100-Gbit/s bitwise logic," *Opt. Lett.*, vol. 23, pp. 1271–1273, 1998.

[16] A. Hamie, A. Sharaiha, M. Guegan, and B. Pucel, "All-optical logic NOR gate using two-cascaded semiconductor optical amplifiers," *IEEE Photon. Technol. Lett.*, vol. 14, pp. 1439–1441, 2002.

[17] S. Adachi, "GaAs, AlAs, and $Al_x Ga_{1-x}As$: Material parameters for use in research and device applications," *J. Appl. Phys.*, vol. 58, pp. R1–R29, 1985.

[18] M. Premaratne, D. Nesic, and G. P. Agrawal, "Pulse amplification and gain recovery in semiconductor optical amplifiers: A systematic analytical approach," *IEEE J. Lightw. Technol.*, vol. 26, pp. 1653–1660, 2008.

[19] A. E. Siegman, *Lasers*. University Science Books, 1986.

[20] P. W. Milonni and J. H. Eberly, *Lasers*, 2nd ed. Wiley, 2010.

[21] L. M. Frantz and J. S. Nodvik, "Theory of pulse propagation in a laser amplifier," *J. Appl. Phys.*, vol. 34, pp. 2346–2349, 1963.

[22] A. E. Siegman, "Design considerations for laser pulse amplifiers," *J. Appl. Phys.*, vol. 35, pp. 460–461, 1964.

[23] P. G. Kryukov and V. S. Letokhov, "Propagation of a light pulse in resonantly amplifying (absorbing) medium," *Sov. Phys. Usp.*, vol. 12, pp. 641–672, 1970.

[24] B. S. G. Pillai, M. Premaratne, D. Abramson, *et al.*, "Analytical characterization of optical pulse propagation in polarization sensitive semiconductor optical amplifiers," *IEEE J. Quantum Electron.*, vol. 42, pp. 1062–1077, 2006.

[25] A. J. Jerri, *Introduction to Integral Equations with Applications*, 2nd ed. Wiley, 1999.

[26] J. Mark and J. Mork, "Subpicosecond gain dynamics in InGaAsP optical amplifiers: Experiment and theory," *Appl. Phys. Lett.*, vol. 61, pp. 2281–2283, 1992.

[27] K. L. Hall, G. Lenz, E. P. Ippen, U. Koren, and G. Raybon, "Carrier heating and spectral hole burning in strained-layer quantum-well laser amplifiers at 1.5 μm," *Appl. Phys. Lett.*, vol. 61, pp. 2512–2514, 1992.

[28] H. K. Tsang, P. A. Snow, I. E. Day, *et al.*, "All-optical modulation with ultra-fast recovery at low pump energies in passive InGaAs/InGaAsP multiquantum well waveguides," *Appl. Phys. Lett.*, vol. 62, pp. 1451–1453, 1993.

[29] M. Asada and Y. Suematsu, "Density matrix theory of semiconductor lasers with relaxation broadening model: Gain and gain-suppression in semiconductor lasers," *IEEE J. Quantum Electron.*, vol. QE-21, pp. 434–441, 1985.

[30] A. Uskov, J. Mork, and J. Mark, "Wave mixing in semiconductor laser amplifiers due to carrier heating and spectral-hole burning," *IEEE J. Quantum Electron.*, vol. 30, pp. 1769–1781, 1994.

[31] K. Henneberger, F. Herzel, S. W. Koch, R. Binder, A. E. Paul, and D. Scott, "Spectral hole burning and gain saturation in short-cavity semiconductor lasers," *Phys. Rev. A*, vol. 45, pp. 1853–1859, 1992.

[32] H. Haug and S. W. Koch, *Quantum Theory of the Optical and Electronic Properties of Semiconductors*, 4th ed. World Scientific Publishing, 2004.

[33] M. Asada, A. Kameyama, and Y. Suematsu, "Gain and intervalence band absorption in quantum-well lasers," *IEEE J. Quantum Electron.*, vol. QE-20, pp. 745–753, 1984.

[34] J. Mork and J. Mark, "Carrier heating in InGaAsP laser amplifiers due to two-photon absorption," *Appl. Phys. Lett.*, vol. 64, pp. 2206–2208, 1994.

[35] R. Binder, D. Scott, A. E. Paul, M. Lindberg, K. Henneberger, and S. W. Koch, "Carrier–carrier scattering and optical dephasing in highly excited semiconductors," *Phys. Rev. A*, vol. 45, pp. 1107–1115, 1992.

7

Raman amplifiers

Light gets scattered when it encounters an obstacle or inhomogeneity even on a microscopic scale. A well-known example is the blue color of the sky, resulting from Rayleigh scattering of light by molecules in the air. Such redirection of energy can be used to amplify signals by taking power from a "pump" wave co-propagating with the signal in an appropriate optical medium. An example of this is provided by Raman scattering. Having said that, it is important to realize that scattering does not always occur when light interacts with a material [1]. In some cases, photons get absorbed in the medium, and their energy is eventually dissipated as heat. In other cases, the absorbed light may be re-emitted after a relatively short time delay in the form of a less energetic photon [2], a process known as fluorescence. If fluorescence takes place after a considerable delay, the same process is called phosphorescence [3].

For a photon to get absorbed by a material, its energy must correspond to the energy required by the atoms or molecules of that material to make a transition from one energy level to a higher energy level. In contrast, the scattering of photons from a material can take place without such a requirement. However, if the energy of the incident photon is close to an allowed energy transition, significant enhancement of scattering can occur. This type of enhanced scattering, often called *resonance scattering*, has characteristics significantly different from "normal" scattering [4]. If the energy of a scattered photon is exactly equal to the incident photon energy, the scattering event is termed elastic scattering. Examples of elastic scattering include *Mie scattering* and *Rayleigh scattering*. The former is observed from dielectric objects (such as biological cells) whose size is large compared to the wavelength of the incident light [5]. However, in general, scattered radiation consists of many other pairs of frequencies of the form $\omega + \Omega_n$ and $\omega - \Omega_n$, where ω is the frequency of the incident radiation and Ω_n ($n = 1, 2, \cdots$) are frequency shifts produced by molecules of the scattering material. Their numerical values depend on the specific rotational, vibrational, and electronic energy levels involved in the scattering process. Such inelastic scattering is called Raman scattering. Raman amplifiers make use of stimulated Raman scattering (SRS) to amplify an optical signal. After discussing the general aspects of Raman scattering in Section 7.1, we focus in Section 7.2 on SRS in optical fibers. Section 7.3 provides details on how this process can be used to make fiber-based Raman amplifiers. Silicon-based Raman amplifiers are covered in Section 7.4.

7.1 Raman effect

The Raman effect was discovered in 1928 by two Indian scientists, C. V. Raman and K. S. Krishnan [6]. It has proven to be extremely useful for a variety of spectroscopic applications in physics, chemistry, and biology. The reason is easy to understand: spontaneous Raman scattering has its origin in the vibrational states of molecules, and reduces the frequency of incoming light by a precise amount associated with a specific molecule. Moreover, it can occur for any frequency of the incident light because of it its nonresonant nature. However, it is important to recognize that the incident photon must be able to interact with one of the vibrational energy levels of a molecule to initiate Raman scattering. If a material allows such an interaction, it is called *Raman-active*. There exists a large number of Raman-active materials, and we consider two of them, silica and silicon, in later sections. In this section we discuss the general aspects of Raman scattering that apply for any Raman-active material.

7.1.1 Spontaneous Raman scattering

Figure 7.1 shows the origin of spontaneous Raman scattering using the vibrational states associated with a specific molecule. The energy of the incident photon is not in resonance with any electronic transition. In quantum mechanics, such a situation is

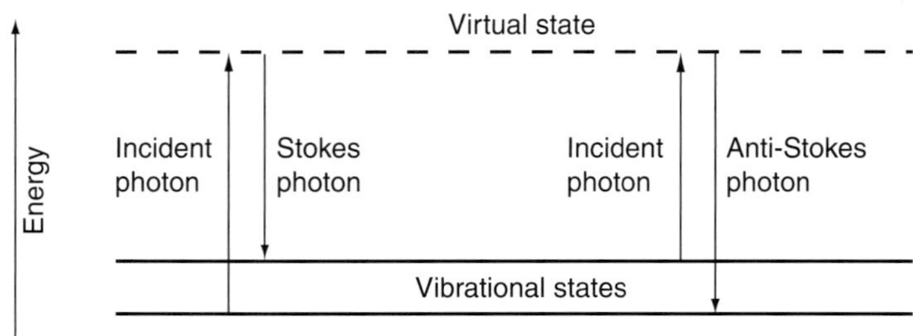

Figure 7.1 Schematic illustration of Stokes and anti-Stokes Raman scattering with the assistance of a virtual energy level.

represented by means of a virtual state, as shown by the dashed line in Figure 7.1. The molecule makes a transition to the virtual electronic state by absorbing the incident photon, but it stays in this virtual state only for the relatively short time permitted by Heisenberg's uncertainty principle (typically < 1 fs) before it emits a photon to end up in one of the vibrational states associated with the ground state. In the case of spontaneous Raman scattering, the molecule ends up in the first vibrational state, and the photon frequency is reduced by a specific amount. The frequency shift experienced by the incident light during spontaneous Raman scattering is called the *Raman shift*, and corresponds to a specific vibrational resonance associated with the Raman-active medium [7]. Another process, which increases the frequency of the incident photon, can also occur with a smaller probability. As a result, for each vibrational resonance, two Raman bands appear in the optical spectrum on opposite sides of the incident frequency. These are classified as follows:

Stokes band A Raman band with frequencies less than the incident radiation is called a Stokes band (see Figure 7.1). During its creation, incident photons have lost energy to the material's molecules, which end up in an excited vibrational state. The stored energy is eventually dissipated as heat.

Anti-Stokes band A Raman band with frequencies greater than the incident radiation is called an anti-Stokes band (see Figure 7.1). During its creation, incident photons receive energy from the material's molecules. This can only happen if a molecule is initially in an excited vibrational state before the incident photon arrives. After the scattering event, the molecule ends up in the lower energy state.

The strength of spontaneous Raman scattering is directly proportional to the number of molecules occupying different vibrational states of the Raman-active material. At room temperature, the population of vibrationally excited states is typically small but not negligible. In thermal equilibrium, the occupancy of vibrational states is governed by the Boltzmann distribution, which favors low-lying energy states over higher energy states at a given temperature. For this reason, each Stokes band dominates over the corresponding anti-Stokes band in thermal equilibrium.

7.1.2 Stimulated Raman scattering

Consider a Raman-active material pumped by an intense optical beam at a certain frequency. If a second optical beam whose frequency is lower by exactly the Raman shift associated with the material's molecules is also launched into this medium, the presence of the second beam should stimulate the emission of Stokes photons from the Raman-active molecules (similar to the case of stimulated emission). This type of interaction, stimulated Raman scattering, discovered in 1962 by Woodbury and Ng [8], is known as (SRS). One can understand this process from Figure 7.1 as follows. Owing to the presence of the pump field, molecules in the electronic ground state are excited to a virtual energy state with higher energy. The presence of another external field matching the Stokes frequency stimulates each excited molecule in this virtual state to emit a Stokes photon such that it ends up in a vibrational state of the ground state [7]. In contrast to conventional gain media (such as semiconductor optical amplifiers or doped fiber amplifiers), in which gain occurs only when a population inversion exists, SRS takes place in Raman-active media without any population inversion. Except for the presence of a pump beam, no additional preparation is necessary for SRS to occur. As expected, if the external Stokes field is switched off, the SRS process falls back to the traditional spontaneous Raman scattering.

On physical grounds one should expect the growth rate of the Stokes field to depend on the intensities of both the pump and the Stokes fields, and this is indeed the case [7]. For this reason, SRS can be categorized as a nonlinear process. In contrast, spontaneous Raman scattering depends linearly only on the intensity of the pump beam. The most striking aspect of the SRS process is that laser light at virtually any wavelength can be used for pumping as long as it lies within the transparency region of the Raman medium. Clearly, SRS-based amplification has a wider applicability than amplification mechanisms that rely on the excited electronic states of a gain medium. It is also interesting to note that if the medium has an external beam coinciding in frequency with the anti-Stokes frequency, it is possible to achieve stimulated scattering of photons at the anti-Stokes frequency. However, this process is considerably less efficient than the normal SRS process because it also depends on how many molecules exist in the excited vibrational state [7].

Figure 7.2 shows schematically several types of Raman scattering that may occur in a Raman-active material when two beams with frequencies $\omega_a > \omega_b$ are incident on it. During the interaction of the two beams within the material, optical phonons

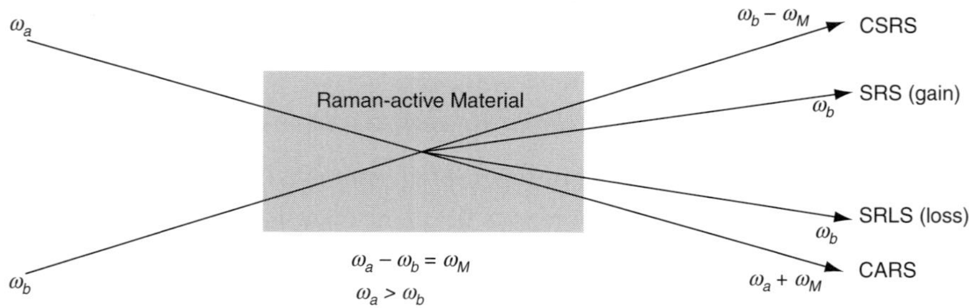

Figure 7.2 Different Raman processes; SRS and SRLS, CARS, CSRS, due to interaction of light with a Raman-active material.

of a specific frequency ω_M are created, where ω_M is a characteristic vibrational frequency associated with the molecules of the medium. The Stokes beam at frequency ω_b is amplified through SRS, while the other beam at frequency ω_a is attenuated. In addition, two other beams may be created from processes known as coherent Stokes Raman scattering (CSRS) and coherent anti-Stokes Raman scattering (CARS). CSRS and CARS differ from conventional SRS in several respects [9], the most important being that their intensities depend on the square of the pump intensity and the square of the number of molecules participating in the scattering process. CARS is generally described as a four-wave mixing (FWM) process enhanced through the Raman effect [10]. CSRS and CARS have less relevance for optical amplification and thus fall outside the scope of this book.

7.2 Raman gain spectrum of optical fibers

Since optical fibers are made of silica glass consisting of randomly distributed SiO_2 molecules, they are a good candidate for Raman amplification by SRS [11]. Stolen and Ippen [12] were the first to observe, in 1973, Raman amplification in optical fibers. However, until the late 1990s, not much interest was shown in the use of Raman amplification in telecommunications systems. One reason was that the erbium-doped fiber amplifiers (EDFAs) discussed in Chapter 5 were serving this purpose reasonably well. Another reason was the unavailability of the high-power semiconductor lasers required for pumping Raman amplifiers. Owing to the need to expand the wavelength-division multiplexed (WDM) systems beyond the traditional telecommunications band near 1550 nm and the inability of EDFAs to operate much beyond their resonance near 1530 nm, Raman-amplifier technology was adopted for modern WDM systems after the year 2000.

The Raman effect in silica arises from vibrational modes of SiO_4 tetrahedra, in which corner oxygen atoms are shared by adjacent units [14]. Figure 7.3 displays the Raman gain spectrum of a silica fiber when both the pump and the signal waves are linearly copolarized. It shows that the Raman gain spans a bandwidth in excess of 40 THz with a dominant peak close to 440 cm^{-1} or 13.2 THz. In addition to the dependence of the Raman gain in Figure 7.3 on the difference in optical frequencies between the pump and signal, Raman gain also depends on the pump wavelength and the relative polarizations of the pump and signal. In particular, it nearly vanishes when the two are orthogonally polarized.

In spite of its complex nature, the Raman gain spectrum can be well approximated using a basis consisting of a finite set of Gaussian functions convolved with Lorentzian functions [15]. The reason behind this expansion-basis choice is that the dynamics of each vibrational mode can be approximated with that of a

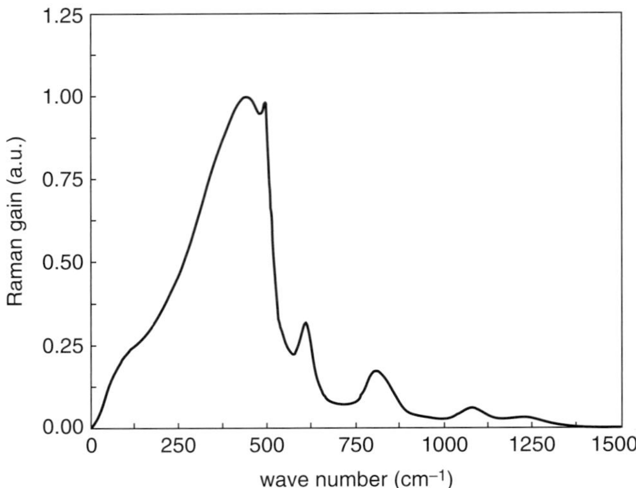

Figure 7.3 Raman gain spectrum of a fused silica fiber with copolarized pump and signal beams. (After Ref. [13]; © OSA 1989)

damped harmonic oscillator with a Lorentzian frequency response. If the oscillation frequencies of a large number of such vibrational modes are represented by a Gaussian distribution, the spectrum of the ensemble becomes a convolution of this Gaussian with individual Lorentzian response functions [16]. It turns out [15] that the inclusion of just 13 vibrational modes provides a reasonable approximation to the experimentally measured Raman impulse response $h_R(t)$ of optical fibers, and hence to the Raman spectrum, which is related to the imaginary part of the Fourier transform of $h_R(t)$. More specifically, the Raman impulse response vanishes for $t < 0$ to ensure causality, and for $t \geq 0$ has the form

$$h_R(t) = \sum_{n=1}^{13} \frac{A_n}{\omega_n} \exp(-\gamma_n t) \exp(-\Gamma_n^2 t^2/4) \sin(\omega_n t), \qquad (7.1)$$

where A_n is the amplitude of the nth vibrational mode, ω_n is its frequency, and γ_n is its decay rate. The parameter Γ_n in the Gaussian function represents the bandwidth associated with the nth vibrational mode.

The Raman impulse response of optical fibers has been measured experimentally [13]. Figure 7.4 shows how well the experimental data can be fit using Eq. (7.1). Table 7.1 provides the values of the parameters used for the 13 vibrational modes, with the notation $f_n = \omega_n/(2\pi c)$, $B_n = A_n/\omega_n$, $\delta_n = \gamma_n/(\pi c)$, and $\Delta_n = \Gamma_n/(\pi c)$. The inset in Figure 7.4 compares the corresponding Raman gain spectra obtained by taking the Fourier transform of $h_R(t)$. It is evident that the approximation in Eq. (7.1) is excellent throughout the frequency range of interest.

Table 7.1. *Parameter values used to fit the Raman response of silica fibers* [15]

n	f_n (cm^{-1})	B_n	Δ_n (cm^{-1})	δ_n (cm^{-1})
1	56.25	1.00	52.10	17.37
2	100.00	11.40	110.42	38.81
3	231.25	36.67	175.00	58.33
4	362.50	67.67	162.50	54.17
5	463.00	74.00	135.33	45.11
6	497.00	4.50	24.50	8.17
7	611.50	6.80	41.50	13.83
8	691.67	4.60	155.00	51.67
9	793.67	4.20	59.50	19.83
10	835.50	4.50	64.30	21.43
11	930.00	2.70	150.00	50.00
12	1 080.00	3.10	91.00	30.33
13	1 215.00	3.00	160.00	53.33

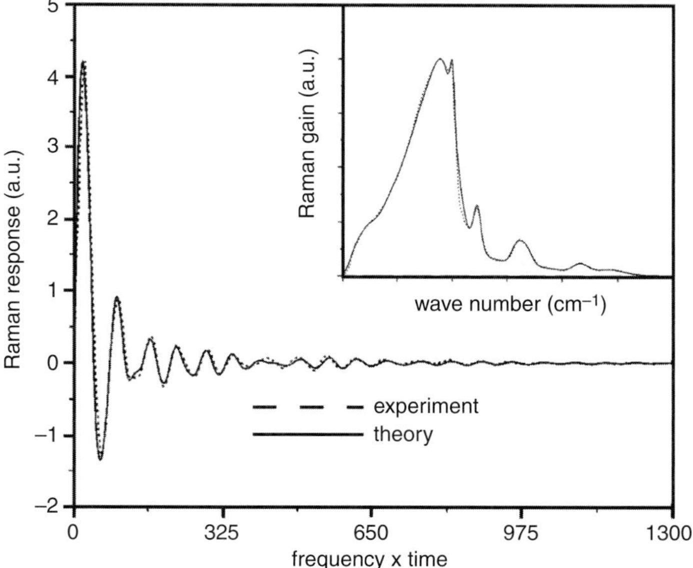

Figure 7.4 Experimentally measured Raman impulse response (solid curve) and a theoretical fit based on Eq. (7.1). The inset shows the corresponding fit to the Raman gain spectrum. (After Ref. [15]; © OSA 2002)

The model in Eq. (7.1) is too simplistic to account for the polarization dependence of the Raman gain. In general, the Raman gain is anisotropic. This can be seen by considering the third-order nonlinear response of silica, whose most general form is given by [17]

$$R_{ijkl}^{(3)}(\tau) = \frac{(1 - f_R)}{3}\delta(\tau)(\delta_{ij}\delta_{kl} + \delta_{ik}\delta_{jl} + \delta_{il}\delta_{jk})$$
$$+ f_R h_a(\tau)\delta_{ij}\delta_{kl} + \frac{f_R}{2}h_b(\tau)(\delta_{ik}\delta_{jl} + \delta_{il}\delta_{jk}), \tag{7.2}$$

where the subscripts take values x, y, and z. The first term, containing $\delta(\tau)$, results from the nearly instantaneous electronic response, but the other two terms have their origins in molecular vibrations. The functions $h_a(\tau)$ and $h_b(\tau)$ represent the isotropic and anisotropic parts of the Raman response function, respectively, and f_R represents the relative contribution of molecular vibrations to the total nonlinear response (determined in practice by fitting the experimental data). The often-used scalar form of the nonlinear Raman response can be obtained by setting i, j, k, and l equal to x or y, and is given by

$$R_{xxxx}^{(3)}(\tau) = R_{yyyy}^{(3)}(\tau) = (1 - f_R)\delta(\tau) + f_R h_a(\tau) + f_R h_b(\tau). \tag{7.3}$$

This form corresponds to launching linearly polarized light along a principal axis of a polarization-maintaining fiber.

Noting that the isotropic part of the Raman response stems predominantly from the symmetric stretching motion of the the bridging oxygen atom in the Si–O–Si bond, a Lorentzian oscillator model can be used to approximate h_a as [17]

$$h_a(\tau) = f_a \frac{\tau_1}{\tau_1^2 + \tau_2^2} \exp\left(-\frac{\tau}{\tau_2}\right) \sin\left(\frac{\tau}{\tau_1}\right), \tag{7.4}$$

where $\tau_1 = 12.2\,\text{fs}$, $\tau_2 = 32\,\text{fs}$, and $f_a = 0.75$ represents the fractional contribution of h_a to the total copolarized Raman response.

To model $h_b(\tau)$, we need to take into account the low-frequency behavior of the Raman gain spectrum. In addition, one needs to consider the bond-bending motion and strong intermediate-range correlations between neighboring bonds in glass media [18]. It was found by Lin and Agrawal [17] empirically that the following function provides a good approximation to $h_b(\tau)$:

$$h_b(\tau) = f_b\left(\frac{2\tau_b - \tau}{\tau_b^2}\right)\exp\left(-\frac{\tau}{\tau_b}\right) + f_c\frac{\tau_1}{\tau_1^2 + \tau_2^2}\exp\left(-\frac{\tau}{\tau_2}\right)\sin\left(\frac{\tau}{\tau_1}\right), \tag{7.5}$$

where f_b and f_c are fractional contributions such that $f_a + f_b + f_c = 1$. The three new parameters appearing in this equation are found to have values $\tau_b = 96\,\text{fs}$, $f_b = 0.21$, and $f_c = 0.04$ to match the dominant peak in the experimentally measured Raman response of silica. Figure 7.5 shows the predicted Raman gain spectrum (solid curve) together with with the experimental data in the case in which

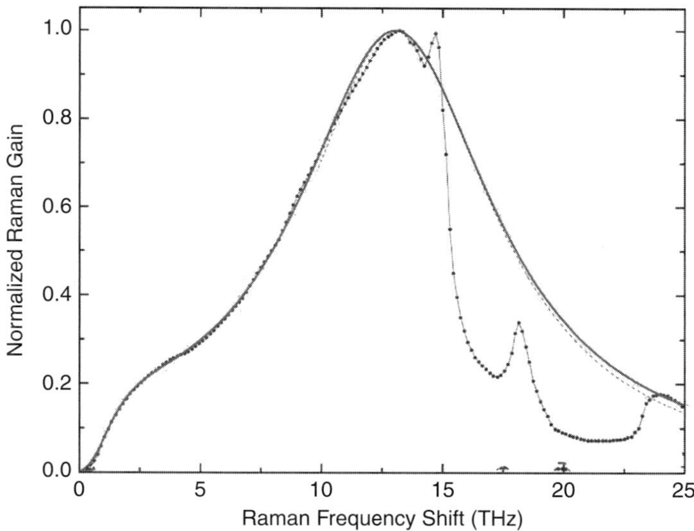

Figure 7.5 The theoretical fit (solid line) to the measured Raman gain spectrum (dotted line) calculated using Eq. (7.3) (After Ref. [17]; © OSA 2006)

the Stokes radiation is copolarized with the pump. The fit is quite good in the low-frequency region of the gain spectrum extending up to the gain peak, but it fails in the frequency region beyond 15 THz because it is based on the response of a single vibrational mode. In practice, the high-frequency response becomes relevant only for ultrashort pump pulses.

The Raman gain spectrum in Figure 7.5 results from the interaction of copolarized pump and signal waves. The case in which the pump and signal waves are orthogonally polarized is also interesting. The Raman gain for this configuration can also be calculated from Eq. (7.2). Figure 7.6 compares the Raman gain spectra for the copolarized and orthogonally polarized cases. It shows that the Raman gain is reduced by more than a factor of 10 near the peak of the Raman curve when the pump and signal are orthogonally polarized. One can use these polarization properties to make a Raman amplifier that is insensitive to polarization by using a polarization-diversity scheme for pumping [11].

Another aspect that we have not considered so far is the dependence of the Raman gain on the fiber design. Clearly, Raman interaction can be enhanced at a given pump power by reducing the diameter of the fiber core to enhance the local intensity. As an example, Figure 7.7 shows the measured Raman gain efficiency [19], defined as the ratio g_R/A_{eff} where A_{eff} is the effective mode area, for standard single-mode fiber (SMF), dispersion-shifted fiber (DSF), and dispersion-compensating fiber (DCF). The Raman gain efficiency depends to some extent on the composition of the fiber core but the most important factor is the size of the fiber core. The DCF

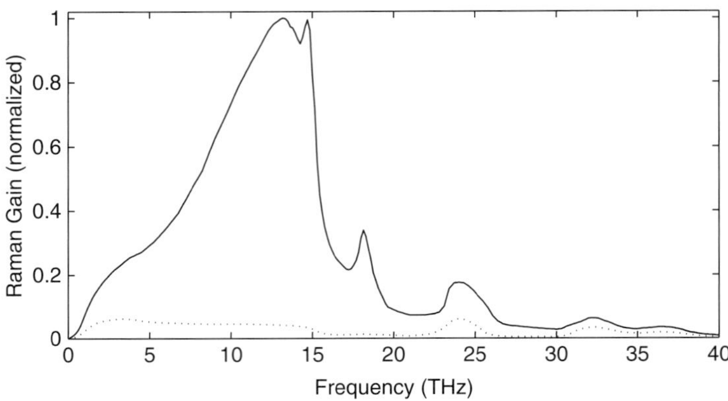

Figure 7.6 Normalized Raman-gain coefficient when pump and signal waves are copolarized (solid curve) or orthogonally polarized (dotted curve). (After Ref. [12]; © OSA 2006)

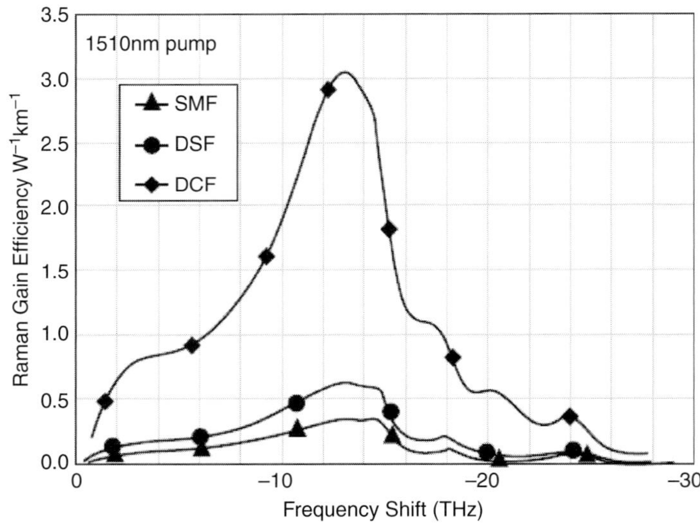

Figure 7.7 Raman gain efficiency for three different fibers. (After Ref. [19]; © OSA 2000)

curve in Figure 7.7 shows significantly higher Raman gain because a DCF employs a relatively narrow core size together with the dopants to change its dispersive properties.

To summarize, Raman gain depends primarily on the frequency difference between the pump and signal waves but not on their absolute frequencies. This is the result of energy conservation, requiring that the energy of the optical phonons participating in the SRS process matches the energy difference between the pump and signal photons. Momentum conservation is also required for such interactions, but this is easily achieved in practice because optical phonons associated with vibrating molecules have a wide range of momenta. An implicit consequence of this feature is that Raman gain is independent of the relative propagation directions of the pump and signal waves. As a result, a Raman amplifier can be pumped in the forward or backward direction with similar performance. However, as we saw in Figure 7.6, Raman gain depends on the relative polarization of the pump and signal waves, and attains its highest value when the waves are copolarized.

7.3 Fiber Raman amplifiers

A very useful application of an optical fiber is to exploit SRS, achieve the Raman gain by pumping it suitably, and use it as a Raman amplifier. Figure 7.8 shows schematically how a fiber can be used as a Raman amplifier. It depicts three potential pumping schemes, obtained when the pump propagates (a) in the direction of the signal (forward pumping), (b) in the opposite direction (backward pumping), or (c) in both directions (bidirectional pumping). We analyze the performance of fiber Raman amplifiers in these three cases by considering the cases of a CW signal and a pulse signal separately.

7.3.1 CW operation of a Raman amplifier

We consider the general case of bidirectional pumping and assume that both pumps as well the signal being amplified are in the form of CW waves. If $I_f(z)$ and $I_b(z)$ represent, respectively, the forward and backward pump intensities at a distance z from the front end of the fiber and $I_s(z)$ is the signal intensity at that distance, the interaction between the pumps and the signal is governed by the following set of three coupled equations [20]:

$$\frac{\partial I_s}{\partial z} = -\alpha_s I_s + g_R(I_f + I_b)I_s, \tag{7.6a}$$

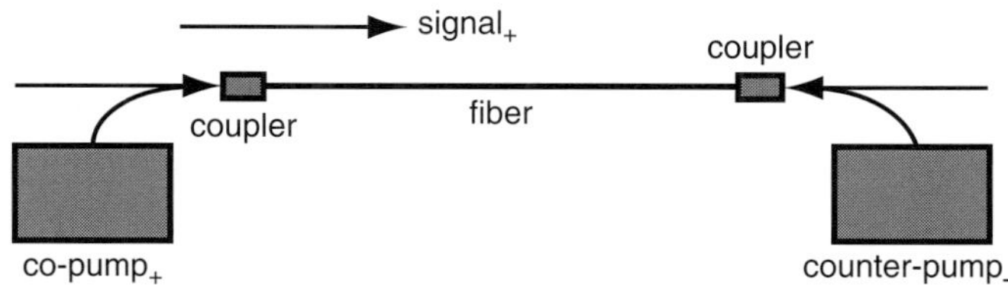

Figure 7.8 Schematic of a fiber Raman amplifier with bidirectional pumping. The amplifier can also be pumped in one direction only, by eliminating the pump in the other direction.

$$\frac{\partial I_f}{\partial z} = -\alpha_s I_f - \frac{\omega_p}{\omega_s} g_R I_f I_s, \tag{7.6b}$$

$$-\frac{\partial I_b}{\partial z} = -\alpha_s I_b - \frac{\omega_p}{\omega_s} g_R I_b I_s, \tag{7.6c}$$

where g_R is the Raman-gain coefficient at the difference frequency $\Omega = \omega_p - \omega_s$, ω_s and ω_p are the signal and pump frequencies, respectively, and α_s and α_p account for fiber losses at these frequencies. The case of forward pumping can be studied by setting $I_b = 0$. Similarly, the case of backward pumping can be studied by setting $I_f = 0$. The latter case is sometimes preferred because it can reduce the impact of pump noise on the signal.

Consider first the forward-pumping case and set $I_b = 0$ in Eqs. (7.6). A simple analytic solution is possible if we ignore the the second term in Eq. (7.6b), which corresponds to pump depletion, by assuming that the pump remains much more intense than the signal throughout the amplifier length. The pump intensity then varies as $I_f(z) = I_0 \exp(-\alpha_p z)$, where I_0 is the input pump intensity. Substituting this form into Eq. (7.6a), we obtain

$$\frac{\partial I_s}{\partial z} = -\alpha_s I_s - g_R I_0 \exp(-\alpha_p z) I_s. \tag{7.7}$$

This equation can be easily integrated and provides the following expression for the signal intensity at the output of a Raman amplifier of length L:

$$I_s(L) = I_s(0) \exp\left(g_R I_0 L_{\text{eff}} - \alpha_s L\right), \tag{7.8}$$

where the effective length of the amplifier is defined as

$$L_{\text{eff}} = \frac{1 - \exp(\alpha_p L)}{\alpha_p}. \tag{7.9}$$

This effective length is shorter than the actual length L because absorption experienced by the pump in the amplification medium reduces the interaction length of the SRS process.

The solution of Eqs. (7.6) shows that the signal power grows exponentially along the amplifier length. The same exponential dependence is found when the amplifier is backward pumped. The solution is more complicated in the case of bidirectional pumping but can be easily obtained. Figure 7.9 shows the evolution of signal power in a 100-km-long Raman amplifier pumped bidirectionally such that the original signal power of 1 mW is recovered at the output end in spite of 0.2 dB km^{-1} loss in the fiber. The percentage of pump power launched in the forward direction is varied from 0% to 100%. In all cases, the total pump power is chosen such that the Raman gain is just sufficient to compensate for fiber losses.

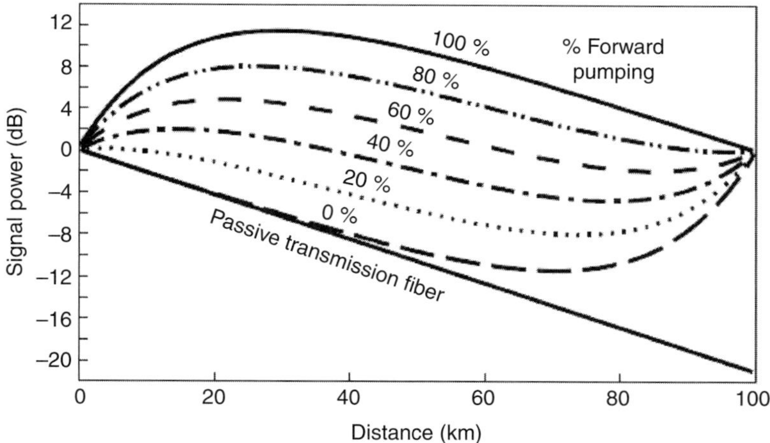

Figure 7.9 Evolution of signal power in a bidirectionally pumped, 100-km-long Raman amplifier as the contribution of forward pumping is varied from 0 to 100%. (After Ref. [20]; © Springer 2003)

7.3.2 Amplification of ultrashort pulses

In this section we consider amplification of ultrashort pulses in a fiber Raman amplifier. As discussed in Section 5.5, pulse amplification in doped-fiber amplifiers is governed by a generalized nonlinear Schrödinger (NLS) equation, modified suitably to account for the optical gain. In the case of Raman amplification, this NLS equation requires further modification, and takes the following form [11]:

$$
\begin{aligned}
\frac{\partial A}{\partial z} &+ \frac{1}{2}\left(\alpha(\omega_0) + j\alpha_1\frac{\partial}{\partial t}\right)A + \frac{j\beta_2}{2}\frac{\partial^2 A}{\partial T^2} - \frac{\beta_3}{6}\frac{\partial^3 A}{\partial T^3} \\
&= j\left(\gamma(\omega_0) + j\gamma_1\frac{\partial}{\partial t}\right)\left(A(z,t)\int_0^\infty R^{(3)}_{xxxx}(t')|A(z,t-t')|^2 dt'\right),
\end{aligned}
\tag{7.10}
$$

where $A(z,t)$ is the slowly varying envelope of the electric field at a reference frequency ω_0, α is the loss parameter, β_2 and β_3 are the second- and third-order dispersion parameters, and γ is the nonlinear parameter, all evaluated at ω_0. The frequency dependence of the loss and nonlinear parameters is included through the derivatives $\alpha_1 = d\alpha/d\omega$ and $\gamma_1 = d\gamma/d\omega$, both evaluated at ω_0. In practice, γ_1 can be approximated by γ/ω_0, if the effective mode area of the fiber is nearly the same at the pump and signal wavelengths.

In the case of Raman amplifiers, the total electric field contains both the signal and the pump fields, i.e., $E = Ae^{-i\omega_0 t} \equiv A_s e^{-i\omega_s t} + A_p e^{-i\omega_p t}$, where ω_s and ω_p are the carrier frequencies of the signal and pump pulses, respectively. If we choose $\omega_0 = \omega_s$, we obtain $A = A_s + A_p e^{-i\Omega t}$, where $\Omega = \omega_p - \omega_s$. Since $\Omega/(2\pi)$ is close to the Raman shift of 13.2 THz for silica fibers, a direct solution of Eq. (7.10) is time-consuming because it requires a temporal step size < 10 fs to cover the entire bandwidth. The computational time can be significantly reduced by recasting the pulse-amplification problem as a coupled set of equations for the signal amplitude A_s and the pump amplitude A_p. Using the preceding form of A in Eq. (7.10), we obtain the following set of two coupled equations [21]:

$$\frac{\partial A_s}{\partial z} + \frac{\alpha_s}{2} A_s + \frac{j\beta_{2s}}{2} \frac{\partial^2 A_s}{\partial T^2} - \frac{\beta_{3s}}{6} \frac{\partial^3 A_s}{\partial T^3}$$

$$= j\gamma_s \left(1 + \frac{j}{\omega_s} \frac{\partial}{\partial t}\right) \left[(1 - f_R)(|A_s|^2 + 2|A_p|^2)A_s\right.$$

$$+ f_R A_s \int_0^\infty [h_a(t') + h_b(t')][|A_s(z, t - t')|^2 + |A_p(z, t - t')|^2]dt'$$

$$\left. + f_R A_p \int_0^\infty [h_a(t') + h_b(t')]A_s(z, t - t')A_p^*(z, t - t')e^{-j\Omega t'}dt'\right],$$

$$(7.11)$$

$$\frac{\partial A_p}{\partial z} + \frac{\alpha_p}{2} A_p + (\beta_{1p} - \beta_{1s})\frac{\partial A_p}{\partial t} + \frac{j\beta_{2p}}{2} \frac{\partial^2 A_p}{\partial T^2} - \frac{\beta_{3p}}{6} \frac{\partial^3 A_p}{\partial T^3}$$

$$= j\gamma_p \left(1 + \frac{j}{\omega_p} \frac{\partial}{\partial t}\right) \left[(1 - f_R)(|A_p|^2 + 2|A_s|^2)A_p\right.$$

$$+ f_R A_p \int_0^\infty [h_a(t') + h_b(t')][|A_s(z, t - t')|^2 + |A_p(z, t - t')|^2]dt'$$

$$\left. + f_R A_s \int_0^\infty [h_a(t') + h_b(t')]A_p(z, t - t')A_s^*(z, t - t')e^{-j\Omega t'}dt'\right],$$

$$(7.12)$$

where the subscripts s and p on the parameters α, β, and γ indicate whether they are evaluated at ω_s or ω_p.

These equations can be simplified considerably for picosecond pump and signal pulses whose widths are much larger than the Raman response time (about 60 fs). The final result can be written in the following compact form [11]:

$$\frac{\partial A_s}{\partial z} + \frac{\alpha_s}{2} A_s + \frac{j\beta_{2s}}{2} \frac{\partial^2 A_s}{\partial T^2}$$

$$= j\gamma_s[|A_s|^2 + (2 - f_R)|A_p|^2]A_p + \frac{g_R}{2}|A_p|^2 A_s, \qquad (7.13\text{a})$$

$$\frac{\partial A_p}{\partial z} + \frac{\alpha_p}{2} A_p + d\frac{\partial A_p}{\partial T} + \frac{j\beta_{2p}}{2} \frac{\partial^2 A_p}{\partial T^2}$$

$$= j\gamma_p[|A_p|^2 + (2 - f_R)|A_s|^2]A_p - \frac{g_R\omega_p}{2\omega_s}|A_s|^2 A_p, \qquad (7.13\text{b})$$

where $d = \beta_{1p} - \beta_{1s}$ represents the group-velocity mismatch and g_R is the Raman gain at the signal frequency.

Amplification of picosecond pulses in Raman amplifiers requires a numerical solution of the preceding two equations. Before solving them, it is useful to introduce four length scales that determine the relative importance of various terms in these equations. For a pump pulse of duration T_0 and peak power P_0, these are defined as follows:

Dispersion length $L_D = T_0^2/|\beta_{2p}|$ is the length scale over which the effects of group-velocity dispersion become important. Typically, $L_D > 1$ km for $T_0 = 5$ ps but becomes ~ 1 m for a femtosecond pulse.

Walk-off length $L_W = T_0/|\beta_{1p} - \beta_{1s}|$ is the length scale over which the walk-off between pump and signal resulting from their different group velocities becomes important. Typically $L_W \approx 1$ m for $T_0 < 1$ ps.

Nonlinear length $L_{NL} = 1/(\gamma_p P_0)$ is the length scale over which nonlinear effects such as self- and cross-phase modulation become important. Typically $L_{NL} \sim 1$ km for $P_0 = 100$ mW.

Raman gain length $L_G = 1/(g_p P_0)$, where g_p is the Raman-gain coefficient at the pump frequency, is the length scale over which Raman gain becomes important. Typically $L_G \sim 1$ km for $P_0 > 100$ mW.

It is important to note that the shortest length among the four length scales plays the dominant role during pulse propagation.

To illustrate the amplification of ultrashort pulses in a fiber Raman amplifier, it is instructive to look at a specific example. Figure 7.10 shows the evolution of the pump and signal pulses in the normal-GVD regime over three walk-off lengths when the pump pulse and fiber parameters are such that $L_D/L_W = 1000$, $L_W/L_{NL} = 24$, and $L_W/L_G = 12$. The pump pulse is assumed to have a Gaussian shape initially. The input signal pulse is also Gaussian with the same width but its peak power is quite small initially ($P_s = 2 \times 10^{-7}$ W at $z = 0$). The signal pulse starts to grow exponentially close to the input end, but its growth slows down because of the walk-off effects between the pump and the signal. In fact, energy transfer from the pump pulse stops after $z = 3L_W$ as the two pulses are then physically separated because of their group-velocity mismatch. Since the signal pulse moves faster than the pump pulse in the normal-GVD regime, the energy for Raman amplification comes from the leading edge of the pump pulse. This is apparent near $z = 2L_W$, where energy transfer has led to a two-peak structure in the pump pulse as a result of pump depletion. The hole near the leading edge corresponds exactly to the location of the signal pulse. The small peak near the leading edge disappears with further propagation as the signal pulse walks through it. The pump pulse at $z = 3L_W$ is asymmetric in shape and appears narrower than the input pulse as it consists of the

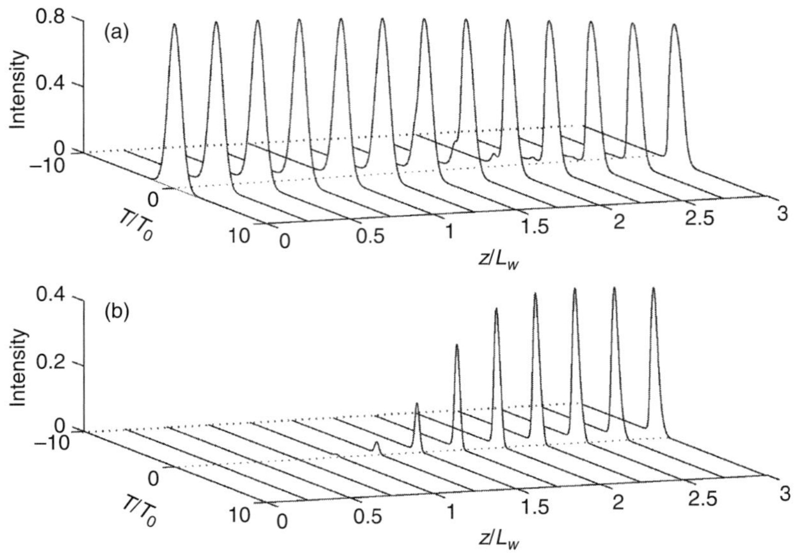

Figure 7.10 Evolution of pump (top) and signal (bottom) pulses over three walk-off lengths when $L_D/L_W = 1000$, $L_W/L_{NL} = 24$, and $L_W/L_G = 12$. (After Ref. [11]; © Elsevier 2007)

trailing portion of the input pulse. The signal pulse is also narrower than the input pulse and is asymmetric with a sharp leading edge.

The preceding results show that the transfer of energy from a pump pulse to a signal pulse of comparable width is hindered considerably by the walk-off effects resulting from a mismatch between their group velocities. This problem is reduced considerably when pump pulses are much wider than the signal pulses, and disappears for a CW pump. Indeed, fiber Raman amplifiers employ a CW pump when they are used for amplifying picosecond signal pulses in telecommunications systems.

7.4 Silicon Raman amplifiers

Silicon is called the material where the "extraordinary is made ordinary" [22]. Silicon continues to dominate the microelectronics industry because of its remarkable electrical, chemical, thermal, and mechanical properties that enable mass-scale manufacturing methods [23]. In contrast to this, no single material or technology plays a dominant role in the photonics area, a fact that creates a plethora of interfacing and manufacturing problems. The use of silicon for photonic integrated circuits is attracting attention because it capitalizes on the success of the silicon revolution [22–24]. Silicon-based optical devices have the potential for providing a monolithically integrated optoelectronic platform. However, as mentioned in Section 6.1, silicon is an indirect bandgap material, and this feature makes light emission in silicon quite inefficient. To make matters worse, nonradiative recombination rates in silicon are much higher than radiative ones.

Because of the indirect bandgap of silicon, it is unlikely that electrical pumping can be used to achieve the population inversions required for optical gain. Even if a population inversion is achieved, nonlinear absorption mechanisms such as two-photon absorption in silicon will make it difficult to obtain a net optical gain. However, it has been known for years that the Raman-gain coefficient in silicon is higher than that of a silica fiber by a large factor. At the same time, the effective mode area is about 100 times smaller in typical silicon waveguides, known as nanowires, because of their relatively small dimensions, resulting in considerable enhancement of pump intensity at a given pump power. For these reasons, the use of the Raman effect in silicon waveguides has attracted considerable attention. Spontaneous Raman scattering was observed in 2002 using such a silicon nanowire [25]. Soon afterward, SRS was employed to achieve Raman gain in such waveguides [26]. Since then, the SRS process has been utilized to make silicon-based Raman amplifiers [26–33] as well as Raman lasers [34–37]. Since silicon nanowires exhibit a strong nonlinear response [38–41], they have been used for many other applications. For example, four-wave mixing has been used to make parametric amplifiers acting as broadband wavelength converters [17,42–47]. Such amplifiers are discussed in Chapter 8.

7.4.1 Coupled pump and signal equations

Similarly to the case of a fiber Raman amplifier, we need to find the coupled set of two equations governing pulse amplification in silicon Raman amplifiers. The form of these equations is similar to those in Eqs. (7.13), but they must be generalized to include new features that are unavoidable in silicon devices. The most important among these is two-photon absorption (TPA), which occurs whenever the energies of pump and signal photons exceed one-half of the bandgap. Since the bandgap of silicon is near 1.1 eV, this is the case in the wavelength region near 1550 nm, where photon energies are close to 0.8 eV. Another complication is that the electron–hole pairs produced by TPA can create a substantial density of free carriers, which not only absorb light (free-carrier absorption) but also change the refractive index within the silicon waveguide region (free-carrier dispersion). A third issue that needs attention is that the Raman gain spectrum of silicon is so narrow compared with optical fibers (only about 100 GHz wide) that the Raman response time is close to 3 ps. Taking into account all these three issues, one obtains the following coupled NLS-type equations that govern the evolution of the pump and signal pulses in a silicon waveguide [48]:

$$\frac{\partial A_s}{\partial z} + \beta_{1s}\frac{\partial A_s}{\partial t} + \frac{j\beta_{2s}}{2}\frac{\partial^2 A_s}{\partial t^2} = -\frac{\alpha_s}{2}A_s + j(\gamma_s|A_s|^2 + 2\gamma_{sp}|A_p|^2)A_s$$

$$-\frac{\sigma_s}{2}(1 + j\mu_s)NA_s + j\gamma_s A_p \int_{-\infty}^{t} h_R(t - t')A_p^*(z, t')A_s(z, t')\,e^{-j\Omega(t-t')}dt',$$

$$(7.14a)$$

$$\frac{\partial A_p}{\partial z} + \beta_{1p}\frac{\partial A_p}{\partial t} + \frac{j\beta_{2p}}{2}\frac{\partial^2 A_p}{\partial t^2} = -\frac{\alpha_p}{2}A_p + j(\gamma_p|A_p|^2 + 2\gamma_{ps}|A_s|^2)A_p$$

$$-\frac{\sigma_p}{2}(1 + j\mu_p)NA_p + j\gamma_p A_s \int_{-\infty}^{t} h_R(t - t')A_s^*(z, t')A_p(z, t')\,e^{j\Omega(t-t')}dt'.$$

$$(7.14b)$$

The first term on the right side of Eqs. (7.14) represents linear losses, while the second term accounts for SPM, cross-phase modulation, and TPA through the nonlinear parameters defined as $\gamma_m = (\omega_m/c)n_2 + j\beta_T(\omega_m)/2$ $(m = p, s)$, where $\beta_T(\omega)$ is the TPA coefficient of silicon. In addition, we have two other nonlinear parameters, defined as $\gamma_{ps} = b_x n_2 + j(\omega_p/\omega_s)\beta_T/2$, where b_x has its origin in the anisotropic nature of third-order susceptibility for silicon [48]. The third term on the right side of Eqs. (7.14) represents the impact of TPA-generated free carriers on the pump and signal pulses through the parameter $\sigma_m = \sigma_r(\lambda_m/\lambda_r)^2$ $(m = p, s)$, where, $\sigma_r = 1.45 \times 10^{-21}$ m^2 is the FCA coefficient at a reference wavelength $\lambda_r = 1550$ nm. The dimensionless parameter $\mu_m = 2k_m\sigma_n/\sigma_r$, with $\sigma_n = 5.3 \times 10^{-27}$ m^3, accounts for free-carrier-induced changes in the refractive index. It plays the same role as the linewidth enhancement factor in SOAs and its value is about 7.5 for silicon amplifiers.

The integrals in Eqs. (7.14) account for SRS whose magnitude depends on the pump-signal frequency detuning, $\Omega = \omega_p - \omega_s$, and the Raman response function $h_R(t)$. Using the classical oscillator model of SRS [49], this function for silicon is given by

$$h_R(t) = \frac{\Omega_R^2}{\Omega_0}\sin(\Omega_0 t)\exp(-\Gamma_R t), \qquad (7.15)$$

where $\Omega_0 = (\Omega_R^2 - \Gamma_R^2)^{1/2}$, Γ_R is the bandwidth of the Lorentzian-shaped Raman gain spectrum in silicon (about 105 GHz), and Ω_R is the Raman shift (about 15.6 THz for silicon). The parameters γ_p and γ_s depend on the Raman gain and are defined as

$$\gamma_p = \frac{g_R \Gamma_R}{\Omega_R}, \tag{7.16a}$$

$$\gamma_s = \frac{\omega_s}{\omega_p} \gamma_p, \tag{7.16b}$$

where g_R is the Raman-gain coefficient.

Similarly to the case of SOAs, we need a rate equation for the density $N(z, t)$ of free carriers before we can solve Eqs. (7.14). This equation should include all mechanisms by which free carriers can be generated and all channels through which they may recombine, including radiative recombination, thermal diffusion, and nonradiative recombination at the defects on waveguide interfaces. It is common to lump the impact of all recombination channels in a single parameter τ_c, called the effective carrier lifetime. With this simplification, the carrier rate equation becomes [48]

$$\frac{\partial N}{\partial t} = -\frac{N}{\tau_c} + \rho_p |A_p|^4 + \rho_s |A_s|^4 + \rho_{ps} |A_p A_s|^2, \tag{7.17}$$

where the carrier-generation parameters are defined as

$$\rho_p = \frac{\beta_T}{2\hbar\omega_p}, \quad \rho_s = \frac{\beta_T}{2\hbar\omega_s}, \quad \rho_{ps} = \frac{2\beta_T}{\hbar\omega_s}. \tag{7.18}$$

Owing to the presence of cross-coupling terms, Eqs. (7.14) and (7.17) cannot be solved analytically. Nevertheless, several approximate solutions that shed light on the operation of silicon Raman amplifiers can be derived in few cases of practical interest [50]. We focus on these solutions in what follows.

7.4.2 CW operation of silicon Raman amplifiers

In the CW regime, the envelopes of the pump and signal fields do not change with time. As a consequence, all time derivatives in Eqs. (7.14) vanish. The integrations in the Raman term can be performed by noting that

$$\int_{-\infty}^{t} h_R(t - t') e^{\pm i\Omega(t-t')} dt' = \frac{\Omega_R^2}{\Omega_R^2 - \Omega^2 \pm 2i\Gamma_R\Omega}. \tag{7.19}$$

In addition, the free-carrier density can be obtained from Eq. (7.17) by setting $\partial N/\partial t = 0$ and is given by

$$N(z) = \tau_c \left(\rho_p |A_p|^4 + \rho_s |A_s|^4 + \rho_{ps} |A_p A_s|^2 \right). \tag{7.20}$$

Using this result in Eqs. (7.14) and introducing the intensities of pump and signal (Stokes) waves, $I_p(z) = |A_p(z)|^2$ and $I_s(z) = |A_s(z)|^2$, we obtain [50]

$$\frac{dI_p}{dz} = -\alpha_p I_p - \beta_{Tp} I_p^2 - \zeta_{ps} I_p I_s - \sigma_p \tau_c (\rho_p I_p^2 + \rho_s I_s^2 + \rho_{ps} I_p I_s) I_p, \tag{7.21a}$$

$$\frac{dI_s}{dz} = -\alpha_s I_s - \beta_{Ts} I_s^2 - \zeta_{sp} I_s I_p - \sigma_s \tau_c (\rho_p I_p^2 + \rho_s I_s^2 + \rho_{ps} I_p I_s) I_s, \tag{7.21b}$$

where we have introduced two new quantities,

$$\zeta_{ps} = 2\beta_T \frac{\omega_p}{\omega_s} + \frac{4g_R \Gamma_R^2 \Omega_R \Omega}{(\Omega_R^2 - \Omega^2)^2 + 4\Gamma_R^2 \Omega^2}, \tag{7.22a}$$

$$\zeta_{sp} = 2\beta_T - \frac{4g_R \Gamma_R^2 \Omega_R \Omega (\omega_s/\omega_p)}{(\Omega_R^2 - \Omega^2)^2 + 4\Gamma_R^2 \Omega^2}. \tag{7.22b}$$

To solve Eqs. (7.21) analytically, we make some reasonable simplifications. First, we assume that linear losses are equal at the pump and signal wavelengths, i.e., $\alpha_p = \alpha_s \equiv \alpha$. Second, we discard the second terms on the right side of Eqs. (7.21) since losses from TPA are typically much smaller than losses from FCA in the case of CW pumping [48]. Third, noting that $\Omega_R \ll \omega_p$ or ω_s, we make two rough approximations, $\sigma_p \approx \sigma_s$ and $\rho_p \approx \rho_s \approx \rho_{ps}/4$. Fourth, noting that $g_R \gg \beta_{Tps}$, we set $\zeta_{ps} \approx |\zeta_{sp}| \equiv \gamma$. With these simplifications, the coupled intensity equations (7.21) become

$$\frac{dI_p}{dz} \approx -\alpha I_p - \kappa (I_p^2 + 4I_p I_s + I_s^2) I_p - \gamma I_s I_p, \tag{7.23a}$$

$$\frac{dI_s}{dz} \approx -\alpha I_s - \kappa (I_p^2 + 4I_p I_s + I_s^2) I_s + \gamma I_p I_s, \tag{7.23b}$$

where $\kappa = \tau_c \sigma_s \rho_s$ is an effective TPA parameter. These equations cannot yet be solved in an analytic form, so, we make one more simplification. It consists of replacing the quantity in the parenthesis of Eqs. (7.23) with $(I_p + I_s)^2$ and amounts to replacing 4 with 2 in the cross-TPA term that corresponds to simultaneous absorbtion of one photon from the pump and another from the signal. Numerical simulations used to judge the error introduced with this replacement show that it is a reasonable approximation. After this change, we obtain the following set of two coupled equations:

$$\frac{dI_p}{dz} \approx -\alpha I_p - \kappa (I_p + I_s)^2 I_p - \gamma I_s I_p, \tag{7.24a}$$

$$\frac{dI_s}{dz} \approx -\alpha I_s - \kappa (I_p + I_s)^2 I_s + \gamma I_p I_s. \tag{7.24b}$$

An important point to note is that, even though we have made several approximations in arriving at Eqs. (7.24), they still contain all the physics of Eqs. (7.23) and should predict all the features associated with the FCA and SRS processes qualitatively.

Equations (7.24) can be solved analytically by noting that the total intensity $I(z) = I_p(z) + I_s(z)$ satisfies the Bernoulli equation (see Ref. [51]) whose solution is given by

$$I(z) = \frac{I_0 \exp(-\alpha z)}{\sqrt{1 + \kappa\, I_0^2 L_{\mathrm{eff}}(2z)}}, \tag{7.25}$$

where $I_0 = I_{p0} + I_{s0}$ is the total input intensity, with $I_{p0} = I_p(0)$ and $I_{s0} = I_s(0)$. By substituting $I_p(z) = I(z) - I_s(z)$ into Eq. (7.24b), we can solve this equation as well. The resulting solution is given by

$$I_s(z) = \frac{I(z)}{1 + (I_0/I_{s0}) \exp[-\gamma I_0 \mathscr{L}_{\mathrm{eff}}(z)]}, \tag{7.26a}$$

$$I_p(z) = I(z) - I_s(z), \tag{7.26b}$$

where the generalized effective length of the silicon waveguide is defined as

$$\mathscr{L}_{\mathrm{eff}}(z) = \frac{f(0) - f(z)}{I_0 \sqrt{\alpha \kappa}}, \tag{7.27a}$$

$$f(z) = \tan^{-1}\left[\sqrt{\frac{\kappa}{\alpha}}\, I(z)\right]. \tag{7.27b}$$

In the limit in which FCA becomes negligible ($I_0 \to 0$ or $\kappa \to 0$), it is easy to show that

$$\mathscr{L}_{\mathrm{eff}}(z) \to L_{\mathrm{eff}}(z) = (1 - e^{-\alpha z})/\alpha, \tag{7.28}$$

and we recover the usual definition of the effective length.

Equation (7.26a) shows that changes in the signal intensity result from two sources with different physical origins. The steadily decreasing function $I(z)$ results from linear losses and FCA, whereas the denominator of Eq. (7.26a) arises from SRS and exhibits a saturable character. The structure of the denominator shows that the generalized effective length has a direct influence on the signal gain. If the pump intensity is so large that the condition $I_0/I_{s0} \gg \exp[\gamma I_0 \mathscr{L}_{\mathrm{eff}}(z)]$ remains satisfied for all z, we can neglect 1 in the denominator of Eq. (7.26a). In this limit, we recover the exponential growth of the signal resulting from Raman amplification and obtain

$$I_s(z) = I_{s0} \exp[\gamma I_0 \mathscr{L}_{\mathrm{eff}}(z)] \frac{I(z)}{I(0)}. \tag{7.29}$$

In the other extreme limit, $I_0/I_{s0} \ll \exp[\gamma I_0 \mathscr{L}_{\mathrm{eff}}(z)]$, we can replace the denominator in Eq. (7.26a) with 1 and obtain $I_s(z) = I(z)$, indicating that the signal

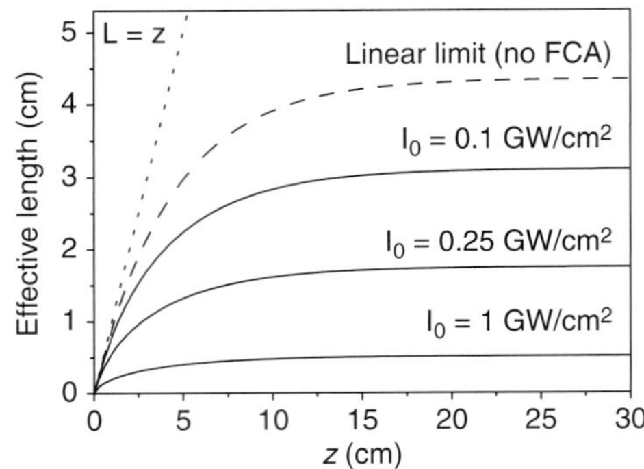

Figure 7.11 $\mathscr{L}_{\mathrm{eff}}(z)$ as a function of propagation distance for different input intensities I_0. The dashed curve shows the linear-loss limit. The dotted line corresponds to the lossless case. (After Ref. [50]; © IEEE 2010)

intensity approaches the total intensity $I(z)$. This is the limit in which the pump is nearly depleted because of an efficient transfer of pump power to the signal.

The influence of FCA on CW Raman amplification manifests in two ways. First, FCA leads to an overall attenuation of the signal, as indicated by the denominator in Eq. (7.25). Second, it leads to a decrease in the generalized effective length compared with L_{eff} and, what is more important, makes it intensity-dependent. The influence of total input intensity on the generalized effective length is illustrated by solid curves in Figure 7.11. Clearly, one can increase $\mathscr{L}_{\mathrm{eff}}(z)$ substantially by decreasing I_0. This is somewhat counterintuitive because one tends to increase pump power to increase the signal gain. The effective length can also be increased by reducing τ_c because κ scales linearly with τ_c. This approach is pursued often for silicon Raman amplifiers because it allows for significant amplification of the signal.

The approximate solution (7.26) contains errors resulting from the simplifications that were used to obtain Eq. (7.23) and the terms that we discarded in Eq. (7.24). We can reduce the errors of the second type by introducing corrective multipliers. Assuming a corrected solution of the form

$$I_{s,\mathrm{corr}}(z) = \xi(z)I_s(z), \tag{7.30a}$$

$$I_{p,\mathrm{corr}}(z) = \xi(z)I_p(z), \tag{7.30b}$$

we find from Eq. (7.23) that the corrective multiplier is

$$\xi(z) = \left(1 + 2\kappa \int_0^z I_p(z')I_s(z')\,dz'\right)^{-1/2}, \tag{7.31}$$

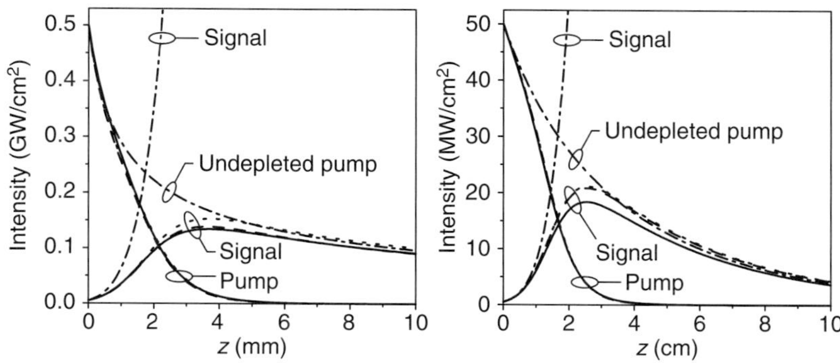

Figure 7.12 Evolution of pump and signal intensities along the amplifier length obtained numerically (solid curves) and predicted analytically (dashed curves). Dotted curves show the uncorrected solution in Eq. (7.26). Dash-dotted curves show the solution when the pump is assumed to remain undepleted. The left and right panels correspond to $I_0 = 0.5$ GW cm^{-2} and 0.05 GW cm^{-2}, respectively. (After Ref. [50]; © IEEE 2010)

where $I_p(z)$ and $I_s(z)$ are given in Eq. (7.26). Once corrected with this multiplier, the analytical solution in Eq. (7.23) becomes quite close to the numerical solution of Eq. (7.21).

The validity range of this analytic solution can be estimated by asking when the TPA terms in Eq. (7.21) become small compared with the linear loss and FCA terms. This requirement leads to the condition $I_0 \ll \alpha/\beta$ or $I_0 \gg \beta/\kappa$. For $\alpha = 1$ dB/cm, $\beta = 0.5$ cm/GW, and λ_s near 1600 nm, we obtain $\alpha/\beta \approx 0.5$ GW cm^{-2} and $\beta/\kappa \approx 0.01$ GW cm^{-2}. Thus, the result in Eq. (7.26) is a good approximate solution of Eq. (7.21) for input pump intensities in the range 0.02 to 0.5 GW cm^{-2}. Figure 7.12 compares the analytical solution with the numerical solution (solid curves) for input powers of $I_0 = 0.5$ GW cm^{-2} (left) and 0.05 GW cm^{-2} (right) using $I_{s0} = 0.01 I_0$. The other parameter values are: $\alpha = 1$ dB cm^{-1}, $\beta = 0.5$ cm GW^{-1}, $\tau_c = 1$ ns, $g_R = 76$ cm GW^{-1}, $\lambda_p = 1550$ nm, and $\lambda_s = 1686$ nm. The corrected and uncorrected solutions are shown by the dashed and dotted curves, respectively. For reference, the solution that corresponds to the undepleted-pump approximation is shown by the dash-dotted curves. It is evident that this approximation becomes invalid within a distance of 2 mm inside a silicon waveguide and should be used with caution.

7.4.3 Amplification of picosecond pulses

We now consider the situation in which a silicon Raman amplifier, pumped with a CW beam, is used to amplify a train of signal pulses whose carrier frequency differs from the pump frequency by the Raman shift of 15.6 THz. Each signal pulse, as it is amplified through SRS, is also affected by several nonlinear processes including SPM, XPM, and TPA. All of these phenomena are included in Eqs. (7.14), which must be solved numerically. Since such an approach requires extensive computational resources and also does not provide much physical insight, one may ask if an approximate solution can be obtained with a different approach. It turns out that the variational technique allows one to derive a set of relatively simple first-order differential equations [52] that describe the evolution of pulse parameters such as amplitude, phase, width, and chirp. A numerical solution of these equations can be obtained much faster than one for the original equations, and it also provides considerable physical insight for designing silicon Raman amplifiers

Variational technique method

The calculus of variations was developed into a full mathematical theory by Euler around 1744 and extended further by Lagrange [53]. Although developed originally for mechanical systems, it can also be used for solving optics problems [54]. Let us assume that the problem under consideration can be described using n generalized coordinates, written in vector form as $\boldsymbol{q} = (q_1, q_2, \cdots, q_n)$. These coordinates change with distance z and time τ as the system evolves. One may thus speak of generalized velocities $\boldsymbol{q}_\tau$ and $\boldsymbol{q}_z$, defined as

$$\boldsymbol{q}_\tau = \left(\frac{\partial q_1}{\partial \tau}, \frac{\partial q_2}{\partial \tau}, \cdots, \frac{\partial q_n}{\partial \tau} \right), \tag{7.32a}$$

$$\boldsymbol{q}_z = \left(\frac{\partial q_1}{\partial z}, \frac{\partial q_2}{\partial z}, \cdots, \frac{\partial q_n}{\partial z} \right). \tag{7.32b}$$

Consider first a conservative system with no dissipation. The dynamic evolution of such a system can be studied by using a generic variational principle known as the principle of least action (or Hamiltonian principle):

$$\delta \left(\iint L(\boldsymbol{q}, \boldsymbol{q}_\tau, \boldsymbol{q}_z, \tau, z) \mathrm{d}\tau \mathrm{d}z \right) = 0, \tag{7.33}$$

where L is called the Lagrangian density of the system, and its integral z and τ over the variables is called the *action*. The operator δ represents the variation of the action taken over both z and τ. The precise meaning of Eq. (7.33) is that the action takes an extreme value (a minimum, maximum, or saddle point) for the space–time trajectory associated with the system's motion. Equation (7.33) leads to the Euler–Lagrange equation,

$$\frac{\partial}{\partial \tau} \frac{\partial L}{\partial \boldsymbol{q}_\tau} + \frac{\partial}{\partial z} \frac{\partial L}{\partial \boldsymbol{q}_z} - \frac{\partial L}{\partial \boldsymbol{q}} = 0, \tag{7.34}$$

where the derivative with respect to a vector $s \equiv (s_1, s_2, \cdots, s_n)$ is defined as

$$\frac{\partial L}{\partial s} = \left(\frac{\partial L}{\partial s_1}, \frac{\partial L}{\partial s_2}, \cdots, \frac{\partial L}{\partial s_n} \right). \tag{7.35}$$

The variational technique has been extended to lossy systems by including the effects of dissipation through a function called the Rayleigh dissipation function, denoted by R [52,55,56]. The generalized Euler–Lagrange equation then takes the form

$$\frac{\partial}{\partial \tau} \frac{\partial L}{\partial q_\tau} + \frac{\partial}{\partial z} \frac{\partial L}{\partial q_z} - \frac{\partial L}{\partial q} + \frac{\partial R}{\partial q_z} + \frac{\partial R}{\partial q_\tau} = 0. \tag{7.36}$$

Inherent in this formalism is the assumption that losses are a function only of the generalized velocities. Even though this assumption is restrictive, it is adequate to handle pulse propagation through silicon amplifiers. The interested reader can find more details in Ref. [56].

We now apply this general approach to the propagation of optical pulses in a nonlinear medium. If $u(z, \tau)$ represents the slowly varying amplitude of a pulse propagating through a silicon amplifier, both u and u^* should be treated as generalized coordinates. The Lagrangian L of the propagation problem depends on the following variables:

$$L \equiv L \left(z, \tau, u(z, \tau), u^*(z, \tau), \frac{\partial u}{\partial z}, \frac{\partial u}{\partial \tau}, \frac{\partial u^*}{\partial z}, \frac{\partial u^*}{\partial \tau} \right). \tag{7.37}$$

We assume that the functional form of the pulse shape at $z = 0$ is known in terms of N specific parameters such that $u(z, \tau) = f(b_1, b_2, \cdots, b_N, \tau)$. Now comes the chief limitation of the variational technique. It consists of assuming that the pulse evolves in such a fashion that its N parameters change with z but the overall shape of the pulse does not change. We can express this requirement by writing $u(z, \tau)$ as [57]

$$u(z, \tau) = f(b_1(z), b_2(z), \cdots, b_N(z), \tau). \tag{7.38}$$

Notice that the N pulse parameters change with z but not with τ.

Since we are only interested in the evolution of pulse parameters with z, the problem can be simplified considerably by carrying out time integration in Eq. (7.33) and reducing the dimensionality of the problem by one. The resulting reduced Lagrangian L_g and the Rayleigh dissipation function are obtained using

$$L_g = \int L \, d\tau, \qquad R_g = \int R \, d\tau. \tag{7.39}$$

By applying the variational principle to the reduced Lagrangian L_g, one can obtain the following reduced Euler–Lagrange equation:

$$\frac{\partial}{\partial z}\frac{\partial L_g}{\partial \boldsymbol{b}_z} - \frac{\partial L_g}{\partial \boldsymbol{b}} + \frac{\partial R_g}{\partial \boldsymbol{b}_z} = 0, \tag{7.40}$$

where $\boldsymbol{b} = (b_1, b_2, \cdots, b_N)$. We use this equation to analyze the amplification of Gaussian pulses in a silicon Raman amplifier.

Variational method for Gaussian pulses

The pulse amplification problem requires the solution of a set of three coupled equations, Eqs. (7.14) and (7.17). However, we can simplify these equations by noting that the pump is much stronger than the signal in practice. Thus, free carriers in the silicon amplifier are generated predominantly by the CW pump. If we discard all terms depending on the signal amplitude A_s in Eq. (7.17), we obtain

$$\frac{\partial N}{\partial t} \approx -\frac{N}{\tau_c} + \rho_p |A_p|^4. \tag{7.41}$$

In the case of a CW pump, N does not vary with time, and we can use the steady-state solution

$$N(z) = \rho_p \tau_c |A_p(z)|^4. \tag{7.42}$$

Even though the free-carrier density is constant over time under CW pumping, it changes with the propagation distance z because the pump power varies with z.

Consider next the pump equation, Eq. (7.14a). For a CW pump, we can neglect the dispersion terms in this equation. The integral in this equation can also be performed analytically. Using $h_R(t)$ from Eq. (7.15) and γ_s from Eq. (7.16), the last term, denoted by Y_R, can be written as

$$Y_R = j\gamma_s |A_p(z)|^2 \int_0^\infty h(t') A_s(z, t - t')\, e^{j\Omega_{spt'}}\, dt'$$

$$\approx \frac{g_R \Gamma_R}{2} |A_p(z)|^2 \int_0^\infty e^{j(\Omega - \Omega_R)t' - \Gamma_R t'} A_s(z, t - t')\, dt', \tag{7.43}$$

where we have assumed $\Omega \approx \Omega_R$ and neglected the nonresonant term in the Raman response function. Taking the Fourier transform of this expression (see Section 1.1.2) and assuming $|\omega + \Omega - \Omega_R| \ll \Gamma_R$, we obtain

$$\widetilde{Y}_R = \mathscr{F}_{t+}\{Y_R\}(\omega) \approx \frac{g_R}{2}\frac{|A_p(z)|^2 A_s(z, \omega)}{1 - j(\omega + \Omega - \Omega_R)/\Gamma_R}. \tag{7.44}$$

This equation shows explicitly the frequency dependence of the Raman susceptibility. It leads to a Lorentzian profile for the Raman gain with the bandwidth Γ_R/π.

To proceed further, we use the condition $\Omega \approx \Omega_R$ (a prerequisite for Raman amplification to take place), expand the result in a Taylor series assuming $\omega \ll \Gamma_R$, retain the first three terms, and take the inverse Fourier transform. The result is

$$Y_R \approx \frac{g_R}{2}|A_p(z)|^2\left[A_s(z,t) + \frac{1}{\Gamma_R}\frac{\partial A_s}{\partial t} + \frac{1}{\Gamma_R^2}\frac{\partial^2 A_s}{\partial t^2}\right]. \tag{7.45}$$

The first-derivative term provides a small correction to the group velocity and can be neglected in practice. With these simplifications, the pump Eq. (7.14a) takes the following simple form:

$$\frac{dA_p}{dz} = -\frac{\alpha_p}{2}A_p - \frac{\kappa_p}{2}(1+j\mu_p)|A_p|^4 A_p + \frac{j}{2}(2k_p n_2 + j\beta_T(\omega_p))|A_p|^2 A_p, \tag{7.46}$$

where $\kappa_p = \tau_c \sigma_p \rho_p$ and $k_p = \omega_p/c$.

The signal equation, Eq. (7.14b), can also be simplified using Y_R from Eq. (7.45). Introducing the retarded time $\tau = t - \beta_{1s}z$ and using $N(z)$ from Eq. (7.42), it can be written as

$$\frac{\partial A_s}{\partial z} + \frac{j\beta_{2s}}{2}\frac{\partial^2 A_s}{\partial \tau^2} = -\frac{\alpha_s}{2}A_s - \frac{\kappa_{sp}}{2}(1+j\mu_s)|A_p|^4 A_s,$$
$$+ jk_s n_2(|A_s|^2 + 2|A_p|^2)A_s + \frac{g_R}{2}\left(1 + T_2^2\frac{\partial^2}{\partial \tau^2}\right)|A_p|^2 A_s, \tag{7.47}$$

where $\kappa_{sp} = \tau_c \sigma_s \rho_p$, $k_s = \omega_s/c$, and $T_2 = 1/\Gamma_R$. To solve this equation with the variational technique, we need the Lagrangian density L and the Rayleigh dissipation function R associated with Eq. (7.47). These are found to be [52]

$$L = \frac{1}{2}\left(A_s\frac{\partial A_s^*}{\partial z} - A_s^*\frac{\partial A_s}{\partial z}\right) + \frac{j\beta_{2s}}{2}\left|\frac{\partial A_s}{\partial \tau}\right|^2 - \frac{j\mu_s\kappa_{sp}}{2}I_p^2|A_s|^2$$
$$+ \frac{j}{2}k_s n_2(|A_s|^2 + 4I_p)|A_s|^2, \tag{7.48}$$

$$R = \frac{1}{2}[\alpha_s + \beta(|A_s|^2 + 2I_p) - g_R I_p + \kappa_{sp}I_p^2]\left(A_s\frac{\partial A_s^*}{\partial z} - A_s^*\frac{\partial A_s}{\partial z}\right)$$
$$- \frac{g_R}{2}T_2^2 I_p\left(\frac{\partial^2 A_s}{\partial \tau^2}\frac{\partial A_s^*}{\partial z} - \frac{\partial^2 A_s^*}{\partial \tau^2}\frac{\partial A_s}{\partial z}\right). \tag{7.49}$$

We assume that the signal pulse launched into the silicon Raman amplifier has a Gaussian shape and maintains this shape during Raman amplification, even though its parameters change and evolve with z. Thus, the signal field has the form

$$A_s(z, \tau) = \sqrt{I_s(z)} \exp\left\{ -[1 - jc_s(z)] \frac{\tau^2}{2T_s^2(z)} + j\varphi_s(z) \right\}, \tag{7.50}$$

where I_s, c_s, T_s, and φ_s represent, respectively, the peak intensity, frequency chirp, width, and phase of the signal pulse. Using the reduced Euler–Lagrange equation (7.40) with $\mathbf{q} = (I_s, c_s, T_s, \varphi_s)$, we obtain the following set of four equations for the four pulse parameters [52]:

$$\frac{dI_s}{dz} = -(\alpha_s + 2\beta I_p)I_s + g_R\left(1 - \frac{T_2^2}{T_s^2}\right)I_p I_s + \beta_{2s}\frac{c_s}{T_s^2}I_s - \frac{5\beta}{4\sqrt{2}}I_s^2 - \kappa_{sp}I_p^2 I_s, \tag{7.51a}$$

$$\frac{dT_s}{dz} = -\beta_{2s}\frac{c_s}{T_s} + \frac{\beta}{4\sqrt{2}}T_s I_s + \frac{g_R}{2}\frac{T_2^2}{T_s}(1 - c_s^2)I_p, \tag{7.51b}$$

$$\frac{dc_s}{dz} = -\frac{\beta_{2s}}{T_s^2}(1 + c_s^2) - \frac{k_s n_2}{\sqrt{2}}I_s + \frac{\beta}{2\sqrt{2}}c_s I_s - g_R\frac{T_2^2}{T_s^2}(1 + c_s^2)c_s I_p, \tag{7.51c}$$

$$\frac{d\varphi_s}{dz} = \frac{\beta_{2s}}{2T_s^2} + \frac{5k_s n_2}{4\sqrt{2}}I_s - \frac{\mu_s \kappa_{sp}}{2}I_p^2 + \left(2k_s n_2 + \frac{g_R}{2}\frac{T_2^2}{T_s^2}c_s\right)I_p. \tag{7.51d}$$

These equations determine the evolution of the signal pulse parameters during Raman amplification. By looking at their structure, one can understand physically which nonlinear phenomenon affects which pulse parameters. For example, the presence of the parameter T_2 in Eqs. (7.51a) to (7.51c) shows explicitly that a finite bandwidth of the Raman gain affects not only pulse amplitude but also pulse width and chirp and thus plays a significant role when input pulse width is comparable to T_2. Since $T_2 = 3$ ps for silicon, such effects become negligible for $T_s > 30$ ps but cannot be neglected for pulse widths close to 10 ps or less.

In practice, the evolution of pulse intensity I_s along the amplifier length is of primary interest as it governs the extent of Raman amplification. Typically, the input pump intensity is much higher than that of the signal. As a result, in the vicinity of the waveguide input, TPA caused by the signal is relatively small and we can safely drop the I_s^2 term in Eq. (7.51a). The amplification of the signal is possible only if $dI_s/dz > 0$, or

$$\kappa_{sp}I_p^2 - \left[g_R\left(1 - \frac{T_2^2}{T_s^2}\right) - 2\beta\right]I_p + \alpha_s - \beta_{2s}\frac{c_s}{T_s^2} < 0. \tag{7.52}$$

If $\alpha_s T_s^2 > \beta_{2s} c_s$, this quadratic equation and the requirement that pump intensity be a real positive quantity lead to the following limitation on the carrier lifetime [52]:

$$\tau_c < \tau_{\text{th}} \equiv \frac{\left[g_R(1 - T_2^2/T_s^2) - 2\beta\right]^2}{4\sigma_s p_p(\alpha_s - \beta_{2s} c_s / T_s^2)}. \tag{7.53}$$

Equation (7.53) shows that the carrier lifetime should be less than a threshold value, τ_{th}, before CW pumping can be used for Raman amplification of a signal pulse. Notice that both the GVD and the pulse chirp affect τ_{th} through the sign of $\beta_{2s} c_s$. In particular, τ_{th} is reduced when $\beta_{2s} c_s$ becomes negative. On the other hand, if $\beta_{2s} c_s > 0$, the threshold value can be made relatively high. Note also that the restriction on τ_c disappears altogether for $\beta_{2s} c_s > \alpha_s T_s^2$. These conclusions are quite important for Raman amplification of picosecond pulses in silicon waveguides and show why analytic tools are essential for understanding nonlinear phenomena in such devices.

As an example, we consider the amplification of 10 ps Gaussian pulses in a 1-cm-long SOI waveguide and solve the three coupled equations (7.51a)–(7.51c) numerically using $\alpha_s = 1 \text{ dB cm}^{-1}$, $\beta = 0.5 \text{ cm GW}^{-1}$, $n_2 = 6 \times 10^{-18} \text{ m}^2 \text{ W}^{-1}$, $g_R = 76 \text{ cm GW}^{-1}$, $\tau_c = 10 \text{ ns}$, $\beta_{2s} = 20 \text{ ps}^2 \text{ m}^{-1}$, $\mu_s = 7.5$, $\lambda_s = 1550 \text{ nm}$, $T_s = 10 \text{ ps}$, $T_2 = 3 \text{ ps}$, $I_p(0) = 250 \text{ MW cm}^{-2}$, and $I_s(0) = 0.25 \text{ MW cm}^{-2}$. Figure 7.13 shows the amplification factor, $G_s = P_s(L)/P_s(0)$, as a function of input pump power for $T_2 = 0$ (solid curves) and $T_2 = 3 \text{ ps}$ (dashed curves) for three values of the carrier lifetime τ_c. The Raman gain bandwidth corresponds to

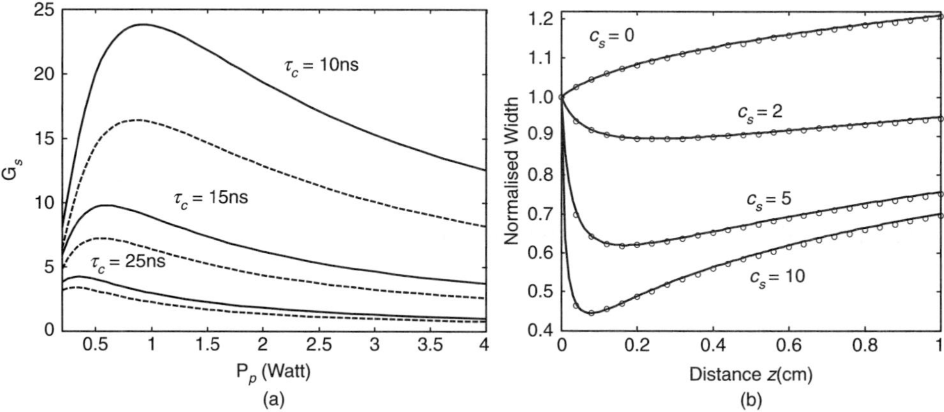

Figure 7.13 (a) Amplification factor as a function of launched pump power for three values of τ_c. Solid curves correspond to $T_2 = 0$ whereas $T_2 = 3 \text{ ps}$ for dashed curves. (b) Evolution of pulse width with z for several values of initial chirping. Open circles represent the corresponding values obtained numerically. (After Ref. [52]; © OSA 2009)

the realistic value of 105 GHz in the latter case. In all cases, an increase in the pump power does not always enhance the output signal power, even decreasing it after a certain value of input pump power. This feature can be attributed to an increase in TPA, and the resulting FCA, with increasing pump powers. The important point to note is that G_s for 10 ps pulses is considerably reduced when $T_2 = 3$ ps because their spectral width is comparable to Raman-gain bandwidth. The signal gain is also affected by the carrier lifetime and increases rapidly as τ_c decreases. Figure 7.13(b) shows how the width of amplified pulses changes along the waveguide. Although unchirped input pulses always broaden because of gain dispersion, chirped pulses go through an initial compression stage. The width reduction occurs because the central part of a linearly chirped pulse experiences more amplification than its pedestals. Numerical results, shown by circles and obtained by solving the full model, support the narrowing of chirped pulses.

Direct-integration method for arbitrarily shaped pulses

A major assumption of the variational method is that the shape of a pulse does not change during its amplification, even though its amplitude, width, and chirp evolve within an SOA. This is a very restrictive assumption because, in reality, pulse shape may deviate considerably from the incident shape in any amplifier. For this reason, we look at another solution method with wider applicability than the variational formalism. However, we must approximate the original propagation equations because, in their complete form, they are not analytically tractable. For this purpose, we need to introduce reasonable assumptions that are applicable in most practical situations.

Consider the generalized NLS equation (7.14a) governing signal propagation in a silicon Raman amplifier. If we assume that the pump remains much stronger than the signal throughout the SOA length, we can approximate $\gamma_s |A_s|^2 + 2\gamma_{sp} |A_p|^2$ by $2\gamma_{sp} |A_p|^2$. With the same approximation, we can use the result given in Eq. (7.42) for the free-carrier density $N(z)$ in the case of a CW pump. Using these simplifications, Eq. (7.14a) takes the form

$$\frac{\partial A_s}{\partial z} - j\mathscr{F}_{\omega+}^{-1}\{\beta(\omega)\}(t)A_s = -\frac{\alpha_s}{2}A_s - \frac{\sigma_s \rho_p \tau_c}{2}(1 + j\mu_s)|A_p(z)|^4 A_s$$

$$+ 2j\gamma_{sp}|A_p|^2 A_s + j\gamma_s A_p \int_{-\infty}^{t} h_R(t - t') A_p^*(z, t') A_s(z, t') e^{-j\Omega(t - t')} dt',$$

$$(7.54)$$

where $\beta(\omega)$ is the propagation constant of the waveguide. We have also assumed that the pump wavelength is chosen such that the peak of the signal spectrum coincides with the Raman-gain peak ($\Omega = \Omega_R$).

We now use the results given in Eqs. (7.43) and (7.44) to obtain

$$j\gamma_s A_p \int_{-\infty}^{t} h_R(t-t')A_p^*(z,t')\widetilde{A}_s(z,t')e^{-j\Omega(t-t')}dt'$$

$$\approx \frac{g_R\Gamma_R}{2}|A_p(z)|^2 \int_{0}^{\infty} e^{-j\Gamma_R t'} A_s(z, t-t')\,dt' \qquad (7.55)$$

$$\approx \mathscr{F}_{\omega+}^{-1}\left\{\frac{g_R}{2}\frac{|A_p(z)|^2 A_s(z,\omega)}{1-j\omega/\Gamma_R}\right\}(t),$$

where we use the notation $\widetilde{A}_s(z,\omega) = \mathscr{F}_{t+}\{A_s\}(\omega)$. Since A_s is the only time-dependent variable in Eq. (7.54), it is easy to take the Fourier transform of this equation to obtain the following linear differential equation for the signal spectrum:

$$\frac{\partial\widetilde{A}_s(z,\omega)}{\partial z} = \left[p_0 + p_1|A_p(z)|^2 + p_2|A_p(z)|^4\right]\widetilde{A}_s(z,\omega), \qquad (7.56)$$

where the parameters p_0, p_1, and p_2 are defined as

$$p_0 = j\beta(\omega) - \frac{\alpha_s}{2}, \qquad (7.57a)$$

$$p_1 = \frac{1}{2}\frac{g_R}{1-j\omega/\Gamma_R} + 2j\gamma_{sp}, \qquad (7.57b)$$

$$p_2 = -\frac{1}{2}\sigma_s\rho_p\tau_c(1+j\mu_s). \qquad (7.57c)$$

Equation (7.56) can be easily integrated to obtain the evolution of the signal spectrum along the amplifier length. The result is given by

$$\widetilde{A}_s(z,\omega) = \widetilde{A}_s(0,\omega)\exp\left(p_0 z + p_1 \int_0^z |A_p(z)|^2 dz + p_2 \int_0^z |A_p(z)|^4 dz\right). \qquad (7.58)$$

To evaluate this expression, we need to know the spatial evolution of the pump intensity $|A_p(z)|^2$ along the amplifier length. The pump intensity can be calculated by carrying out the following two steps. First, we multiply Eq. (7.46) by $A_p^*(z)$ to get

$$A_p^*(z)\frac{dA_p(z)}{dz} = -\frac{\alpha_p}{2}|A_p(z)|^2 - \frac{\kappa_p}{2}(1+j\mu_p)|A_p(z)|^6$$

$$+ \frac{j}{2}[2k_p n_2 + j\beta_T(\omega_p)]|A_p(z)|^4. \qquad (7.59)$$

Second, we take the complex conjugate of Eq. (7.46) and multiply the resulting equation by $A_p(z)$ to get

$$A_p(z)\frac{dA_p^*(z)}{dz} = -\frac{\alpha_p}{2}|A_p(z)|^2 - \frac{\kappa_p}{2}(1 - j\mu_p)|A_p(z)|^6 + \\ -\frac{j}{2}[2k_p n_2 - j\beta_T(\omega_p)]|A_p(z)|^4. \tag{7.60}$$

We now add the preceding two equations to obtain the following equation for the spatial evolution of the CW pump intensity along the silicon amplifier:

$$\frac{dI_p(z)}{dz} = -\alpha_p I_p(z) - \kappa_p I_p^3(z) - \beta_T(\omega_p)I_p^2(z), \tag{7.61}$$

where $I_p(z) = |A_p(z)|^2$. This equation has an analytical solution depending on the sign of the parameter $\Delta = 4\alpha_p\kappa_p - \beta_T^2(\omega_p)$. When $\Delta > 0$, the solution is given by

$$\left[\frac{I_p^2(z)}{I_p^2(0)}\right]\left[\frac{\alpha_p I_p(0) + \kappa_p I_p^3(0) + \beta_T(\omega_p)I_p^2(0)}{\alpha_p I_p(z) + \kappa_p I_p^3(z) + \beta_T(\omega_p)I_p^2(z)}\right]e^{2\alpha_p z} \\ = \exp\left[2b\left(\tan^{-1}[F(z)] - \tan^{-1}[F(0)]\right)\right], \tag{7.62}$$

where the function $F(z)$ and the constant b are defined as

$$F(z) = [2\kappa_p I_p(z) + \beta_T(\omega_p)]/\sqrt{\Delta}, \qquad b = \beta_T(\omega_p)/\sqrt{\Delta}. \tag{7.63}$$

When $\Delta < 0$, the right-hand side (RHS) of this solution changes to

$$\text{RHS} = \left(\frac{F(z) - 1}{F(z) + 1}\right)^b \left(\frac{F(0) - 1}{F(0) + 1}\right)^b. \tag{7.64}$$

The solution of the preceding transcendental equation provides the pump intensity $I_p(z)$ at any distance z. The use of this pump intensity in Eq. (7.58) gives us the signal spectrum $\tilde{A}_s(z, \omega)$. Taking its Fourier transform, we can study the evolution of the signal envelope $A_s(z, t)$ for an arbitrary pulse shape. Clearly, this direct-integration method is superior to the variational technique because it imposes no restriction on the shape of the signal pulse as it is amplified in the silicon Raman amplifier. The only assumption it makes is that the amplifier is pumped by an intense CW pump beam. The situation changes considerably when a pump pulse is used to amplify a signal pulse of comparable with. This configuration has attracted attention recently [58, 59]. More specifically, it was found that the SPM-induced spectral broadening of pump pulses affects the signal gain considerably because it leads to saturation of the Raman gain. A theoretical understanding of this saturation behavior requires a numerical solution of the coupled NLS equations given as Eqs. (7.14a) and (7.14b).

References

[1] M. Born and E. Wolf, *Principles of Optics*, 7th ed. Cambridge University Press, 1999.

[2] L. Barron, *Molecular Light Scattering and Optical Activity*, 2nd ed. Cambridge University Press, 2004.

[3] D. M. Hercules, ed., *Fluorescence and Phosphorescence Analysis: Principles and Applications*. Interscience, 1966.

[4] A. A. Kokhanovsky, *Optics of Light Scattering Media: Problems and Solutions*, 2nd ed. Springer, 2001.

[5] C. F. Bohren and D. R. Huffman, *Absorption and Scattering of Light by Small Particles*. Wiley InterScience, 1983.

[6] C. V. Raman and K. S. Krishnan, "A new type of secondary radiation," *Nature*, vol. 121, pp. 501–502, 1928.

[7] D. A. Long, *The Raman Effect: A Unified Treatment of the Theory of Raman Scattering by Molecules*. Wiley, 2002.

[8] E. J. Woodbury and W. K. Ng, "Ruby laser operation in the near IR," *Proc. Inst. Radio Eng.*, vol. 50, pp. 2367–2368, 1962.

[9] R. L. Sutherland, D. G. McLean, and S. Kirkpatrick, *Handbook of Nonlinear Optics*. CRC Press, 2003.

[10] B. Bobbs and C. Warne, "Raman-resonant four-wave mixing and energy transfer," *J. Opt. Soc. Am. B*, vol. 7, pp. 234–238, 1990.

[11] G. P. Agrawal, *Nonlinear Fiber Optics*, 4th ed. Academic Press, 2007.

[12] R. H. Stolen and E. P. Ippen, "Raman gain in glass optical waveguides," *Appl. Phys. Lett.,*, vol. 22, pp. 276–281, 1973.

[13] R. H. Stolen, J. P. Gordon, W. J. Tomlinson, and H. A. Haus, "Raman response function of silica-core fibers," *J. Opt. Soc. Am. B*, vol. 6, pp. 1159–1166, 1989.

[14] J. A. Buck, *Fundamentals of Optical Fibers*. Wiley InterScience, 1995.

[15] D. Hollenbeck and C. Cantrell, "Multiple-vibrational-mode model for fiber-optic Raman gain spectrum and response function," *J. Opt. Soc. Am. B*, vol. 19, pp. 2886–2892, 2002.

[16] A. C. G. Mitchell and M. W. Zemansky, *Resonance Radiation and Excited Atoms*. Cambridge University Press, 1971.

[17] Q. Lin and G. P. Agrawal, "Raman response function for silica fibers," *Opt. Lett.*, vol. 31, pp. 3086–3088, 2006.

[18] F. L. Galeener, A. J. Leadbetter, and M. W. Stringfellow, "Comparison of the neutron, Raman, and infrared vibrational spectra of vitreous SiO_2, GeO_2, and BeF_2," *Phys. Rev. B*, vol. 27, pp. 1052–1078, 1983.

[19] S. Namiki and Y. Emori, "Broadband Raman amplifiers: Design and practice" In A. Mecozzi, M. Shimizu, and J. Zyskind, eds., *Optical Amplifiers and Their Applications*, vol. 44 of OSA. Trends in Optics and Photonics, paper OMB2. Optical Society of America, 2000.

[20] J. Bromage, P. J. Winzer, and R. J. Essiambre, *Raman Amplifiers for Telecommunications* pp. 491–567. Springer, 2003.

[21] C. Finot, "Influence of the pumping configuration on the generation of optical similaritons in optical fibers," *Opt. Commun.*, vol. 249, pp. 553–561, 2005.

[22] H. J. Leamy and J. H. Wernick, "Semiconductor silicon: The extraordinary made ordinary," *MRS Bull.*, vol. 22, pp. 47–55, 1997.

[23] L. Pavesi and D. L. Lockwood, eds., *Silicon Photonics*. Springer, 2004.

[24] R. A. Soref, "Silicon-based optoelectronics," *Proc. IEEE*, vol. 81, pp. 1687–1706, 1993.

[25] R. Claps, D. Dimitropoulos, Y. Han, and B. Jalali, "Observation of Raman emission in silicon waveguides at 1.54 μm," *Opt. Express*, vol. 10, pp. 1305–1313, 2002.

[26] R. Claps, D. Dimitropoulos, V. Raghunathan, Y. Han, and B. Jalali, "Observation of stimulated Raman amplification in silicon waveguides," *Opt. Express*, vol. 11, pp. 1731–1739, 2003.

[27] T. K. Liang and H. K. Tsang, "Role of free carriers from two-photon absorption in Raman amplification in silicon-on-insulator waveguides," *Appl. Phys. Lett.*, vol. 84, pp. 2745–2747, 2004.

[28] R. Claps, V. Raghunathan, D. Dimitropoulos, and B. Jalali, "Influence of nonlinear absorption on Raman amplification in silicon-on-insulator waveguides," *Opt. Express*, vol. 12, pp. 2774–2780, 2004.

[29] O. Boyraz and B. Jalali, "Demonstration of 11 dB fiber-to-fiber gain in a silicon Raman amplifier," *IEICE Electron. Express*, vol. 1, pp. 429–434, 2004.

[30] A. Liu, H. Rong, M. Paniccia, O. Cohen, and D. Hak, "Net optical gain in a low loss silicon-on-insulator waveguide by stimulated Raman scattering," *Opt. Express*, vol. 12, pp. 4261–4268, 2004.

[31] R. Espinola, J. Dadap, R. Osgood, S. J. McNab, and Y. A. Vlasov, "Raman amplification in ultrasmall silicon-on-insulator wire waveguides," *Opt. Express*, vol. 12, pp. 3713–3718, 2004.

[32] R. Jones, H. Rong, A. Liu, *et al.*, "Net continuous wave optical gain in a low loss silicon-on-insulator waveguide by stimulated Raman scattering," *Opt. Express*, vol. 13, pp. 519–525, 2005.

[33] S. Fathpour, K. K. Tsia, and B. Jalali, "Energy harvesting in silicon Raman amplifiers," *Appl. Phys. Lett.*, vol. 89, p. 061109 (3 pages), 2006.

[34] M. Krause, H. Renner, and E. Brinkmeyer, "Analysis of Raman lasing characteristics in silicon-on-insulator waveguides," *Opt. Express*, vol. 12, pp. 5703–5710, 2004.

[35] H. Rong, A. Liu, R. Jones, *et al.*, "An all-silicon Raman laser," *Nature*, vol. 433, pp. 292–294, 2005.

[36] O. Boyraz and B. Jalali, "Demonstration of directly modulated silicon Raman laser," *Opt. Express*, vol. 13, pp. 796–800, 2005.

[37] M. Krause, H. Renner, and E. Brinkmeyer, "Efficient Raman lasing in tapered silicon waveguides," *Spectroscopy*, vol. 21, pp. 26–32, 2006.

[38] B. Jalali, V. Raghunathan, D. Dimitropoulos, and O. Boyraz, "Raman-based silicon photonics," *IEEE J. Sel. Top. Quantum Electron.*, vol. 12, pp. 412–421, 2006.

[39] M. A. Foster, A. C. Turner, M. Lipson, and A. L. Gaeta, "Nonlinear optics in photonic nanowires," *Opt. Express*, vol. 16, pp. 1300–1320, 2008.

[40] O. Boyraz, P. Koonath, V. Raghunathan, and B. Jalali, "All optical switching and continuum generation in silicon waveguides," *Opt. Express*, vol. 12, pp. 4094–4102, 2004.

[41] B. Jalali, S. Yegnanarayanan, T. Yoon, T. Yoshimoto, I. Rendina, and F. Coppinger, "Advances in silicon-on-insulator optoelectronics," *IEEE J. Sel. Top. Quantum Electron.*, vol. 4, pp. 938–947, 1998.

[42] R. Claps, V. Raghunathan, D. Dimitropoulos, and B. Jalali, "Anti-Stokes Raman conversion in silicon waveguides," *Opt. Express*, vol. 11, pp. 2862–2872, 2003.

[43] D. Dimitropoulos, V. Raghunathan, R. Claps, and B. Jalali, "Phase-matching and nonlinear optical processes in silicon waveguides," *Opt. Express*, vol. 12, pp. 149–160, 2004.

[44] V. Raghunathan, R. Claps, D. Dimitropoulos, and B. Jalali, "Parametric Raman wavelength conversion in scaled silicon waveguides," *J. Lightw. Technol.*, vol. 23, pp. 2094–2012, 2005.

[45] R. L. Espinola, J. I. Dadap, R. M. Osgood, S. J. McNab, and Y. A. Vlasov, "C-band wavelength conversion in silicon photonic wire waveguides," *Opt. Express*, vol. 13, pp. 4341–4349, 2005.

[46] H. Fukuda, K. Yamada, T. Shoji, *et al.*, "Four-wave mixing in silicon wire waveguides," *Opt. Express*, vol. 13, pp. 4629–4637, 2005.

[47] H. Rong, Y. H. Kuo, A. Liu, M. Paniccia, and O. Cohen, "High efficiency wavelength conversion of 10 Gb/s data in silicon waveguides," *Opt. Express*, vol. 14, pp. 1182–1188, 2006.

[48] Q. Lin, O. J. Painter, and G. P. Agrawal, "Nonlinear optical phenomena in silicon waveguides: Modelling and applications," *Opt. Express.*, vol. 15, pp. 16 604–16 644, 2007.

[49] R. W. Boyd, *Nonlinear Optics*, 3rd ed. Academic Press, 2008.

[50] I. D. Rukhlenko, M. Premaratne, and G. P. Agrawal, "Nonlinear silicon photonics: Analytical tools," *IEEE J. Sel. Topics Quantum Electron.*, vol. 16, pp. 200–215, 2010.

[51] A. D. Polyanin, *Handbook of Exact Solutions for Ordinary Differential Equations.* Chapman and Hall/CRC, 2003.

[52] S. Roy, S. K. Bhadra, and G. P. Agrawal, "Raman amplification of optical pulses in silicon waveguides: Effects of finite gain bandwidth, pulse width and chirp," *J. Opt. Soc. Am. B*, vol. 26, pp. 17–25, 2009.

[53] W. W. R. Ball, *A Short Account of the History of Mathematics.* Martino, 2004.

[54] B. A. Malomed, "Variational methods in nonlinear fiber optics and related fields." In *Progress in Optics*, E. Wolf, ed., vol. 43, pp. 71–191. Elsevier, 2002.

[55] H. Goldstein, C. P. Poole, and J. L. Safko, *Classical Mechanics.* Addison-Wesley, 2001.

[56] F. Riewe, "Nonconservative Lagrangian and Hamiltonian mechanics," *Phys. Rev. E*, vol. 53, pp. 1890–1899, 1996.

[57] S. C. Cerda, S. B. Cavalcanti, and J. M. Hickmann, "A variational approach of nonlinear dissipative pulse propagation," *Eur. Phys. J. D.*, vol. 1, pp. 313–316, 1998.

[58] D. R. Solli, P. Koonath, and B. Jalali, "Broadband Raman amplification in silicon," *Appl. Phys. Lett.*, vol. 93, p. 191105, 2008.

[59] F. Kroeger, A. Ryasnyanskiy, A. Baron, N. Dubreuil, P. Delaye, R. Frey, G. Roosen, and D. P. Peyrade, "Saturation of Raman amplification by self-phase modulation in silicon nanowaveguides," *Appl. Phys. Lett.*, vol. 96, p. 241102, 2010.

8

Optical parametric amplifiers

Optical parametric amplifiers constitute a category of amplifiers whose operation is based on a physical process that is quite distinct from other amplifiers discussed in the preceding chapters. The major difference is that an atomic population is not transferred to any excited state of the system, in the sense that the initial and final quantum-mechanical states of the atoms or molecules of the medium remain unchanged [1]. In contrast, molecules of the gain medium end up in an excited vibrational state, in the case of Raman amplifiers. Similarly, atoms are transferred from an excited electronic state to a lower energy state in the case of fiber amplifiers and semiconductor optical amplifiers.

Optical parametric amplification is a nonlinear process in which energy is transferred from a pump wave to the signal being amplified. This process was first used to make optical amplifiers during the 1960s and has proved quite useful for practical applications. In Section 8.1 we present the basic physics behind parametric amplification, and then focus on the phase-matching requirement in Section 8.2. In Section 8.3 fiber-based parametric amplifiers are covered in detail, because of the technological importance of such amplifiers. Section 8.4 is devoted to parametric amplification in birefringent crystals, while Section 8.5 focuses on how phase matching is accomplished in birefringent fibers.

8.1 Physics behind parametric amplification

From a classical point of view, a parametric process has its origin in the nonlinear nature of an optical material, manifested though the second- and third-order susceptibilities, $\chi^{(2)}$ and $\chi^{(3)}$, respectively [2, 3]. In the case of $\chi^{(2)}$, parametric amplification is synonymous with difference-frequency generation. To illustrate this case, consider the scheme shown in Figure 8.1, where an input field at frequency ω_p, called the pump, and another input field at frequency ω_s, called the signal, interact in a medium exhibiting significant $\chi^{(2)}$ nonlinearity. We assume

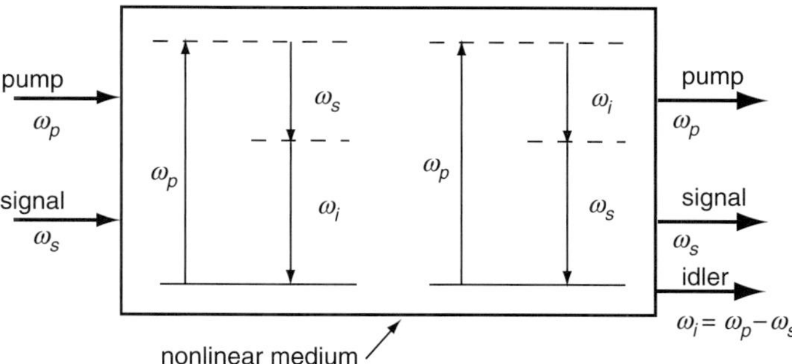

Figure 8.1 Optical parametric amplification viewed as difference-frequency generation. Energy-level diagrams show how absorption of a pump photon creates two photons at the signal and idler frequencies. Dashed lines show virtual states involved in this parametric process. Both the signal and the idler are amplified at the output at the expense of the pump.

$\omega_p > \omega_s$ so that pump photons have more energy than signal photons. Under certain conditions, discussed later, the interaction of the pump and the signal in the nonlinear medium leads to the creation of another optical field, called the *idler*, whose frequency $\omega_i = \omega_p - \omega_s$ corresponds exactly to the difference between the two incident frequencies. This process can be understood as the annihilation of a pump photon, followed by the simultaneous creation of two photons at the signal and idler frequencies [2–4].

Quantum-mechanically, the creation of an idler wave during the parametric interaction of two incident waves can be described as a process in which an atom or molecule of the nonlinear medium absorbs one pump photon to make a transition from its ground state to a virtual high-energy state (the upper dashed line in the energy-level diagrams of Figure 8.1). Since a virtual state survives for a very short time dictated by Heisenberg's uncertainty principle, the atom decays to the ground state almost immediately. Although this decay can take place spontaneously by a process known as spontaneous parametric down-conversion, it is stimulated by signal photons such that a two-stage emission process creates one photon at the signal frequency and another at the idler frequency. Once idler photons are created, they can also stimulate the decay of the virtual state through the second process shown in Figure 8.1. The net result is a continuous transfer of pump power to the signal and idler waves all along the length of a nonlinear medium, leading to amplification of the signal at the expense of the pump [5]. The optical gain experienced by the signal by this process is called the *parametric gain*. It is important to stress that virtual states exist only momentarily in this picture. The time interval Δt for which a virtual state can survive is roughly $\hbar/\Delta E$, where ΔE is the energy difference between the virtual state and the ground state [2]. Noting that $\Delta E = \hbar \omega_p$ for the upper virtual state in Figure 8.1, $\Delta t \approx 1/\omega_p$ is below 1 fs in the visible regime.

In general, regardless of the number of waves interacting with one another, any gain process in which the initial and final states of the material are identical is called a parametric-gain process [4, 5]. Since no energy is transferred to the gain medium during a parametric-gain process, the susceptibility associated with such a process is a real-valued function [2, 3]. In contrast to this, nonparametric gain mechanisms discussed earlier in the chapters on fiber amplifiers and semiconductor optical amplifiers employ transitions between distinct energy levels and require a population inversion in which more atoms are in the excited state than in the ground state [6,7]. Since energy stored in the gain medium can decay through other channels (such as spontaneous emission), a nonparametric gain medium always exhibits a susceptibility with an imaginary part [2].

As we shall see later, mathematical description of the parametric response of a medium is carried out by invoking changes in the medium's refractive index and then relating those changes to phases of the individual waves. Therefore, the refractive index of the parametric medium plays a central role in this chapter, and its frequency dependence leads to the so-called phase-matching condition (discussed in the next section). It is interesting to note that the real part of the refractive index associated with any dielectric medium represents parametric or nonresonant processes, while its imaginary part represents nonparametric or resonant processes [3]. As we saw in Chapter 2, these processes are tightly coupled to one another through the Kramers–Kronig relations in both linear and nonlinear regimes [8, 9].

Before considering the phase-matching condition, we clarify some terminology issues. The parametric-gain process in Figure 8.1 is often referred to as three-wave mixing because it involves three waves: pump, signal, and idler [2, 3, 48]. If the frequencies of the signal and idler are distinct, the process is identified as nondegenerate three-wave mixing. If both the frequencies and polarization states of the signal and idler coincide, the process is called degenerate. In the degenerate case, the signal frequency has to be exactly one-half the pump frequency. If only the pump is launched at the input end, signal and idler waves can still be generated, through a process called spontaneous parametric down-conversion, provided the phase-matching condition is satisfied for some specific frequencies.

If a nonlinear medium does not exhibit second-order susceptibility, parametric amplification requires the use of third-order susceptibility, $\chi^{(3)}$. In this case, two pumps and a signal are launched into the medium, and the idler frequency is given by $\omega_i = \omega_{p1} + \omega_{p2} - \omega_s$. In the energy-level diagram of Figure 8.1, the transition to the virtual excited state now involves the simultaneous absorption of two pump photons. Since four distinct waves are involved, this parametric-gain process is referred to as four-wave mixing (FWM). In practice, two photons can come from a single pump beam at the frequency ω_p. In this case, the parametric process is known as degenerate FWM or single-pump FWM, and the idler frequency is given by $\omega_i = 2\omega_p - \omega_s$. If the pump and signal frequencies also coincide ($\omega_s = \omega_p$), such a parametric process becomes a fully degenerate FWM process in which all four waves have the same frequency.

8.2 Phase-matching condition

Parametric processes are sensitive to the phases of the optical waves involved; that is, energy transfer from the pump to the signal and idler waves requires a definite phase relationship among the waves involved in the parametric process, and this is expressed mathematically as a phase-matching condition [2,4,10]. This condition depends inherently on the pump and signal frequencies and makes it clear that the parametric gain required for signal amplification cannot be achieved for arbitrary pump and signal wavelengths. As phase matching is required for efficient parametric interaction in a nonlinear medium, the gain bandwidth is largely determined by the phase-matching bandwidth, which itself depends on the characteristics of the medium involved. In short, efficient parametric interaction can only take place in a certain frequency window and is highly dependent on the dispersive properties of the medium, or on the nature of wave-guiding in the case of an optical waveguide.

From a physical standpoint, phase matching has its origin in the requirement for momentum conservation. In the quantum-mechanical picture, we can think of a three-wave mixing process as the breakup of a pump photon of energy $\hbar\omega_p$ into two photons of energies $\hbar\omega_s$ and $\hbar\omega_i$. Energy is clearly conserved if $\omega_p = \omega_s + \omega_i$. However, momentum conservation requires that the corresponding wave vectors (also called propagation vectors) satisfy the phase-matching condition [2,4,10]

$$k_p = k_s + k_i. \tag{8.1}$$

If several pump waves are involved in the parametric process, the phase-matching condition can be easily extend to account for them. For example, if two pumps with propagation vectors k_{p1} and k_{p2} are involved in a nondegenerate FWM process, the phase-matching condition (8.1) becomes

$$k_{p1} + k_{p2} = k_s + k_i. \tag{8.2}$$

In most cases, the phase-matching condition cannot be satisfied precisely, for a variety of reasons. Therefore, it is common to define a phase-mismatch parameter Δk to characterize deviations from the ideal matching condition, as follows:

$$\Delta k = \begin{cases} k_s + k_i - k_p, & \text{for three-wave mixing,} \\ k_s + k_i - k_{p1} - k_{p2}, & \text{for four-wave mixing.} \end{cases} \tag{8.3}$$

When $\Delta k \neq 0$, the contributions to signal and idler get out of phase and interfere destructively after a distance L_{coh}, commonly known as the coherence length of the interaction process and defined as

$$L_{\text{coh}} = \frac{2\pi}{|\Delta k|}. \tag{8.4}$$

In a nondispersive, isotropic medium, the phase-matching condition is satisfied automatically in the case of collinear interaction because the propagation constant, $k = (n\omega/c) \propto \omega$, is directly proportional to ω when n does not depend on frequency. However, it is not possible to find materials for which refractive index stays constant over a wide frequency range. In the case of a normally dispersive medium where the refractive index $n(\omega)$ increases with frequency, $n(\omega_p)$ exceeds in value both $n(\omega_s)$ and $n(\omega_i)$. Clearly, it is impossible to satisfy the phase-matching condition in Eq. (8.1) for a normally dispersive material. However, if the material has anomalous dispersive properties (i.e., refractive index decreases in magnitude with increasing frequency), then it is possible to satisfy the phase-matching condition with the right choice of frequencies.

A technique that is used commonly for phase matching makes use of the birefringence properties of the nonlinear medium. In this case, the refractive index of the medium depends on the state of polarization of the incident beams and it becomes possible to obtain phase matching even in a normally dispersive medium. We discuss this phase-matching technique in some detail in Section 8.4, where we discuss specific examples of parametric amplification in a birefringent medium.

8.3 Four-wave mixing in optical fibers

Optical fibers do not exhibit nonlinear effects governed by the second-order susceptibility $\chi^{(2)}$. For this reason, FWM governed by the third-order susceptibility $\chi^{(3)}$ is employed for parametric amplification [10]. Even though the magnitude of $\chi^{(3)}$ for silica fibers is relatively small compared with many other materials, relatively long propagation lengths of pump and signal waves in the form of well-confined fiber modes make possible the use of optical fibers as parametric amplifiers. In this section we first discuss fiber-optic parametric amplifiers pumped with a single laser and then consider briefly the case of dual-wavelength pumps.

8.3.1 Coupled amplitude equations

We assume that the CW pump and signal waves at frequencies ω_p and ω_s are launched into a single-mode fiber. Their parametric interaction through FWM creates the idler wave at the frequency $\omega_i = 2\omega_p - \omega_s$. The total electric field at any point inside the fiber is given by

$$E(\mathbf{r}, t) = E_p(\mathbf{r}, t) + E_s(\mathbf{r}, t) + E_r(\mathbf{r}, t). \tag{8.5}$$

Since each wave propagates as a fundamental mode inside the fiber, it is useful to write their electric fields in the form

$$E_p(\mathbf{r}, t) = \mathrm{Re}[F(x, y)A_p(z)\exp(j\beta_p z - j\omega_p t)], \tag{8.6a}$$

$$E_s(\mathbf{r}, t) = \mathrm{Re}[F(x, y)A_s(z)\exp(j\beta_s z - j\omega_s t)], \tag{8.6b}$$

$$E_i(\mathbf{r}, t) = \mathrm{Re}[F(x, y)A_i(z)\exp(j\beta_i z - j\omega_i t)], \tag{8.6c}$$

where $A_p(z)$, $A_s(z)$, and $A_i(z)$ are the slowly varying amplitudes, and β_p, β_s, and β_i are the propagation constants, for the pump, signal, and idler waves, respectively. For simplicity, the mode distribution $F(x, y)$ is assumed to be the same for all three waves. This is justified in the case of FWM because the three frequencies are not too far apart from one another.

Substituting these expressions into Maxwell's equations, including the third-order nonlinear effects through $\chi^{(3)}$, and making the slowly-varying-envelope approximation, it is possible to derive the following set of three coupled equations [10, 11]:

$$\frac{dA_p}{dz} + \frac{\alpha_p}{2}A_p = j\gamma\left(|A_p|^2 + 2|A_s|^2 + 2|A_i|^2\right)A_p + 2j\gamma A_s A_i A_p^* \exp(j\Delta\beta z), \tag{8.7a}$$

$$\frac{dA_s}{dz} + \frac{\alpha_s}{2}A_s = j\gamma\left(|A_s|^2 + 2|A_p|^2 + 2|A_i|^2\right)A_s + j\gamma A_i^* A_p^2 \exp(-j\Delta\beta z), \tag{8.7b}$$

$$\frac{dA_i}{dz} + \frac{\alpha_i}{2}A_i = j\gamma\left(|A_i|^2 + 2|A_s|^2 + 2|A_p|^2\right)A_i + j\gamma A_s^* A_p^2 \exp(-j\Delta\beta z), \tag{8.7c}$$

where $\Delta\beta = \beta_s + \beta_i - 2\beta_p$ is the propagation-constant mismatch, γ is the nonlinear coefficient of the fiber (related linearly to $\chi^{(3)}$), and α_p, α_s, and α_i are loss coefficients of the pump, signal, and idler waves, respectively.

In Eqs. (8.7), multiple nonlinear terms on the right side are responsible for different nonlinear effects. The first term represents the self-phase modulation (SPM) phenomenon. The next two terms represent the cross-phase modulation (XPM) between two waves. The last term in each equation represents the FWM term that is responsible for energy transfer among the three waves through their parametric interaction. In these equations, the fields A_q ($q = p, s, i$) are normalized such that $|A_q|^2$ provides the optical power in watts for the corresponding wave.

Following Refs. [12] and [13], it is possible to recast these equations in terms of mode powers and their phases. Such a formulation not only provides insight into the nature of energy transfer among the three waves but also leads to the correct phase-matching condition under high-power conditions. Using $A_q = \sqrt{P_q}\exp(\theta_q)$, with $q = p, s, i$, we can write Eqs. (8.7) in the following equivalent form:

$$\frac{dP_p}{dz} = -\alpha_p A_p - 4\gamma \left(P_p^2 P_s P_i\right)^{1/2} \sin\theta, \tag{8.8a}$$

$$\frac{dP_s}{dz} = -\alpha_s A_s + 2\gamma \left(P_p^2 P_s P_i\right)^{1/2} \sin\theta, \tag{8.8b}$$

$$\frac{dP_i}{dz} = -\alpha_i A_i + 2\gamma \left(P_p^2 P_s P_i\right)^{1/2} \sin\theta, \tag{8.8c}$$

$$\frac{d\theta}{dz} = \Delta\beta + \gamma(2P_p - P_s - P_i)$$

$$+ \gamma \left[P_p(P_i/P_s)^{1/2} + P_p(P_s/P_i)^{1/2} - 2(P_s P_i)^{1/2}\right]\cos(\theta), \tag{8.8d}$$

where $\theta(z) = \Delta\beta z + \theta_s(z) + \theta_i(z) - 2\theta_p(z)$.

8.3.2 Impact of nonlinear effects on phase matching

The preceding set of four equations requires four initial conditions at $z = 0$. Three of them, $P_p(0) = P_{p0}$, $P_s(0) = P_{s0}$, and $P_i(0) = 0$, follow from the fact that only the pump and the signal are launched into the fiber. However, the fourth one, for $\theta(0)$, is not immediately obvious. It turns out that θ has an initial value of $\pi/2$ at $z = 0$ [13]. To see why this must be so, we note from the idler equation in the set (8.7) that after a short distance δz, the idler field is given by $A_i = j\gamma A_p^2 A_s^* \exp(-j\Delta\beta z)\delta z$. Since A_i should vanish in the limit $\delta z \to 0$, the phase of the idler is set such that $\theta_i = 2\theta_p - \theta_s - \Delta\beta z + \pi/2$, or $\theta = \pi/2$, at $z = 0$.

If θ stays at its initial value of $\pi/2$, power transfer from the pump to the signal (and to the idler) will continue all along the fiber. As seen from Eq. (8.8d), θ can stay at its initial value only if the right side of this equation vanishes. This is possible only if the condition

$$\Delta\beta + \gamma\left(2P_p - P_s - P_i\right) = 0 \tag{8.9}$$

remains satisfied along the fiber length. Physically, in the presence of SPM and XPM, the phase-matching condition $\Delta\beta = 0$ is modified and replaced with Eq. (8.9). In practice, it is hard to satisfy this condition exactly for all distances, and it is useful to introduce a net phase mismatch [13, 14],

$$\kappa = \Delta\beta + \gamma\left(2P_p - P_s - P_i\right). \tag{8.10}$$

A scalar form is employed here because κ points in the z direction.

Further simplification can be introduced by assuming that the parametric amplifier operates in a regime where the undepleted-pump approximation is valid [10]. This can happen in practice when the input pump power is so much larger than the signal power that it remains nearly unchanged even when the signal and idler are amplified inside the fiber. Under such conditions, we can neglect P_s and P_i in comparison to $2P_p$ in Eq. (8.10) and obtain

$$\kappa \approx \Delta\beta + 2\gamma P_p. \tag{8.11}$$

It is possible to obtain an approximate analytic expression for $\Delta\beta = 2\beta_p - \beta_s - \beta_i$ in terms of the fiber's dispersion parameters. For this purpose, we note that $\beta_s \equiv \beta(\omega_s)$ and $\beta_i \equiv \beta(\omega_i)$ can be expanded around the pump frequency ω_p as

$$\beta(\omega_s) = \beta(\omega_p) + \beta_1(\omega_p)(\omega_s - \omega_p) + \tfrac{1}{2}\beta_2(\omega_p)(\omega_s - \omega_p)^2 + \cdots, \tag{8.12a}$$

$$\beta(\omega_i) = \beta(\omega_p) + \beta_1(\omega_p)(\omega_i - \omega_p) + \tfrac{1}{2}\beta_2(\omega_p)(\omega_i - \omega_p)^2 + \cdots, \tag{8.12b}$$

where $\beta_m = d^m\beta/d\omega_m$. Introducing $\Omega = \omega_s - \omega_p$ as a detuning of the signal from the pump and noting that $\omega_i = 2\omega_p - \omega_s$, we obtain the simple expression [10]

$$\Delta\beta \approx \beta_2(\omega_p)\Omega^2 + \frac{\beta_4(\omega_p)}{12}\Omega_s^4 + \cdots, \tag{8.13}$$

where only the even-order dispersion terms appear. Often, the first term in this series dominates. Assuming that to be the case, the net phase mismatch in Eq. (8.11) is given by

$$\kappa \approx \beta_2(\omega_p)\Omega_s^2 + 2\gamma P_p. \tag{8.14}$$

8.3.3 Parametric gain

We now obtain an expression for the parametric gain in the approximation that the pump remains undepleted all along the fiber length. For this purpose, it is better to solve the set (8.7) of three amplitude equations. We further assume that the fiber length is short enough (< 1 km) that losses can be neglected for all three waves. With these simplifications, the pump equation can be solved easily to obtain the pump power $A_p = \sqrt{P_p}\exp(2jP_p z)$, where P_p is the input pump power, which we have assumed to remain constant along the fiber. Since the pump modulates its own phase through SPM, we include its phase changes. The signal and idler equations then take the following simplified form:

$$\frac{dA_s}{dz} = 2j\gamma P_p A_s + j\gamma P_p A_i^* \exp(-j\Delta\beta z + j2\gamma P_p z), \tag{8.15a}$$

$$\frac{dA_i^*}{dz} = -2j\gamma P_p A_i^* - j\gamma P_p A_s \exp(j\Delta\beta z - j2\gamma P_p z), \tag{8.15b}$$

where we have kept only the dominant pump-induced XPM terms. Since A_s couples to A_i^*, the idler equation is written for A_i^* rather than A_i. This feature is referred to as phase conjugation and has many important applications. For example, it can be used for dispersion compensation in fiber-optic communications systems [15].

The preceding set of two first-order differential equations can be solved analytically. To solve them, we first make the substitution [14]

$$A_s = B_s \exp(2j\gamma P_p z), \qquad A_i^* = B_i^* \exp(-2j\gamma P_p z), \qquad (8.16)$$

and obtain the following simplified coupled equations:

$$\frac{dB_s}{dz} = j\gamma P_p B_i^* \exp(-j\kappa z), \qquad (8.17a)$$

$$\frac{dB_i^*}{dz} = -j\gamma P_p B_s \exp(j\kappa z). \qquad (8.17b)$$

These equations clearly show the role played by the net phase mismatch κ. We now assume a solution of the form

$$B_s = C \exp(gz - j\kappa z/2), \qquad B_i^* = D \exp(gz + j\kappa z/2), \qquad (8.18)$$

where C and D are constants, and obtain a set of two linear coupled equations of the form

$$\begin{bmatrix} g - j\kappa/2 & -j\gamma P_p \\ j\gamma P_p & g + j\kappa/2 \end{bmatrix} \begin{bmatrix} C \\ D \end{bmatrix} = 0. \qquad (8.19)$$

These equations have a nontrivial solution only if the determinant of the coefficient matrix vanishes. This requirement leads to

$$g = \pm\sqrt{(\gamma P_p)^2 - (\kappa/2)^2}. \qquad (8.20)$$

This relation shows that g is real when $2\gamma P_p$ exceeds κ. Since both the signal and the idler are amplified under such conditions, g is referred to as the parametric-gain coefficient.

Since g has two possible values, a general solution of Eq. (8.17) is of the form

$$B_s(z) = (C_1 e^{gz} + C_2 e^{-gz})e^{-j\kappa z/2}, \qquad (8.21a)$$

$$B_i^*(z) = (D_1 e^{gz} + D_2 e^{-gz})e^{j\kappa z/2}, \qquad (8.21b)$$

where the four constants are determined from the boundary conditions. In a typical experimental situation, no idler wave exists at the fiber input end. Thus, $B_s(0) \neq 0$ but $dB_s/dz = 0$. The constants C_1 and C_2 can be found using these boundary conditions. Similarly the use of $B_i^*(0) = 0$ and $dB_i^*/dz = -j\gamma P_p B_s(0)$ provides

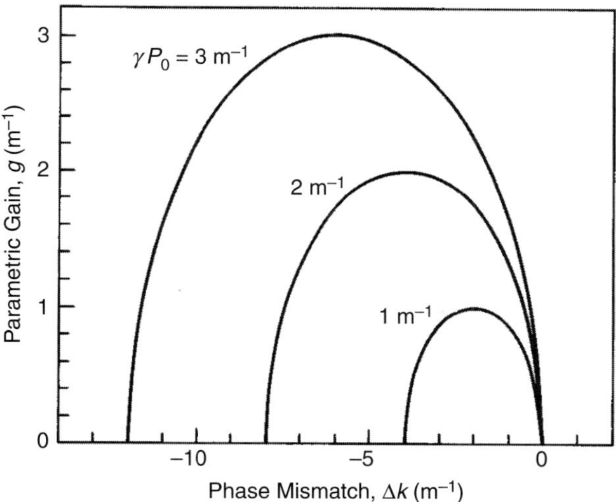

Figure 8.2 Variation of parametric-gain coefficient g with $\Delta\beta$ for several pump powers P_p. (After Ref. [10]; © Elsevier 2007)

us with the constants D_1 and D_2. Using them, we obtain the final solution in the form [14]

$$B_s(z) = B_s(0)[\cosh(gz) - (j\kappa/2g)\sinh(gz)]e^{-j\kappa z/2}, \tag{8.22a}$$

$$B_i^*(z) = -j(\gamma P_p/g)B_s(0)\sinh(gz)e^{j\kappa z/2}. \tag{8.22b}$$

The preceding solution shows that both the signal and the idler grow along the fiber length if g is real. This is possible only if $2\gamma P_p$ exceeds $\kappa < 2\gamma P_p$, or $\Delta\beta < 4\gamma P_p$ if we use Eq. (8.11). Figure 8.2 shows variations of g with $\Delta\beta$ for several values of γP_p. The maximum value of g occurs when the net phase mismatch vanishes ($\kappa = 0$). Shifts in the gain peak with peak power result from the power dependence of κ.

8.3.4 Signal gain in parametric amplifiers

The general solution in Eq. (8.22) can be used to calculate the signal and idler powers using $P_q = |B_q|^2$ ($q = s, i$). The result can be written in the form

$$P_s(z) = P_s(0)\left[1 + (\gamma P_p/g)^2 \sinh^2(gz)\right], \tag{8.23a}$$

$$P_i(z) = P_s(0)(\gamma P_p/g)^2 \sinh^2(gz). \tag{8.23b}$$

Equation (8.23b) shows that the idler wave is generated almost immediately after the input signal is launched into the fiber. Its power increases as z^2 initially, but both the signal and idler grow exponentially after a distance such that $gz > 1$. As the idler is amplified together with the signal all along the fiber, it can build up to nearly the same level as the signal at the amplifier output. In practical terms, the same parametric process can be used to amplify a weak signal *and* to generate simultaneously a new wave at the idler frequency. The idler wave mimics all features of the input signal except that its phase is reversed (or conjugated). Among other things, such phase conjugation can be used for dispersion compensation and wavelength conversion in a WDM system [10].

Further insight can be gained by considering the factor by which the signal is amplified in an amplifier of length L. This signal gain, $G_s = P_s(L)/P_s(0)$, can be obtained from Eq. (8.23a) in the form

$$G_s = 1 + (\gamma P_p/g)^2 \sinh^2(gL) = 1 + \sinh^2(gL)/(gL_{NL})^2, \qquad (8.24)$$

where $L_{NL} = (\gamma P_p)^{-1}$ is the nonlinear length. Consider now the impact of net phase mismatch κ on the signal gain. If phase matching is perfect ($\kappa = 0$), and noting that $g = \gamma P_p$, the amplifier gain can be written as

$$G_s = \cos^2(\gamma P_p L) \approx \tfrac{1}{4}\exp(2\gamma P_p L), \qquad (8.25)$$

where the last approximation applies in the limit $\gamma P_p L \gg 1$. This result shows that the signal power increases exponentially with the pump power P_p in the case of perfect phase matching. As κ increases, g decreases, and the signal gain is reduced. When $\kappa = \gamma P_p$, g vanishes and the amplification factor is given by $G_s = 1 + (\gamma P_p L)^2$, i.e., the signal grows only quadratically with an increase in the pump power or the amplifier length.

One may ask what happens when $\kappa > \gamma P_p$, and g becomes imaginary. Using $g = jh$ in Eq. (8.24) we find that the gain is now given by $G_s = 1 + (\gamma P_p/h)^2 \sin^2(hL)$. In the limit $\kappa \gg \gamma P_p$, the gain expression reduces to

$$G_s \approx 1 + (\gamma P_p L)^2 \frac{\sin^2(\kappa z/2)}{(\kappa L/2)^2}. \qquad (8.26)$$

The signal gain is relatively small and increases with pump power as P_p^2 if phase mismatch is relatively large.

As an example, consider a parametric amplifier designed using a 2.5-km-long fiber with its zero-dispersion wavelength λ_0 at 1550 nm, i.e., $\beta_2(\omega_0) = 0$, where $\omega_0 = 2\pi c/\lambda_0$. The amplifier is pumped with 0.5 W of power, resulting in a nonlinear length of 1 km when $\gamma = 2\ \text{W}^{-1}\ \text{km}^{-1}$. Figure 8.3.4(a) shows the gain spectra for several pump wavelengths around 1550 nm. The dispersion parameter at the pump wavelength was obtained using the Taylor series expansion

$$\beta_2(\omega_p) = \beta_3(\omega_0)(\omega_p - \omega_0) + \tfrac{1}{2}\beta_4(\omega_0)(\omega_p - \omega_0)^2, \qquad (8.27)$$

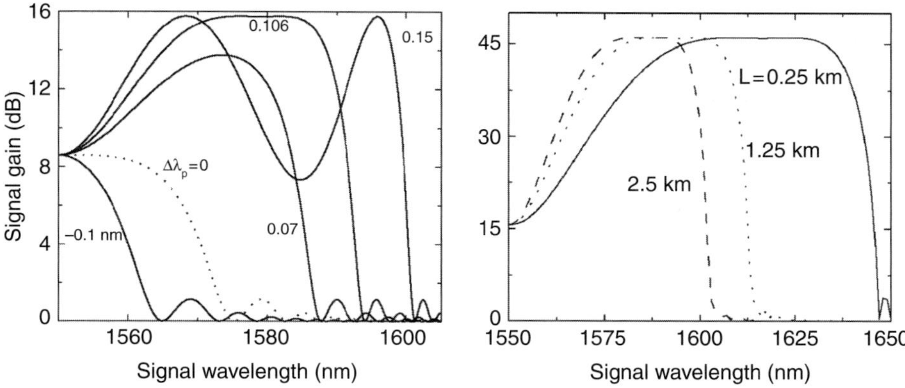

Figure 8.3 (a) Gain spectra of a single-pump fiber-optic parametric amplifier for several values of pump detuning $\Delta\lambda_p$ from the zero-dispersion wavelength. (b) Gain spectra for three different fiber lengths, assuming $\gamma P_0 L = 6$ in each case. The pump wavelength is optimized in each case, and only half of the gain spectrum is shown because of its symmetric nature.

with $\beta_3(\omega_0) = 0.1 \text{ ps}^3 \text{ km}^{-1}$ and $\beta_4(\omega_0 = 10^{-4} \text{ ps}^4 \text{ km}^{-1}$. The dotted curve shows the case $\omega_p = \omega_0$ for which pump wavelength coincides exactly with the zero-dispersion wavelength. The peak gain in this case is about 8 dB. When the pump is detuned by -0.1 nm such that it experiences normal GVD, gain bandwidth is substantially reduced. In contrast, both the peak gain and the bandwidth are enhanced when the pump is detuned toward the anomalous-dispersion side. In this region, the gain spectrum is sensitive to the exact value of pump wavelength. The best situation from a practical standpoint occurs for a pump detuning of 0.106 nm because in that case the gain is nearly constant over a wide spectral region. In all cases, the signal gain drops sharply and exhibits low-amplitude oscillations when the signal detuning from the pump exceeds a critical value. This feature is related to the large-detuning case in which the signal gain is given by Eq. (8.26).

8.3.5 Amplifier bandwidth

An important property of any optical amplifier is the bandwidth over which the amplifier can provide a relatively uniform gain. This amplifier bandwidth, $\Delta\Omega_A$, is usually defined as the range of ω_s over which G_s exceeds 50% of its peak value, and depends on many factors such as fiber length L and pump power P_p. However, a convenient definition of $\Delta\Omega_A$ corresponds to a net phase mismatch of $\kappa = 2\pi/L$. This definition provides a bandwidth slightly larger than the full width at half maximum [14].

As we saw earlier, the signal gain peaks at a signal frequency for which the phase-matching condition is perfectly satisfied ($\kappa = 0$). In the case of a single-mode fiber, κ is given by Eq. (8.11). This equation shows that phase matching cannot occur if the pump frequency lies in the normal-dispersion region of the fiber (unless fourth-order dispersive effects are considered). However, β_2 is negative when the pump frequency lies in the anomalous-dispersion region of the fiber, and $\kappa = 0$ can be achieved for

$$\Omega_s = |\omega_s - \omega_p| = \sqrt{\frac{2\gamma P_p}{|\beta_2(\omega_p)|}}. \tag{8.28}$$

The amplifier bandwidth can now be obtained by finding the frequency shift $\Delta\Omega_A$ for which $\kappa = 2\pi/L$. Again using κ from Eq. (8.11), we obtain the condition

$$\beta_2(\omega_p)(\Omega_s + \Delta\Omega_A)^2 + 2\gamma P_p = 2\pi/L. \tag{8.29}$$

Since typically $\Delta\Omega_A \ll \Omega_s$, we can neglect the $(\Delta\Omega_A)^2$ term in this expression and obtain the simple expression [14]

$$\Delta\Omega_A = \frac{\pi}{|\beta_2(\omega_p)|\Omega_s L} = \frac{\pi}{L}[2\gamma P_p|\beta_2(\omega_p)|]^{-1/2}. \tag{8.30}$$

A more accurate expression of the amplifier bandwidth can be obtained from Eq. (8.24), providing the signal gain as a function of g, which depends on κ as indicated in Eq. (8.20). Its use leads to the following bandwidth expression [14]:

$$\Delta\Omega_A = \frac{1}{|\beta_2(\omega_p)|\Omega_s}\left[\left(\frac{\pi}{L}\right)^2 + (\gamma P_p)^2\right]^{1/2}. \tag{8.31}$$

At relatively low pump powers, this reduces to the bandwidth given in Eq. (8.30). However, at high pump powers, it leads to a bandwidth that scales linearly with the pump power as

$$\Delta\Omega_A \approx \frac{\gamma P_p}{|\beta_2(\omega_p)|\Omega_s}. \tag{8.32}$$

As a rough estimate, the bandwidth is only 160 GHz for a fiber-optic parametric amplifier designed using a dispersion-shifted fiber such that $\gamma = 2$ W^{-1} km^{-1} and $\beta_2(\omega_p) = -1$ ps^2 km^{-1} at the pump wavelength, and the amplifier is pumped with 1 W of CW power. Equation (8.30) shows that at a given pump power, the bandwidth can only be increased considerably by reducing $|\beta_2(\omega_p)|$ and using short fiber lengths.

From a practical standpoint, one wants to maximize both the peak gain and the gain bandwidth at a given pump power P_p. Since the peak gain in Eq. (8.24) scales exponentially with $\gamma P_p L$, it can be increased by increasing the fiber length L. However, from Eq. (8.30), gain bandwidth scales inversely with L. The only solution is to use a fiber as short as possible. Of course, shortening of the fiber length must be accompanied by a corresponding increase in the value of γP_p to maintain the same amount of gain. This behavior is illustrated in Figure 8.3(b), where the gain bandwidth increases considerably when large values of γP_0 are combined with shorter fiber lengths such that $\gamma P_0 L$ remains fixed. The solid curve obtained for a 250-m-long FOPA exhibits a 30-nm-wide region on each side of the pump wavelength over which the gain is nearly flat.

The nonlinear parameter γ can be increased by reducing the effective area of the fiber mode. This is the approach adopted in modern fiber parametric amplifiers, designed using highly nonlinear fibers with $\gamma > 10 \ \mathrm{W}^{-1} \mathrm{km}^{-1}$ and choosing the pump wavelength close to the zero-dispersion wavelength of the fiber so that $|\beta_2(\omega_p)|$ is reduced. However, as $\beta_2(\omega_p)$ is reduced to below $0.1 \ \mathrm{ps}^2 \mathrm{km}^{-1}$, one must consider the impact of higher-order dispersive effects. It turns out that the bandwidth of a fiber-optic parametric amplifier can be increased to beyond 5 THz by suitably optimizing the pump wavelength. A 200 nm gain bandwidth was achieved as early as 2001 by using a 20-m-long fiber with $\gamma = 18 \ \mathrm{W}^{-1} \mathrm{km}^{-1}$ [16]. The required pump power ($\sim$10 W) was large enough that the signal was also amplified by SRS when its wavelength exceeded the pump wavelength.

8.3.6 Dual-pump parametric amplifiers

A shortcoming of single-pump parametric amplifiers is their polarization sensitivity, stemming from the fact that parametric interaction of the four waves participating in the FWM process depends on their relative polarization states, even in an isotropic medium such as a silica fiber. This problem can be solved using two orthogonally polarized pumps. The use of two pumps can also enhance the bandwidth of a parametric amplifier.

It is relatively easy to extend the FWM theory given earlier in this section. The total electric field in Eq. (8.5) should include two pumps by replacing E_p with $E_{p1} + E_{p2}$. As a result, the set (8.7) of three amplitude equations now consists of the four following equations [10]:

$$\frac{dA_{p1}}{dz} + \frac{\alpha_{p1}}{2} A_{p1}$$

$$= j\gamma \left(|A_{p1}|^2 + 2|A_{p2}|^2 + 2|A_s|^2 + 2|A_i|^2 \right) A_1 + 2j\gamma A_s A_i A_{p2}^* e^{j\Delta\beta z},$$

$$\tag{8.33a}$$

$$\frac{dA_{p2}}{dz} + \frac{\alpha_{p2}}{2} A_{p2}$$

$$= j\gamma \left(|A_{p2}|^2 + 2|A_{p1}|^2 + 2|A_s|^2 + 2|A_i|^2 \right) A_2 + 2j\gamma A_s A_i A_{p1}^* e^{j\Delta\beta z},$$

$$\tag{8.33b}$$

$$\frac{dA_s}{dz} + \frac{\alpha_s}{2} A_s$$
$$= j\gamma \left(|A_s|^2 + 2|A_{p1}|^2 + 2|A_{p2}|^2 + 2|A_i|^2 \right) A_s + 2j\gamma A_i^* A_{p1} A_{p2} e^{-j\Delta\beta z},$$
$$(8.33c)$$

$$\frac{dA_i}{dz} + \frac{\alpha_i}{2} A_i$$
$$= j\gamma \left(|A_i|^2 + 2|A_s|^2 + 2|A_{p1}|^2 + 2|A_{p2}|^2 \right) A_i + 2j\gamma A_s^* A_{p1} A_{p2} e^{-j\Delta\beta z},$$
$$(8.33d)$$

where A_{p1} and A_{p2} are the amplitudes and α_{p1} and α_{p2} are the loss coefficients of the two pumps at frequencies ω_{p1} and ω_{p2}, respectively. All other parameters have the same meaning as before, except that the phase mismatch is now defined as

$$\Delta\beta = \beta(\omega_s) + \beta(\omega_i) - \beta(\omega_{p1}) - \beta(\omega_{p2}). \tag{8.34}$$

As before, we can solve these equations analytically when the two pumps are so strong that their deletion can be ignored. Although fiber losses can be included [17], we neglect them so that the results can be compared to the single-pump case. The two pump equations can be solved easily to obtain

$$A_{p1}(z) = \sqrt{P_1} \exp[j\gamma(P_1 + 2P_2)z], \tag{8.35a}$$

$$A_{p2}(z) = \sqrt{P_2} \exp[j\gamma(P_2 + 2P_1)z], \tag{8.35b}$$

where P_1 and P_2 are the incident pump powers. Substituting these expressions into signal and idler wave equations, we obtain

$$\frac{dA_s}{dz} = 2j\gamma(P_1 + P_2)A_s + 2j\gamma\sqrt{P_1 P_2}\exp(j\vartheta)A_i^*, \tag{8.36a}$$

$$\frac{dA_i^*}{dz} = -2j\gamma(P_1 + P_2)A_i^* - 2j\gamma\sqrt{P_1 P_2}\exp(-j\vartheta)A_s, \tag{8.36b}$$

where $\vartheta = \Delta\beta z + 3\gamma(P_1 + P_2)z$.

We can remove the pump-induced XPM terms with the substitution

$$A_s = B_s \exp[2j\gamma(P_1 + P_2)z], \qquad A_i^* = B_i^* \exp[-2j\gamma(P_1 + P_2)z], \tag{8.37}$$

and obtain the following set of two equations:

$$\frac{dB_s}{dz} = j\gamma\sqrt{P_1 P_2}\, B_i^* \exp(-j\kappa z), \tag{8.38a}$$

$$\frac{dB_i^*}{dz} = -j\gamma\sqrt{P_1 P_2}\, B_s \exp(j\kappa z). \tag{8.38b}$$

where κ is the net phase mismatch. These equations are formally identical to Eqs. (8.17) for the single-pump case and can be solved using the same method. In particular, the solution given in Eq. (8.22) remains valid but the parametric gain g is now given by

$$g = \sqrt{4\gamma^2 P_1 P_2 - (\kappa/2)^2}, \qquad \kappa = \Delta\beta + \gamma(P_1 + P_2). \tag{8.39}$$

The signal gain G_s for a parametric amplifier of length L is now found to be

$$G_s = 1 + (2\gamma/g)^2 P_1 P_2 \sinh^2(gL). \tag{8.40}$$

The most commonly used configuration for a dual-pump parametric amplifier employs a relatively large wavelength difference between the two pumps. At the same time, the center frequency ω_c is set close to the zero-dispersion frequency ω_0 of the fiber so that the linear phase mismatch is constant over a broad range of frequencies. To achieve a fairly wide phase-matching range, the two pump wavelengths should be located on opposite sides of the zero-dispersion wavelength in a symmetrical fashion [18]. With this arrangement, κ can be reduced to nearly zero over a wide wavelength range, resulting in a gain spectrum that is nearly flat over this entire range.

The preceding discussion assumes that only the nondegenerate FWM process, $\omega_{p1} + \omega_{p2} \rightarrow \omega_s + \omega_i$, contributes to the parametric gain. In reality, the situation is much more complicated for dual-pump parametric amplifiers because the degenerate FWM process associated with each pump occurs simultaneously and creates two other idlers at frequencies $2\omega_{p1} - \omega_s$ and $2\omega_{p2} - \omega_s$. These three idlers can create several other idler fields through various FWM processes [10]. A complete description of a dual-pump amplifier becomes quite complicated if one includes all underlying FWM processes. Fortunately, the phase-matching conditions associated with these processes are quite different. When the two pumps are located symmetrically far from the zero-dispersion wavelength of the fiber, the dominant contribution to the central flat part of the gain spectrum results from the nondegenerate FWM process $\omega_{p1} + \omega_{p2} \rightarrow \omega_s + \omega_i$.

Figure 8.4 compares the signal gain calculated numerically as a function of signal wavelength using all idlers (solid curve) with that obtained using only the nondegenerate FWM process. Other FWM processes only affect the edges of gain spectrum and reduce the gain bandwidth by 10% to 20%. The device parameters used in this calculation were $L = 0.5$ km, $\gamma = 10$ W^{-1} km^{-1}, $P_1 = P_2 = 0.5$ W, $\beta_3(\omega_0) = 0.1$ ps^3 km^{-1}, $\beta_4(\omega_0) = 10^{-4}$ ps^4 km^{-1}, $\lambda_{p1} = 1502.6$ nm, $\lambda_{p2} = 1600.6$ nm, and $\lambda_0 = 1550$ nm. The central flat part of the gain spectrum with a bandwidth of 80 nm (or 10 THz) results from the dual-pump nature of this parametric amplifier.

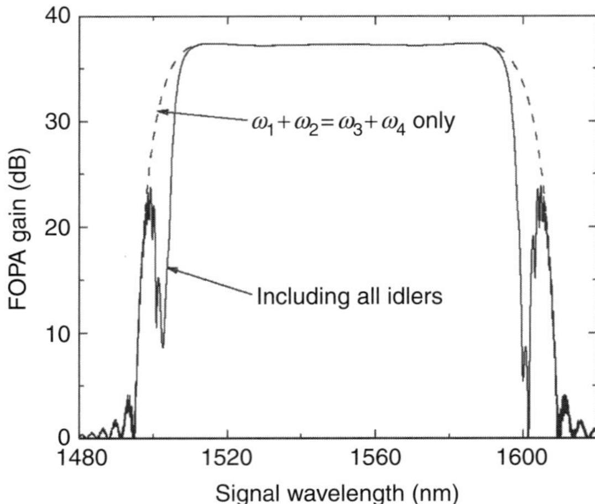

Figure 8.4 Gain spectrum of a dual-pump FOPA when all idlers are included (solid curve). The dashed spectrum is obtained when only a single idler generated through the dominant nondegenerate FWM process is included.

A bandwidth of close to 40 nm was achieved for a dual-pump parametric amplifier in a 2003 experiment [19], and this value was increased beyond 50 nm in later experiments. It is hard to much larger amplifier bandwidths in practice because of a number of detrimental factors. A major limitation stems from the fact that a realistic fiber is far from having a perfect cylindrical core. In practice, the core shape and size may vary along the fiber length in a random fashion as a consequence of the manufacturing process. Such imperfections produce random variations in the zero-dispersion wavelength of the fiber along its length. Since the phase-matching condition depends on this wavelength, and parametric gain is extremely sensitive to dispersion parameters of the fiber, even small changes in the zero-dispersion wavelength (<0.1 nm) produce large changes in the gain spectrum.

8.4 Three-wave mixing in birefringent crystals

The most widely used method of achieving parametric amplification is to employ the birefringence of an anisotropic nonlinear medium for phase matching a three-wave mixing process [2, 10]. In a birefringent medium, depending on the polarization of the propagating light, its refractive index takes different values. There is a direct relationship between the internal structure of a crystal and its birefringent behavior [20]. Birefringence can occur only if the material is anisotropic (directionally dependent), and it does not occur in a centro-symmetric medium such as a silicon crystal with cubic symmetry [21,22]). In this section, we consider parametric amplification in a birefringent crystal (e.g., a potassium dihydrogen phosphate or KDP crystal). To simplify the following discussion, we consider only uniaxial crystals, but the analysis can be easily extended to biaxial or other anisotropic media.

8.4.1 Uniaxial nonlinear crystals

Unlike an isotropic medium, the Poynting vector S does not remain parallel to the propagation vector k of an optical wave in an anisotropic medium. In electrically anisotropic crystals, the electric field E is not parallel to the electrical flux density vector D. However, the angle between D and E equals the angle between k and S [20]. In such a material, it is possible to launch an optical wave in a certain direction such that k is parallel to S. Such a wave is called an ordinary wave (o-wave). All other waves for which this property does not hold are known as extraordinary waves (e-waves).

In general, the dielectric constant of any optical crystal takes the form of a 3×3 matrix whose elements obey the symmetry property $\varepsilon_{ij} = \varepsilon_{ji}$ to satisfy the reciprocal nature of optical propagation. This matrix can be diagonalized along specific coordinate axes known as the principal axes of the crystal. If the diagonal elements are all equal to one another, then the crystal is isotropic. If the two diagonal elements are equal to one another but different from the third one, the crystal is known as uniaxial. If all the diagonal elements are distinct, the crystal is termed biaxial. This classification has a one-to-one correspondence with the index ellipsoid of the wave normals associated with a crystal [20]. The isotropic crystals have ellipsoids in the form of a sphere; the uniaxial crystals have ellipsoids with degeneracy along two axes; and biaxial crystals corresponds to fully nondegenerate ellipsoids.

In a uniaxial crystal, the direction associated with the single distinct diagonal element of the permittivity matrix is called the optical axis. To describe a uniaxial crystal, it is sufficient to define two refractive indices: n_e, the refractive index along the optical axis, and n_o, the refractive index perpendicular to the optical axis (which is the same along any direction in a plane perpendicular to the optical axis). The dielectric constant of such a medium can be written in the form

$$\varepsilon = \varepsilon_0 \begin{bmatrix} n_o^2 & 0 & 0 \\ 0 & n_o^2 & 0 \\ 0 & 0 & n_e^2 \end{bmatrix}. \tag{8.41}$$

The uniaxial crystal is classified as positive uniaxial if $n_o > n_e$; otherwise it is known as negative uniaxial. The magnitude of birefringence of a uniaxial crystal is governed by the difference between the two refractive indices associated with ordinary and extraordinary waves [2].

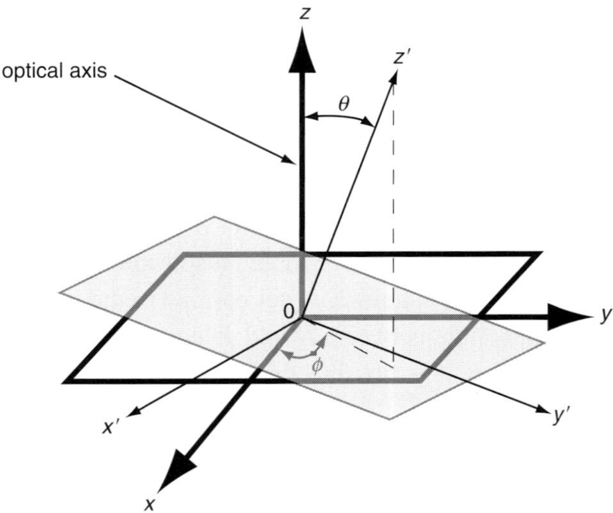

Figure 8.5 Rotation of the optical axis of a uniaxial crystal by an angle θ, transforming the right-handed coordinate system $\{x, y, z\}$ into $\{x', y', z'\}$.

8.4.2 Ordinary and extraordinary waves

An interesting feature of a birefringent crystal is that, when an optical wave is launched into the crystal along a direction making an angle θ with its optical axis, the refractive index $n(\omega, \theta)$ seen by this wave depends on both the frequency of the wave ω and the angle θ. This property can be used to satisfy the phase-matching condition associated with the three-wave mixing process. To calculate $n(\omega, \theta)$ for any angle θ, we need to consider the rotation of the principle coordinate system $\{x, y, z\}$, into a rotated frame $\{x', y', z'\}$ as shown schematically in Figure 8.5.

Note that the diagonal permittivity matrix in Eq. (8.41) is defined relative to a coordinate system $\{x, y, z\}$ in which the optical axis coincides with the z axis (see Figure 8.5). Let us denote the unit vectors along the axes of this coordinate system by $\{\hat{x}, \hat{y}, \hat{z}\}$. When we rotate this coordinate system by an angle θ from the optical axis, the unit vectors in the rotated coordinate system are related to the original ones by [23]

$$\hat{x}' = \hat{x} \sin(\phi) - \hat{y} \cos(\phi), \tag{8.42a}$$

$$\hat{y}' = \hat{x} \cos(\theta) \cos(\phi) + \hat{y} \cos(\theta) \sin(\phi) - \hat{z} \sin(\theta), \tag{8.42b}$$

$$\hat{z}' = \hat{x} \sin(\theta) \cos(\phi) + \hat{y} \sin(\theta) \sin(\phi) + \hat{z} \cos(\theta). \tag{8.42c}$$

It is easy to see from Figure 8.5 that the axis $\hat{x}'$ lies at the intersection of the xy and $x'y'$ planes. The rotation matrix R that takes any vector V with components (V_x, V_y, V_z) in the original coordinate system into its components $(V_{x'}, V_{y'}, V_{z'})$ in the rotated frame can be written as

$$\begin{bmatrix} V_{x'} \\ V_{y'} \\ V_{z'} \end{bmatrix} = R \begin{bmatrix} V_x \\ V_y \\ V_z \end{bmatrix} \equiv \begin{bmatrix} \sin(\phi) & -\cos(\phi) & 0 \\ \cos(\theta)\cos(\phi) & \cos(\theta)\sin(\phi) & -\sin(\theta) \\ \sin(\theta)\cos(\phi) & \sin(\theta)\sin(\phi) & \cos(\theta) \end{bmatrix} \begin{bmatrix} V_x \\ V_y \\ V_z \end{bmatrix}. \tag{8.43}$$

Using this rotation matrix, we can transform the permittivity matrix in Eq. (8.41) into a frame rotated by θ relative to the optical axis. The transformed permittivity matrix ε' has the form

$$\begin{aligned} \varepsilon' &= R\varepsilon_0 \begin{bmatrix} n_o^2 & 0 & 0 \\ 0 & n_o^2 & 0 \\ 0 & 0 & n_e^2 \end{bmatrix} R^{-1} \\ &= \varepsilon_0 \begin{bmatrix} n_o^2 & 0 & 0 \\ 0 & n_o^2\cos^2(\theta) + n_e^2\sin^2(\theta) & (n_o^2 - n_e^2)\sin(\theta)\cos(\theta) \\ 0 & (n_o^2 - n_e^2)\sin(\theta)\cos(\theta) & n_o^2\sin^2(\theta) + n_e^2\cos^2(\theta) \end{bmatrix}. \end{aligned} \tag{8.44}$$

For a wave propagating along the $+\hat{z}'$ direction, its propagation constant k' can be written as $k' = k\hat{z}'$. Noting that D and B are orthogonal to this axis, we use the relations $k' \times E = \omega B$ and $D = \varepsilon' \cdot E$ from Maxwell's equations to obtain

$$\begin{bmatrix} \frac{1}{n_o^2} & 0 \\ 0 & \frac{\cos^2(\theta)}{n_o^2} + \frac{\sin^2(\theta)}{n_e^2} \end{bmatrix} \begin{bmatrix} D_{x'} \\ D_{y'} \end{bmatrix} = \varepsilon_0 \begin{bmatrix} 0 & \frac{\omega}{k} \\ -\frac{\omega}{k} & 0 \end{bmatrix} \begin{bmatrix} B_{x'} \\ B_{y'} \end{bmatrix}. \tag{8.45}$$

Similarly, using the relations $k' \times H = -\omega D$ and $B = \mu_0 H$ from Maxwell's equations, we obtain in the rotated frame

$$\begin{bmatrix} B_{x'} \\ B_{y'} \end{bmatrix} = \mu_0 \begin{bmatrix} 0 & -\frac{\omega}{k} \\ \frac{\omega}{k} & 0 \end{bmatrix} \begin{bmatrix} D_{x'} \\ D_{y'} \end{bmatrix}. \tag{8.46}$$

Combining the two preceding equations, we obtain the matrix relation

$$\begin{bmatrix} \frac{1}{n_o^2} - \varepsilon_0\mu_0\frac{\omega^2}{k^2} & 0 \\ 0 & \frac{\cos^2(\theta)}{n_o^2} + \frac{\sin^2(\theta)}{n_e^2} - \varepsilon_0\mu_0\frac{\omega^2}{k^2} \end{bmatrix} \begin{bmatrix} D_{x'} \\ D_{y'} \end{bmatrix} = 0, \tag{8.47}$$

leading to the following two equations for the two components of D:

$$\left[\frac{1}{n_o^2} - \varepsilon_0\mu_0\frac{\omega^2}{k^2} \right] D_{x'} = 0, \tag{8.48a}$$

$$\left[\frac{\cos^2(\theta)}{n_o^2} + \frac{\sin^2(\theta)}{n_e^2} - \varepsilon_0\mu_0\frac{\omega^2}{k^2} \right] D_{y'} = 0. \tag{8.48b}$$

Table 8.1. *Phase-matching schemes for uniaxial crystals*

Scheme	Crystal type	Pump	Signal	Idler
Type I	positive uniaxial ($n_e > n_o$)	$\perp$	$\parallel$	$\parallel$
Type I	negative uniaxial ($n_e < n_o$)	$\parallel$	$\perp$	$\perp$
Type II	positive uniaxial ($n_e > n_o$)	$\perp$	$\perp$	$\parallel$
Type II	negative uniaxial ($n_e < n_o$)	$\parallel$	$\parallel$	$\perp$

It is clear from these equations that, when $\theta \neq 0$, it is not possible to have a nonzero solution for both $D_{x'}$ and $D_{y'}$ because the bracketed terms cannot vanish simultaneously. Hence, we are left with the following two possibilities:

- $D_{x'} \neq 0$ and $D_{y'} = 0$: this case corresponds to a wave launched perpendicular to the plane containing $\hat{z}$ and $\hat{z}'$. These waves are called ordinary waves, and the refractive index $n_\perp(\omega, \theta)$ is independent of their direction of propagation:

$$\frac{1}{n_\perp^2(\omega, \theta)} = \frac{1}{n_o^2(\omega)}. \tag{8.49}$$

- $D_{y'} \neq 0$ and $D_{x'} = 0$: this case corresponds to a wave launched parallel to the plane containing $\hat{z}$ and $\hat{z}'$. These waves are called extraordinary waves, and the refractive index $n_\parallel(\omega, \theta)$ depends on their direction of propagation:

$$\frac{1}{n_\parallel^2(\omega, \theta)} = \frac{\cos^2(\theta)}{n_o^2(\omega)} + \frac{\sin^2(\theta)}{n_e^2(\omega)}. \tag{8.50}$$

8.4.3 Phase-matching condition

Suppose pump and signal waves are launched into a uniaxial crystal along the directions making angles θ_p and θ_s from the optical axis, respectively. Then, the idler wave is generated in a direction θ_i from the optical axis such that the phase-matching condition is automatically satisfied. From Eq. (8.1), this happens when the refractive indices satisfy the relation

$$\omega_p n(\omega_p, \theta_p) = \omega_s n(\omega_s, \theta_s) + \omega_i n(\omega_i, \theta_i). \tag{8.51}$$

This condition can be satisfied in four different ways, listed in Table 8.1 [24]. Type-I phase matching refers to the situation in which the signal and idler waves have the same state of polarization. In contrast, they are orthogonally polarized in the case of type-II phase matching. The $\parallel$ and $\perp$ planes in Table 8.1 are relative to the plane containing the $\hat{z}$ and $\hat{z}'$ axes in Figure 8.5.

Consider, for example, a negative uniaxial crystal subjected to a type-I scheme (see the second row of Table 8.1). In this case, the signal and idler waves propagate as ordinary waves and do not see any change in the refractive index if their angle of incidence changes because $n_\perp(\omega, \theta)$ is independent of the incident angle θ. Thus, only the refractive index of the pump changes with incident angle. The phase-matching condition Eq. (8.51) for this scenario can be written as

$$\frac{\cos^2(\theta_p)}{n_o^2(\omega_p)} + \frac{\sin^2(\theta_p)}{n_e^2(\omega_p)} = \frac{\omega_p^2}{[\omega_s n_o(\omega_s) + (\omega_p - \omega_s)n_o(\omega_p - \omega_s)]^2}, \qquad (8.52)$$

where we have used $\omega_i = \omega_p - \omega_s$ as required for three-wave mixing. If a real solution for θ_p exists for this equation, then it is possible to achieve type-I phase matching in a negative uniaxial crystal by launching the pump in that direction.

The situation is somewhat more complicated for type-II phase matching in a negative uniaxial crystal (see the fourth row of Table 8.1). Because the idler is an ordinary wave, it does not see any change in refractive index relative to the incident angle θ_i. However, the refractive indices for both the pump and the signal change with their incident angles. Phase matching in this case requires that the following two equations be satisfied simultaneously:

$$\frac{\cos^2(\theta_p)}{n_o^2(\omega_p)} + \frac{\sin^2(\theta_p)}{n_e^2(\omega_p)} = \frac{\omega_p^2}{[\omega_s n_s(\omega_s, \theta_s) + (\omega_p - \omega_s)n_o(\omega_p - \omega_s)]^2}, \qquad (8.53a)$$

$$\frac{\cos^2(\theta_s)}{n_o^2(\omega_s)} + \frac{\sin^2(\theta_s)}{n_e^2(\omega_s)} = \frac{1}{n_s^2(\omega_s, \theta_s)}. \qquad (8.53b)$$

If a real solution exists for both θ_p and θ_s, then it possible to achieve type-II phase matching. It is possible to have situations where such such a solution does not exist in a uniaxial crystal for any choice of input parameters. Once the phase-matching condition is satisfied, it is easy to carry out an analysis of the parametric amplification in uniaxial crystals in a way similar to the theory developed earlier for optical fibers [2].

8.5 Phase matching in birefringent fibers

As mentioned earlier, birefringence vanishes for silica glass because of its isotropic nature. However, when silica is used to make single-mode fibers, it is possible to introduce birefringence intentionally. The reason is that a single-mode fiber is not truly single-mode when the polarization of the light is taken into account. Rather, such a fiber supports two degenerate modes that are polarized in two orthogonal directions. Under ideal conditions (perfect cylindrical symmetry and a stress-free fiber), these two modes have the same effective refractive index and they couple to

one another. Real optical fibers do not have perfect cylindrical symmetry because
of random variations in the core shape along the fiber length, and exhibit weak
fluctuating birefringence that changes the state of polarization in a random fashion
as light propagates through the fiber. This phenomenon, referred to as polarization-
mode dispersion (PMD), has been studied extensively because of its importance
for long-haul lightwave systems [25–27].

8.5.1 Polarization-maintaining fibers

The effects of such random birefringence can be overcome by intentionally introducing a large birefringence during the manufacturing process, either by using
an elliptical core (geometrical birefringence) or by stressing a circular core in a
specific direction (stress-induced birefringence). Such fibers are sometimes called
polarization-maintaining fibers because they have two principal axes along which
the the state of linear polarization of the incident light is maintained. These axes
are called the slow and fast axes, based on the speed at which light polarized along
them travels within the fiber. If n_s and n_f are the effective mode indices along the
slow and fast axes of the fiber, respectively, $n_s > n_f$ because the speed is inversely
proportional to the refractive index. Using these parameters, it is possible to define
the modal birefringence as a dimensionless parameter [28],

$$B_m = n_s - n_f. \tag{8.54}$$

The important point to note is that n_s and n_f play the same roles as n_o and n_e in
the case of birefringent crystals. This makes it possible to use a birefringent fiber
for parametric amplification, provided FWM is employed in place of three-wave
mixing.

When low-power, CW light is launched with its polarization direction oriented
at an angle with respect to the slow (or fast) axis, the polarization state of the CW
light changes along the fiber from linear to elliptic, elliptic to circular, and then back
to linear, in a periodic manner. The distance over which the two modes complete
one such cycle is known as the beat length, defined by [28]

$$L_B = \frac{\lambda}{B_m}, \tag{8.55}$$

where λ is the wavelength of the light launched into the fiber. The beat length can
be as small as 1 cm in high-birefringence fibers with $B_m = 10^{-4}$, but it may exceed
1 m in low-birefringence fibers [10].

Figure 8.6 shows how a large modal birefringence can be induced in a single-
mode fiber by surrounding its 3-μm-radius core by two stress-inducing elements.
The magnitude of the birefringence depends on both the shape and the size of these

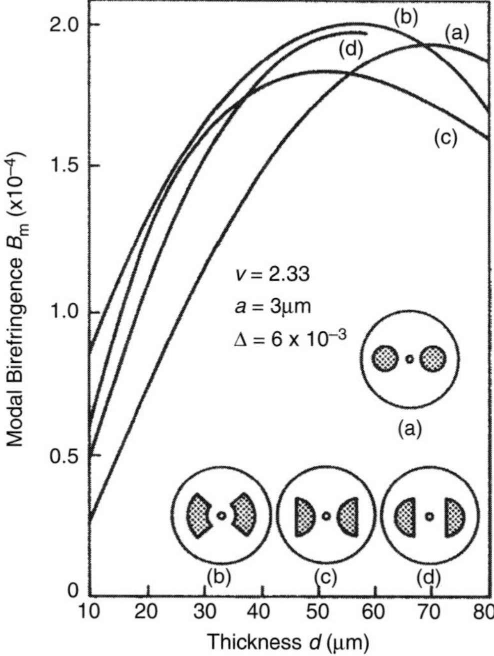

Figure 8.6 Variation of birefringence parameter B_m with thickness of the stress-inducing element, for four different polarization-preserving fibers. Different shapes of the stress-applying elements (shaded region) are shown in the inset. (After Ref. [29]; © IEEE 1986)

elements. The most notable feature is that a value of $B_m \sim 10^{-4}$ can easily be obtained in practice.

To make use of fiber birefringence for parametric amplification, we can follow the FWM theory of Section 8.3, with the difference that the propagation constants associated with the pump, signal, and idler waves have additional contributions resulting from fiber birefringence, i.e.,

$$\beta_m = \beta(\omega_m) + (\omega_m/c)\Delta n_m, \quad (m = p, s, i), \tag{8.56}$$

where the first term results from fiber dispersion and the second term accounts for fiber birefringence. Here Δn_m can be either Δn_s or Δn_f, depending on whether the corresponding optical wave is polarized along the slow or the fast axis of the fiber. The quantities Δn_s and Δn_f represent changes in the effective mode indices resulting from birefringence for the optical fields polarized along the slow and fast axes of the fiber, respectively. Their difference,

$$\delta n = \Delta n_s - \Delta n_f, \tag{8.57}$$

is related to the fiber birefringence parameter B_m.

As seen in Section 8.3, the parametric gain depends on the net phase mismatch κ. As an example, if the pump is polarized along the slow axis, and the signal and the idler are polarized along the fast axis, then the net gain given in Eq. (8.14) has an additional term resulting from fiber birefringence, given by [10]

$$\kappa \approx \beta_2(\omega_p)\Omega_s^2 + 2\gamma P_p/3 + [\Delta n_f(\omega_s + \omega_i) - 2\Delta n_s\omega_p]/c, \tag{8.58}$$

where the nonlinear XPM term has 2/3 in place of 2 because the pump is polarized orthogonally to both the signal and the idler [10]. Noting the energy conservation condition, $\omega_s + \omega_i = 2\omega_p$, we can write κ in the form

$$\kappa \approx \beta_2(\omega_p)\Omega_s^2 + 2\gamma P_p/3 - 2\delta n\omega_p/c, \tag{8.59}$$

where we have used the relation (8.57).

Since the birefringence term is negative, the FWM process can be phase-matched even in the normal-dispersion region of a birefringent fiber. Assuming this to be the case, the frequency shift $\Omega_s = \omega_s - \omega_p$ for which phase matching can occur is found to be

$$\Omega_s = \sqrt{2\delta n\omega_p/(|\beta_2|c)}, \tag{8.60}$$

where we have neglected the nonlinear term because it does not affect the frequency shift significantly. As an example, at a pump wavelength of 0.532 μm, the dispersion parameter has a value of $\beta_2(\omega_p) \approx 60$ ps^2 km^{-1}. If we use $\delta n = 1 \times 10^{-5}$ for the fiber birefringence, the frequency shift $\Omega_s/2\pi$ is $\sim$10 THz. In a 1981 experiment on FWM in birefringent fibers, the frequency shift was indeed in the range of 10 to 30 THz [30]. Furthermore, the measured values of Ω_s agreed well with those estimated from Eq. (8.60).

The use of birefringence for phase matching in single-mode fibers has an added advantage in that the frequency shift Ω_s can be tuned over a considerable range ($\sim$4 THz). Such a tuning is possible because birefringence can be changed by means of external factors such as stress and temperature. In one experiment, the fiber was pressed with a flat plate to apply the stress [31]. The frequency shift Ω_s could be tuned through 4 THz for a stress of 0.3 kg cm^{-1}. In a similar experiment, stress was applied by wrapping the fiber around a cylindrical rod [32]. The frequency shift Ω_s was tuned through 3 THz by changing the rod diameter. Tuning is also possible by varying the temperature, as the built-in stress in birefringent fibers is temperature-dependent. A tuning range of 2.4 THz was demonstrated by heating the fiber up to 700°C [33].

Equation (8.60) for the frequency shift is derived for a specific choice of field polarizations in which the pump is polarized along the slow axis while the signal and idler fields are polarized along the fast axis. Several other combinations can

Table 8.2. *Phase-matched FWM processes in birefringent fibers*[a]

Process	A_1	A_2	A_s	A_i	Frequency shift Ω_s	Condition
I	s	f	s	f	$\delta n/(|\beta_2|c)$	$\beta_2 > 0$
II	s	f	f	s	$\delta n/(|\beta_2|c)$	$\beta_2 < 0$
III	s	s	f	f	$(4\pi\,\delta n/|\beta_2|\lambda_p)^{1/2}$	$\beta_2 > 0$
IV	f	f	s	s	$(4\pi\,\delta n/|\beta_2|\lambda_p)^{1/2}$	$\beta_2 < 0$

[a] The symbols s and f denote, respectively, the direction of polarization along the slow and fast axes of a fiber with birefringence δn; λ_p is the pump wavelength.

be used for phase matching, depending on whether β_2 is positive or negative. The corresponding frequency shifts Ω_s are obtained using $\kappa = 0$ and are given by an expression similar to that in Eq. (8.60). Table 8.2 lists the four phase-matching processes that can occur in birefringent fibers, together with the corresponding frequency shifts [34]. The frequency shifts of the first two processes are smaller, by more than one order of magnitude, then those of the other two processes. All frequency shifts in Table 8.2 are approximate because the frequency dependence of δn has been ignored; its inclusion can reduce them by about 10%. Several other phase-matched processes have been identified [35] but are not generally observed in a silica fiber because of its predominantly isotropic nature.

From a practical standpoint, the four processes shown in Table 8.2 can be divided into two categories. The first two correspond to the case in which pump power is divided between the slow and fast modes. In contrast, the pump field is polarized along a principal axis of the fiber for the remaining two processes. In the first category, the parametric gain is maximum when the pump power is divided equally, by choosing $\theta = 45°$, where θ is the polarization angle measured from the slow axis. Even then, different processes compete with one another because the parametric gain is nearly the same for each of them. In one experiment, parametric interaction occurring as a result of process I was observed by using 15 ps pump pulses from a mode-locked dye laser operating at 585.3 nm [36]. Because of a relatively small group-velocity mismatch among the four waves in this case, process I became dominant compared with the others.

Figure 8.7 shows the spectrum observed at the output of a 20-m-long fiber for an input peak power $\sim$1 kW and a pump-polarization angle $\theta = 44°$. The Stokes and anti-Stokes bands located near ± 4 THz are due to FWM phase-matched by process I. As expected, the Stokes band is polarized along the slow axis, while the anti-Stokes band is polarized along the fast axis. Asymmetric broadening of

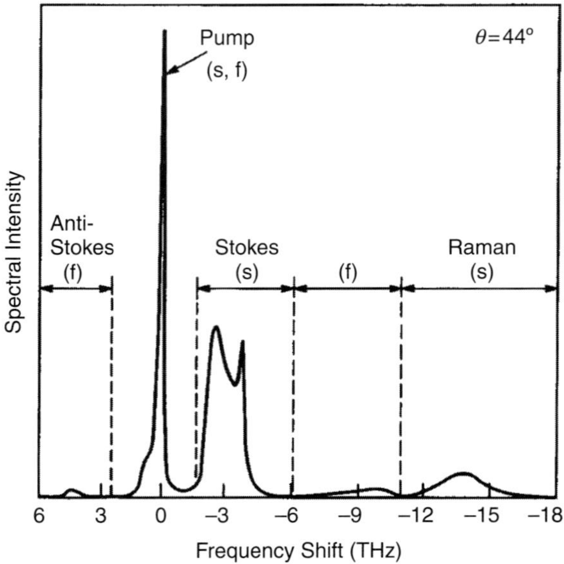

Figure 8.7 Output spectrum showing Stokes and anti-Stokes bands generated in a 20-m-long birefringent fiber when 15 ps pump pulses with a peak power ~1 kW are incident on the fiber. The pump is polarized at 44° from the slow axis. (After Ref. [36]; © OSA 1984)

the pump and Stokes bands results from the combined effects of SPM and XPM. Selective enhancement of the Stokes band is due to the Raman gain. The peak near 13 THz is also due to SRS. It is polarized along the slow axis because the pump component along that axis is slightly more intense for $\theta = 44°$. An increase in θ by 2° flips the polarization of the Raman peak along the fast axis. The small peak near 10 THz results from a nondegenerate FWM process in which the pump and the Stokes band created by it act as two distinct pump waves ($\omega_{p1} \neq \omega_{p2}$) and the Raman band provides a weak signal for the parametric process to occur. Phase matching can occur only if the Raman band is polarized along the slow axis. Indeed, the peak near 10 THz disappeared when θ was increased beyond 45° to flip the polarization of the Raman band.

The magnitude of signal gain G_s is not the same for each of the four FWM processes listed in Table 8.2, and is reduced substantially from the case discussed in Section 8.3 in which all waves are copolarized. The reason is that the strength of the parametric interaction among the four waves depends on their relative states of polarization. For example, when the pump is launched at 45° from the slow or fast axis of the fiber, it excites two distinct orthogonal modes of the fiber. This situation is similar to the dual-pump configuration discussed in Section 8.3.6, even though the two pumps have the same frequency, and we must use the expression for G_s

given in Eq. (8.40). However, if we take into account the fact that γ was reduced by a factor of 3 when the three fields were not copolarized, G_s is reduced considerably. Moreover, the effective value of the parametric gain g in Eq. (8.39) is given by

$$g = \sqrt{4(\gamma^2/9)P_1 P_2 - \kappa^2/4}. \tag{8.61}$$

Clearly, the FWM process becomes much less efficient in a single-mode fiber when its birefringence is employed for phase matching. A full vector theory of FWM in optical fibers was developed in 2004 [37]. It shows that the use of circularly polarized pumps is beneficial in some cases for improving the efficiency of the underlying FWM process. We refer the reader to the discussion in Ref. [10] for further details.

References

[1] B. R. Mollow and R. J. Glauber, "Quantum theory of parametric amplification," *Phys. Rev.*, vol. 160, pp. 1076–1096, 1967.

[2] R. W. Boyd, *Nonlinear Optics*, 3rd ed. Academic Press, 2008.

[3] P. N. Butcher and D. Cotter, *The Elements of Nonlinear Optics*. Cambridge University Press, 2003.

[4] Y. R. Shen, *The Principles of Nonlinear Optics*. Wiley, 1991.

[5] C. L. Tang and L. K. Cheng, *Fundamentals of Optical Parametric Processes and Oscillators*. Harwood Academic Publishers, 1995.

[6] E. Desurvire, *Erbium-Doped Fiber Amplifiers: Principles and Applications*. Wiley, 1994.

[7] A. E. Siegman, *Lasers*. University Science Books, 1986.

[8] K. R. Waters, J. Mobley, and J. G. Miller, "Causality-imposed (Kramers–Kronig) relationships between attenuation and dispersion," *IEEE Trans. Ultrason., Ferroelectr., Freq. Control*, vol. 52, pp. 822–833, 2005.

[9] D. C. Hutchings, M. Sheik-Bahae, D. J. Hagan, and E. W. V. Stryland, "Kramers–Kronig relations in nonlinear optics," *Opt. and Quant. Electron.*, vol. 24, pp. 1–30, 1992.

[10] G. P. Agrawal, *Nonlinear Fiber Optics*, 4th ed. Academic Press, 2007.

[11] G. Cappellini and S. Trillo, "Third-order three-wave mixing in single-mode fibers: exact solutions and spatial instability effects," *J. Opt. Soc. Am. B*, vol. 8, pp. 824–838, 1991.

[12] A. Vatarescu, "Light conversion in nonlinear monomode optical fibers," *J. Lightw. Technol.*, vol. LT-5, pp. 1652–1659, 1987.

[13] K. Inoue and T. Mukai, "Signal wavelength dependence of gain saturation in a fiber optical parametric amplifier," *Opt. Lett.*, vol. 26, pp. 10–12, 2001.

[14] R. H. Stolen and J. E. Bjorkholm, "Parametric amplification and frequency conversion in optical fibers," *IEEE J. Quantum Electron.*, vol. 18, pp. 1062–1072, 1982.

[15] G. P. Agrawal, *Lightwave Technology: Telecommunications Systems*. Wiley-InterScience, 2005.

[16] M. Ho, K. Uesaka, M. Marhic, Y. Akasaka, and L. G. Kazovsky, "200-nm-bandwidth fiber optical amplifier combining parametric and Raman gain," *J. Lightw. Technol.*, vol. 19, pp. 977–981, 2001.

[17] M. Gao, C. Jiang, W. Hu, J. Zhang, and J. Wang, "The effect of phase mismatch on two-pump fiber optical parametrical amplifier," *Opt. Laser Technol.*, vol. 39, pp. 327–332, 2007.

[18] C. J. McKinstrie, S. Radic, and A. R. Chraplyvy, "Parametric amplifiers driven by two pump waves," *IEEE J. Sel. Top. Quantum Electron.*, vol. 8, pp. 538–547, 2002.

[19] S. Radic, C. J. McKinstrie, R. M. Jopson, J. C. Centanni, Q. Lin, and G. P. Agrawal, "Record performance of parametric amplifier constructed with highly nonlinear fibre," *Electron. Lett.*, vol. 39, pp. 838–839, 2003.

[20] M. Born and E. Wolf, *Principles of Optics*, 7th ed. Cambridge University Press, 1999.

[21] G. T. Reed and A. P. Knights, *Silicon Photonics: An Introduction*. Wiley, 2004.

[22] R. S. Jacobsen, K. N. Andersen, P. I. Borel, *et al.*, "Strained silicon as a new electro-optic material," *Nature*, vol. 441, pp. 199–202, 2006.

[23] J. A. Kong, *Electromagnetic Wave Theory*. Wiley, 1986.

[24] J. E. Midwinter and J. Warner, "The effects of phase matching method and of uni-axial crystal symmetry on the polar distribution of second-order non-linear optical polarization," *Brit. J. Appl. Phys.*, vol. 16, pp. 1135–1142, 1965.

[25] C. D. Poole and J. Nagel, *Optical Fiber Telecommunications III*, I. P. Kaminow and T. L. Koch, eds. vol. 3A, chap. 6, Academic Press, 1997.

[26] H. Kogelnik, R. M. Jopson, and L. E. Nelson, *Optical Fiber Telecommunications IV*, I. P. Kaminow and T. Li, eds. vol. 4A, chap. 15, Academic Press, 2002.

[27] J. N. Damask, *Polarization Optics in Telecommunications*. Springer, 2005.

[28] I. P. Kaminow, "Polarization in optical fibers," *IEEE J. Quantum Electron.*, vol. QE-17, pp. 15–22, 1981.

[29] J. Noda, K. Okamoto, and Y. Sasaki, "Polarization-maintaining fibers and their applications," *J. Lightw. Technol.*, vol. LT-4, pp. 1071–1089, 1986.

[30] R. H. Stolen, M. A. Bösch, and C. Lin, "Phase matching in birefringent fibers," *Opt. Lett.*, vol. 6, pp. 213–215, 1981.

[31] K. Kitayama, S. Seikai, and N. Uchida, "Stress-induced frequency turning for stim-ulated four-photon mixing in a birefringement single-mode fiber," *Appl. Phys. Lett.*, vol. 41, pp. 322–324, 1982.

[32] K. Kitayama and M. Ohashi, "Frequency tuning for stimulated four-photon mixing by bending-induced birefringence in a single-mode fiber," *Appl. Phys. Lett.*, vol. 41, pp. 619–621, 1982.

[33] M. Ohashi, K. Kitayama, N. Shibata, and S. Seikai, "Frequency tuning of a stokes wave for stimulated four-photon mixing by temperature-induced birefringence change," *Opt. Lett.*, vol. 10, pp. 77–79, 1985.

[34] P. N. Morgan and J. M. Liu, "Parametric four-photon mixing followed by stimulated Raman scattering with optical pulses in birefringent optical fibers," *IEEE J. Quantum Electron.*, vol. 27, pp. 1011–1021, 1991.

[35] R. K. Jain and K. Stenersen, "Phase-matched four-photon mixing processes in birefringent fibers," *Appl. Phys. B*, vol. 35, pp. 49–57, 1984.

[36] K. Stenersen and R. K. Jain, "Small-stokes-shift frequency conversion in single-mode birefringent fibers," *Opt. Commun.*, vol. 51, pp. 121–126, 1984.

[37] Q. Lin and G. P. Agrawal, "Vector theory of four-wave mixing: Polarization effects in fiber-optic parametric amplifiers," *J. Opt. Soc. Am. B*, vol. 21, pp. 1216–1224, 2004.

9

Gain in optical metamaterials

The recent development of artificially structured optical materials—termed *optical metamaterials*—has led to a variety of interesting optical effects that cannot be observed in naturally occurring materials. Indeed, the prefix "meta" means "beyond" in the Greek language, and thus a metamaterial is a material with properties beyond those of naturally occurring materials. Examples of the novel optical phenomena made possible by the advent of metamaterials include optical magnetism [1, 2], negative refractive index [3, 4], and hyperbolic dispersion [5, 6]. Metamaterials constitute a 21st-century area of engineering science that is not only expanding fundamental knowledge about electromagnetic wave propagation but is also providing new solutions to complex problems in a wide range of disciplines, from data networking to biological imaging. Although metamaterials have attracted public attention, most people see them only in devices such as Harry Potter's cloak of invisibility, or machines like StarCraft's Arbiter, with the ability to make things invisible. Indeed, the research on metamaterials indicates that the invisibility cloak is a real possibility, and might find applications in advanced defence technologies. However, it is worth mentioning other opportunities where such advanced materials can find practical applications. A very important one is the transformation of evanescent waves into propagating waves, enabling one to view subwavelength-scale objects with an optical microscope, thereby surpassing the diffraction limit [7–9]. Other applications include the achievement of optical cloaking [10, 11], improved photovoltaics [12], and so-called nanolasers [13, 14].

In Section 9.1, we focus on various types of available metamaterial structures and their classification. Spasers are seen as viable nano-amplifiers for metamaterial structures. In Section 9.2, we consider two different ways of transforming a standard spaser into an optical amplifier. The dynamic equations governing a spaser-based amplifier are given in this section. In Section 9.3, we look at the amplification of optical signal propagating in metamaterial structures using three-wave mixing. One drawback of the three-wave mixing process is that we have to rely on an external pump medium to alter the magnitude of gain. To overcome this limitation, we consider in Section 9.4 the four-wave mixing interaction in a metamaterial medium containing active doping centers whose density sets the achievable gain in the metamaterial. Section 9.5 is devoted to the important topic of backward self-induced transparency as a way to reduce losses in metamaterial media.

9.1 Classification of metamaterials

Metamaterials, being designable man-made materials, present us with the extraordinary ability to control the path of light within them. Owing to their potential a high refractive index with a positive or negative sign, they can bend light in various directions, or slow its speed, or completely reverse its propagation direction. Obviously, this is the ultimate control of light, and can be applied to any photonic device. The question is how such control is accomplished.

The answer is related to the composite nature of metamaterials on a nanoscale that allows one to synthesize their optical responses. More specifically, metamaterials are made by embedding artificially fabricated inclusions (or inhomogeneities) within the volume of a host material in the case of three-dimensional (3D) metamaterials, or placed on the surface of a host material in the case of two-dimensional (2D) metamaterials. The geometry of these inclusions, their material composition, and the placement sites within the host medium contribute in various degrees to their optical properties. Naturally occurring optical materials react mainly to the electric field of an electromagnetic wave because of the relatively weak interaction of the magnetic field with the material's atoms. While traditional material research employs novel chemical compositions to modify a material's properties, metamaterials are obtained by the nanostructuring of existing materials, thus creating artificial atoms (or meta-atoms) that interact with light. Therefore, optical metamaterials can be engineered in such a way that both the electric and the magnetic fields interact strongly with the structure of a metamaterial (or its meta-atoms), revealing many properties not observed in the natural world [15].

Figure 9.1 shows schematically the general structure of a metamaterial, resulting in the formation of meta-atoms through nanoscale inhomogeneities. Because distances between meta-atoms are on a subwavelength scale, such a structure responds collectively to electromagnetic radiation. The new properties emerge as a result of the interaction of an electromagnetic field with the internal structure of a metamaterial. By configuring the placement of various elements, the optical properties of a metamaterial can be manipulated. For instance, using metallic structural elements, it is possible to create a metamaterial with a strong magnetic response at optical frequencies, even though its individual components are nonmagnetic. The availability

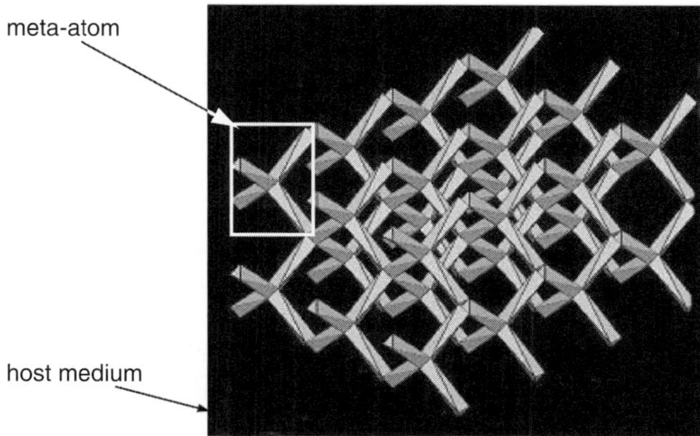

Figure 9.1 Schematic structure of a metamaterial showing the concept of meta-atoms; distances among different elements are of subwavelength size.

of a large number of design parameters provides vast flexibility in controlling the behavior of light propagating in a metamaterial.

Natural crystals have been classified based on their specific atomic configuration (basis) placed in a Bravais lattice (matrix) [16]. Optical metamaterials are the congregate of meta-atoms arranged in a few regular structures. The meta-atom placement structure can be mapped to traditional crystal structures in most cases [17]. Such a mapping provides considerable insight into the properties of metamaterials because of our existing understanding of traditional crystal groups. From this point of view, there is a strong analogy between naturally occurring materials and metamaterials.

There are two fundamentally different ways to create new metameterials. One is to change the structural arrangement of the meta-atoms, and the other is to alter the properties of the meta-atoms. A meta-atom can be made of distinct elements with appropriate electromagnetic properties. Individual meta-atoms can then be arranged to form a lattice which preserves the electromagnetic properties of each sublattice. Classification of metamaterials has already been done for 2D passive metamaterials, using the concept of wallpaper groups [18]. Using this terminology, Figure 9.2 shows the group classification for 2D metamaterial designs in the cases of square and hexagonal unit cells [17]. It also shows all the possible groups resulting from the combination of primitive cells up to $n = 3$. For example, by tiling the entire surface with one of the three square metamaterials (under the $n = 1$ heading), we obtain patterns that belong to the group pm, pmm, or p4m. The $n = 2$ column shows groups with two different metamaterials in one unit cell, tiling the surface with a checkerboard pattern. The $n = 3$ column gives groups that are combinations of three primitive cells spanned by two tiling vectors oriented along $-30°$ and $30°$.

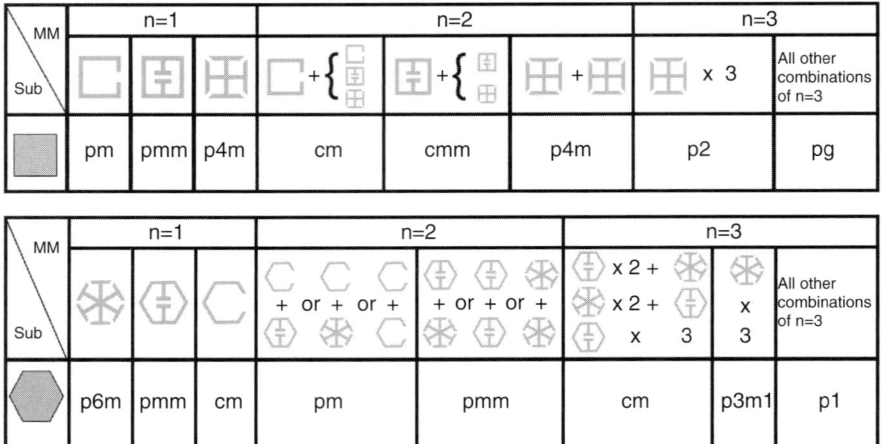

Figure 9.2 Group classification for two dimensional metamaterial designs. (After Ref. [17]; © OSA 2008)

The bottom part of this figure shows groups that correspond to hexagonal primitive cells.

So far, uncharacteristically high losses in optical metamaterials compared with the naturally occurring optical materials have limited their operational bandwidth and have severely constrained their use in much-anticipated applications in the areas of sensing and imaging. Both losses and dispersion affect light propagation in optical metamaterials, and are detrimental to promising applications relying on information transfer over distances exceeding several wavelengths [10, 19, 20]. Clearly, effective methods to overcome these inherent limitations are needed. Losses can be partially attributed to the resonant nature of meta-atoms. Recently proposed methods to control these losses in optical metamaterials include the compensation of losses through parametric amplification [21, 22], gain-assisted dispersion management [23, 24], and electromagnetically induced transparency [25, 26]. These methods can be broadly classified into two groups. In the first group, a gain material is embedded in the metal-dielectric composite making up the meta-material. In the second group, a nonlinear process such as three- or four-wave mixing is employed to achieve parametric gain.

Both schemes are limited in practice, for physical reasons. For example, fundamental laws of physics prohibit having an active metamaterial at all frequencies. This is because any wave propagating through an optical metamaterial must have a positive energy density throughout. However, the dispersive nature of metamaterials suggests that metamaterials must have finite losses at some frequencies to comply with fundamental causality constraints [27]. Therefore, metamaterials cannot be completely lossless, except perhaps at certain discrete frequencies or bands [28]. However, one advantage of optical metamaterials over naturally occurring optical materials is that their optical responses can be tailored to suit specific applications. This kind of engineering demands a fundamental knowledge of how the gain depends on the unit cell (or meta-atom) and how it affects the overall response of a metamaterial [29].

9.2 Schemes for loss compensation in metamaterials

As discussed earlier, losses in metamaterials need to be overcome before these materials can be fully utilized. Many schemes have been proposed in recent years for reducing or compensating for losses in metamaterial structures [14,30–40], ranging from purely simulation-based proposals [35, 39] to real experimental demonstrations [14, 32]. It is instructive to consider some of these schemes briefly, because they provide valuable insights into the way researchers have tried to compensate for intrinsic losses in metamaterials.

9.2.1 Review of loss-compensation schemes

Fu *et al.* showed through numerical simulations that intrinsic losses could be reduced in metamaterials made of binary metals and semiconductor quantum dots [39]. However, it was not possible to show conclusively that such behavior results from the stimulated emission of a plasmonic wave in the structure. They have also considered several other structures, with very similar results [40].

Gordon *et al.* carried out extensive numerical simulations to study whether the use of active, plasmonic-coated, spherical nanoparticles can compensate for intrinsic losses in a metamaterial [35]. The nanoparticles comprised a central core, containing a three-level gain medium, surrounded by a plasmonic metal shell. To investigate the possibility of subwavelength resonant scattering, they used nanospheres around 20–30 nm in size, representative of substructures in a typical optical metamaterial. Their results suggested the possibility of creating a subwavelength laser, known as a spaser (see Section 3.2).

Noginov *et al.* observed experimentally the ability to compensate for losses in a metal by using the optical gain in a dielectric medium consisting of silver particles in a rhodamine 6G dye liquid [30]. Their experiment showed a six-fold enhancement of Rayleigh scattering. Such an enhancement can only be achieved through a strong resonance mode of surface plasmons in such a metamaterial.

Oulton *et al.* [32] achieved experimentally a nanometer-scale plasmonic laser emitting light in the form of an optical beam whose spatial extent was 100 times smaller than the diffraction limit. They used a hybrid plasmonic waveguide consisting of a semiconductor (cadmium sulphide) nanowire, capable of exhibiting high gain when pumped electrically or optically, that was separated from a silver surface by an insulating gap that was only 5 nm wide. Measurements of the emission lifetime showed a broadband enhancement of the nanowire's spontaneous emission rate by up to six times, together with thresholdless lasing.

Noginov *et al.* generated a coherent plasmonic field using the stimulated emission of surface-plasmon polaritons on a nanoparticle of 44 nm diameter made up of a gold core (diameter about 15 nm) surrounded by a dye-doped silica shell [14]. In this experiment, it was possible to completely overcome the loss of localized surface plasmons by introducing optical gain through a spaser operating at a visible wavelength. The internal surface plasmons operated at such a wavelength that out-coupling of surface-plasmon oscillations to photonic modes occurred at a wavelength of 531 nm. Owing to the relatively small size of the gold particle, the Q factor of the cavity was limited by its losses. Moreover, the 44 nm size was not large enough to support a purely photonic mode in the visible wavelengths. Such a nanoparticle can only be excited through resonant energy transfer from excited molecules to surface-plasmon oscillations. This process initiates stimulated emission of surface plasmons in a luminous mode, as originally suggested by Bergman and Stockman [41] in 2003. This work has demonstrated conclusively the ability to create a spaser, fueling a large amount of research activity in this area.

In Section 3.2, we looked at the spaser as a device for compensating losses in plasmonic waves. However, a spaser cannot be directly used as an amplifier because it possess an inherent source of optical feedback. More specifically, surface-plasmon modes in the spaser core exert periodic perturbations on the gain medium, causing feedback which is responsible for saturating the optical gain [34]. Moreover, when a spaser operates in the CW regime, the gain experienced by the surface plasmons should compensate completely for the losses, i.e., the net amplification experienced by a surface plasmon must be zero. Because of this feature, one might naively conclude that a spaser cannot be operated as an amplifier capable of providing net amplification to a surface plasmon interacting with it. However, as Stockman showed recently [34], it is possible to design a spaser-based amplifier using two different strategies:

Dynamic or transient approach makes use of the fact that, even though the spaser has an inherent feedback mechanism that leads to zero net amplification in the steady state, such equilibrium is not reached instantaneously (it needs about 250 fs [34]). Therefore, if the spaser is used in the femtosecond transient regime, just after the population inversion has occurred but before the CW regime is established, it could amplify short optical pulses, making it an effective nanosize amplifier of such pulses.

Bistability approach is based on operating the spaser as a bistable device by placing a saturable absorber within the gain medium. The saturable absorber prevents lasing of the spaser. However, injection of surface plasmons above a certain threshold level set by the saturable absorber and the cavity geometry initiate lasing, thus switching on the device. Such a spaser with CW pumping behaves as an ultrafast nanoscale amplifier if the initial operating point is not too close to the spaser threshold. A dynamic analysis shows that it has the ability to amplify and switch with a characteristic time scale of about 100 fs [34].

9.2.2 Modeling of spaser-based amplifiers

We consider a spaser made up of a metal (or semiconductor inclusions) with permittivity $\varepsilon_\mathrm{m}(\omega)$, embedded in a host medium with permittivity ε_h [41]. The surface-plasmon eigenmodes $\varphi_n(r)$ and corresponding eigenvalues s_n can be calculated by solving the eigenvalue equation [42]

$$\nabla \Theta(r) \nabla \varphi_n(r) = s_n \nabla^2 \varphi_n(r), \tag{9.1a}$$

$$\Theta(r) = \begin{cases} 1 & \text{if } r \text{ in the metal,} \\ 0 & \text{if } r \text{ in the dielectric,} \end{cases} \tag{9.1b}$$

$$\int_V |\nabla^2 \varphi_n(r)|^2 d^3 r = 1, \tag{9.1c}$$

where V is the volume of the spaser medium. The eigenvalues s_n must be real and possess values between 0 and 1.

It is common to assume homogeneous boundary conditions such that both $\Theta(r)$ and $\varphi_n(r)$ are set to zero at the boundaries. The eigenvalues s_n enable us to calculate the frequency ω_n of the nth surface-plasmon mode using the relation [41]

$$s_n = \frac{\varepsilon_\mathrm{h}}{\varepsilon_\mathrm{h} - \varepsilon_\mathrm{m}(\omega_n - j\gamma_n)}, \tag{9.2}$$

where γ_n is the corresponding relaxation rate.

The surface-plasmon eigenmodes provide us with a complete basis for expanding various field variables associated with the spaser. A field quantity of interest is the electric field. It can be easily represented using the surface-plasmon eigenmodes $\varphi_n(r)$ and the corresponding creation and annihilation operators, $\hat{a}_n^\dagger$ and $\hat{a}_n$, and has the form [34]

$$E(r, t) = -\sum_n \sqrt{\frac{4\pi \hbar s_n}{\varepsilon_d s_n'}} [\hat{a}_n(t) + \hat{a}_n^\dagger(t)] \nabla \varphi_n(r), \qquad (9.3)$$

where

$$s_n' = \text{Re} \left[\frac{d}{d\omega} \left(\frac{\varepsilon_h}{\varepsilon_h - \varepsilon_m(\omega)} \right)_{\omega = \omega_n} \right]. \qquad (9.4)$$

Even though, in principle, it is possible to treat all quantities of interest quantum-mechanically, it complicates the analysis and requires extensive numerical computations. In practice, it is much more useful to adopt a semiclassical approach, in which the active medium is treated quantum-mechanically but surface plasmons are assumed to respond quasi-classically. This amounts to replacing the operator $\hat{a}_n$ with a c-number a_n whose time dependence has the form

$$a_n(t) = a_{0n} \exp(-j\omega t), \qquad (9.5)$$

where a_{0n} is a slowly varying amplitude such that $|a_{0n}|^2$ gives the number of coherent surface plasmons in the nth mode. Owing to its deterministic nature, such an approach neglects quantum fluctuations of surface-plasmon modes. However, this is not a significant source of error and can be corrected by introducing fluctuations through appropriately chosen relaxation rates [34].

To treat the active medium quantum-mechanically, we adopt the following model. If $|1\rangle$ and $|2\rangle$ represent the ground and excited states of the chromophore, respectively, we assume that the transition $|2\rangle \rightleftharpoons |1\rangle$ is in resonance with a specific spasing eigenmode. To initiate gain in this two-level system through population inversion, it is essential to include a third level. Similarly to the case of fiber amplifiers discussed in Chapter 5, we assume that this third level is pumped optically or electrically at a rate $\wp$, but that the pumped population decays rapidly to the state $|2\rangle$.

Let $\rho^{(p)}(t)$ be the density matrix of the two-level system associated with the pth chromophore. Using the theory given in Section 4.2, we can write its equation of motion in the form

$$\frac{\partial \rho^{(p)}}{\partial t} = \frac{1}{j\hbar}[\mathbb{H}, \rho^{(p)}], \qquad (9.6)$$

where $\mathbb{H}$ is the Hamiltonian of the spaser in the form [34]

$$\mathbb{H} = \mathbb{H}_g + \hbar \sum_n \omega_n \hat{a}_n^\dagger \hat{a}_n - \sum_p E(r_p) d^{(p)}. \qquad (9.7)$$

Here, r_p is the position of the pth chromophore, $d^{(p)}$ is its dipole moment, and $\mathbb{H}_g$ is the Hamiltonian of the gain medium.

To simplify the equation of motion of the density operator, we introduce the standard rotating-wave approximation (see Section 4.3.4), which gives prominence to the optical field and chromophores. Within this approximation, the off-diagonal elements of the density matrix can be written in the form

$$\rho_{12}^{(p)}(t) = \zeta_{12}^{(p)}(t)\exp(j\omega t), \qquad \rho_{21}^{(p)}(t) = \rho_{12}^{(p)*}(t), \tag{9.8}$$

where $\zeta_{12}^{(p)}(t)$ is the slowly varying amplitude of the nondiagonal element $\rho_{12}^{(p)}(t)$ of the density matrix. Following our derivations of density-matrix equations in chapters 4 and 6, we can write the following two equations for the gain medium of the spaser:

$$\frac{d}{dt}\zeta_{12}^{(p)}(t) = -j(\omega - \omega_{12})\zeta_{12}^{(p)}(t) - \Gamma_{12}\zeta_{12}^{(p)}(t) + jn_{21}^{(p)}(t)\Omega_{12}^{(p)}, \tag{9.9a}$$

$$\frac{d}{dt}n_{21}^{(p)}(t) = -4\mathrm{Im}\left[\zeta_{12}^{(p)}(t)\Omega_{21}^{(p)}\right] - \gamma_2^{(p)}\left[1 + n_{21}^{(p)}(t)\right] + \wp\left[1 - n_{21}^{(p)}(t)\right], \tag{9.9b}$$

where $n_{21}^{(p)} = \rho_{22}^{(p)} - \rho_{11}^{(p)}$ is the population inversion of the spasing transition and Γ_{12} is the polarization relaxation rate.

The decay rate $\gamma_2^{(p)}$ represents spontaneous decay of the population from the state $|2\rangle$ to the state $|1\rangle$ (responsible for spontaneous emission from excited atoms) and is given by

$$\gamma_2^{(p)} = \frac{8\pi s_n}{\gamma_n \varepsilon_d s_n'}\left|\nabla\varphi_n(\mathbf{r}_p)\mathbf{d}_{12}^{(p)}\right|^2 \frac{(\Gamma_{12} + \gamma_n)^2}{(\omega_{12} - \omega_n)^2 + (\Gamma_{12} + \gamma_n)^2}, \tag{9.10}$$

where γ_n is the relaxation rate of the nth surface-plasmon mode of the spaser (see Eq. (9.2)). The Rabi frequency $\Omega_{12}^{(p)}$ for the transitions in the pth chromophore (a real quantity) can be written as

$$\Omega_{12}^{(p)} = \left|\frac{a_{0n}}{\hbar}\sqrt{\frac{4\pi\hbar s_n}{\varepsilon_d s_n'}}\nabla\varphi_n(\mathbf{r}_p)\mathbf{d}_{12}^{(p)}\right|, \tag{9.11}$$

where $\mathbf{d}_{12}^{(p)}$ is the dipole moment corresponding to the transition between states $|1\rangle$ and $|2\rangle$.

The intricate dynamics of the gain medium will contribute to the surface-plasmon field in the spaser medium through the stimulated emission process. As we have replaced the annihilation operator with a c-number, $a_{0n}\exp(-j\omega t)$, the corresponding differential equation, derived using the Hamiltonian $\mathbb{H}$ in Eq. (9.7), is found to be

$$\frac{da_{0n}}{dt} = j(\omega - \omega_n)a_{0n} - \gamma_n a_{0n} + j\sum_p \zeta_{21}^{(p)*}(t)\Omega_{12}^{(p)}. \tag{9.12}$$

This completes the derivation of the dynamic equations governing the operation of a spaser. These equations cannot be solved analytically under transient conditions and thus require numerical integration. The interested reader is referred to Ref. [34] for a detailed numerical investigation of these equations under amplifying conditions.

9.3 Amplification through three-wave mixing

It is possible to utilize the nonlinear process of three-wave mixing to provide parametric amplification to compensate for losses in a metamaterial in negative-index regions [43]. A plethora of resonant conditions are responsible for the unique features of metamaterials. Therefore, if a metamaterial has regions in which the refractive index is negative, it is most likely that the negative values are limited to a certain narrow spectral region. In this section we discuss how parametric amplification can be employed in practice.

9.3.1 Configuration and basic equations for three-wave mixing

In a three-wave mixing process, the interaction of the signal and pump waves at the frequencies ω_1 and ω_3, respectively, generates an idler wave at the difference frequency, $\omega_2 = \omega_3 - \omega_1$. We limit our analysis to metamaterials in which the signal wave experiences a negative index of refraction (i.e., $n(\omega_1) < 0$) but both the pump and idler waves experience a positive refractive index (i.e., $n(\omega_2) > 0$ and $n(\omega_3) > 0$). Here we have assumed that the function $n(\omega)$ gives the refractive index of the metamaterial with the correct sign over the entire frequency range of interest. Suppose the propagation vectors of the signal wave, idler wave, and pump are given by k_1, k_2, and k_3, respectively. If the three vectors point from left to right as shown in Figure 9.3, the idler and pump have energy flows in the direction of their wave vectors, but the signal energy flows from right to left because of $n(\omega_1) < 0$.

Parametric amplification occurs when the energy of a pump photon is used to create a pair of photons simultaneously at the signal and idler frequencies. Although not immediately obvious, it is better to solve Maxwell's equations for the magnetic fields associated with the three waves [43], as in the case of photonic crystals. We introduce the slowly varying part of the magnetic field h_m using

$$H_m(z, t) = \tfrac{1}{2}\{h_m(z, t) \exp[j(k_m z - \omega_m t)] + \text{c. c.}\}, \qquad (9.13)$$

where $m = 1, 2, 3$ for signal, idler, and pump, respectively. The coupled equations governing the nonlinear interaction of the three waves in a $\chi^{(2)}$ medium under the

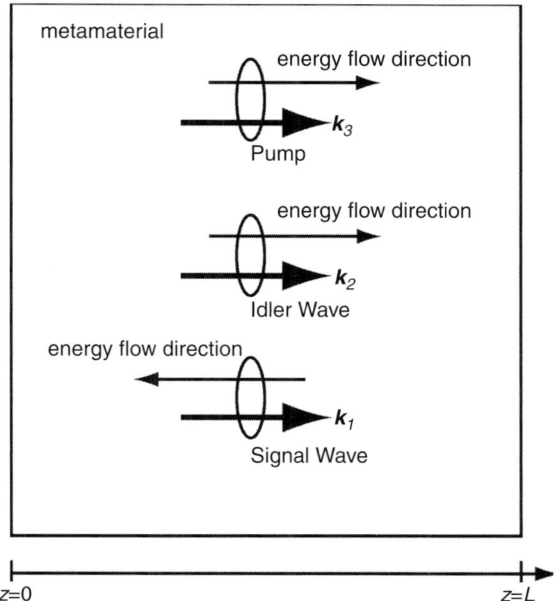

Figure 9.3 Schmematic diagram showing the directions of the propagation vectors for the pump, signal, and idler waves and the corresponding energy-flow directions (related to the Poynting vector) in a metamaterial in which the signal wave experiences a negative refractive index.

slowly-varying-envelope approximation can be written in the form [44]

$$-\frac{\partial h_1}{\partial z} + \frac{1}{\upsilon(\omega_1)}\frac{\partial h_1}{\partial t} = -\frac{1}{2}\alpha(\omega_1)h_1 + j\frac{\omega_1^2\varepsilon(\omega_1)}{c^2 k_1}\chi^{(2)}h_3 h_2^* \exp(j\Delta kz), \quad (9.14a)$$

$$\frac{\partial h_2}{\partial z} + \frac{1}{\upsilon(\omega_2)}\frac{\partial h_2}{\partial t} = -\frac{1}{2}\alpha(\omega_2)h_2 + j\frac{\omega_2^2\varepsilon(\omega_2)}{c^2 k_2}\chi^{(2)}h_3 h_1^* \exp(j\Delta kz), \quad (9.14b)$$

$$\frac{\partial h_3}{\partial z} + \frac{1}{\upsilon(\omega_3)}\frac{\partial h_3}{\partial t} = -\frac{1}{2}\alpha(\omega_3)h_3 + j\frac{\omega_3^2\varepsilon(\omega_3)}{c^2 k_3}\chi^{(2)}h_1 h_2 \exp(j\Delta kz), \quad (9.14c)$$

where $\Delta k = k_3 - k_2 - k_1$ represents the phase mismatch, $\alpha(\omega)$ is the loss coefficient of the medium, $\mu(\omega)$ is the permeability of the metamaterial, and $\upsilon(\omega)$ is the speed of the light in the medium, all at the frequency ω.

9.3.2 Solution with undepleted-pump approximation

In general, the preceding set of equations must be solved numerically. However, we can solve them approximately if we make the so-called undepleted-pump approximation, assuming that the pump does not loose a significant amount of power to the signal during the amplification process. Taking the pump field as approximately constant within the metamaterial and assuming steady-state conditions, we obtain the following reduced set of two coupled equations:

$$-\frac{dh_1}{dz} = -\frac{1}{2}\alpha(\omega_1)h_1 + j\frac{\omega_1^2\varepsilon(\omega_1)}{c^2 k_1}\chi^{(2)}h_3 h_2^* \exp\left(j\Delta kz\right), \qquad (9.15a)$$

$$\frac{dh_2}{dz} = -\frac{1}{2}\alpha(\omega_2)h_2 + j\frac{\omega_2^2\varepsilon(\omega_2)}{c^2 k_2}\chi^{(2)}h_3 h_1^* \exp\left(j\Delta kz\right). \qquad (9.15b)$$

To simplify these equations further, we introduce the normalized amplitudes a_1 and a_2 with the following definition [45]:

$$a_m(\omega_m) = \sqrt{\frac{\eta(\omega_m)}{\omega_m}}h_m \quad (m = 1, 2), \qquad (9.16)$$

where η, evaluated at appropriate frequency, represents the wave impedance of the corresponding medium. The quantities a_1^2 and a_2^2 are proportional to the number of photons at their corresponding frequencies. Substituting the preceding definition into Eq. (9.15), we obtain

$$-\frac{da_1}{dz} = -\frac{1}{2}\alpha(\omega_1)a_1 + j\frac{\omega_1^2\varepsilon(\omega_1)}{c^2 k_1}\chi^{(2)}h_3\sqrt{\frac{\omega_2\eta(\omega_1)}{\omega_1\eta(\omega_2)}}a_2^* \exp\left(j\Delta kz\right), \quad (9.17a)$$

$$\frac{da_2}{dz} = -\frac{1}{2}\alpha(\omega_2)a_2 + j\frac{\omega_2^2\varepsilon(\omega_2)}{c^2 k_2}\chi^{(2)}h_3\sqrt{\frac{\omega_1\eta(\omega_2)}{\omega_2\eta(\omega_1)}}a_1^* \exp\left(j\Delta kz\right). \quad (9.17b)$$

Noting that the ratio k_2/k_1 can can be written as

$$\frac{k_2}{k_1} = \frac{\omega_2}{\omega_1}\sqrt{\frac{\varepsilon(\omega_2)\mu(\omega_2)}{\varepsilon(\omega_1)\mu(\omega_1)}}, \qquad (9.18)$$

it is possible to introduce a factor g in Eqs. (9.15) using

$$\begin{aligned}
g &= \frac{\chi^{(2)}h_3}{c^2}\sqrt{\frac{\omega_1\omega_2}{\eta(\omega_1)\eta(\omega_2)}} \\[2mm]
&\equiv \frac{\omega_1^2\varepsilon(\omega_1)}{c^2 k_1}\chi^{(2)}h_3\sqrt{\frac{\omega_2\eta(\omega_1)}{\omega_1\eta(\omega_2)}} \\[2mm]
&\equiv \frac{\omega_2^2\varepsilon(\omega_2)}{c^2 k_2}\chi^{(2)}h_3\sqrt{\frac{\omega_1\eta(\omega_2)}{\omega_2\eta(\omega_1)}}.
\end{aligned} \qquad (9.19)$$

The parameter g can be identified as the local gain in this amplifier configuration. However, care should be exercised in using this definition because the overall parametric gain depends on g in a more complicated fashion than it does with conventional amplifiers. The use of g in Eqs. (9.17) leads to the following compact form of the two coupled amplitude equations [45]:

$$\frac{da_1}{dz} = \frac{1}{2}\alpha(\omega_1)a_1 - jga_2^* \exp(j\Delta kz), \tag{9.20a}$$

$$\frac{da_2}{dz} = -\frac{1}{2}\alpha(\omega_2)a_2 + jga_1^* \exp(j\Delta kz). \tag{9.20b}$$

Even though our derivation of the coupled amplitude equations was almost identical to a conventional analysis based on the electric field [44], these equations show an asymmetry which is not present in materials with a purely positive refractive index [43]. The signs of the nonlinear term differ because the signal and idler waves propagate in a medium with permittivities of opposite signs. The loss terms also have opposite signs because energy flows in opposite directions in a metamaterial.

Equations (9.20) can be solved by imposing appropriate boundary conditions. Noting that energy flow for the signal occurs from right to left, the boundary condition for the signal must be set at the right-hand end, $z = L$, where L is the thickness of the metamaterial slab. Using the boundary conditions $a_1(z = L) = a_{1L}$ and $a_2(z = 0) = a_{20}$, we obtain the following analytic solution [45]:

$$a_1(z) = F_1 \exp[(\beta_1 + j\Delta k/2)z] + F_2 \exp[(\beta_2 + j\Delta k/2)z], \tag{9.21a}$$

$$a_2(z) = \kappa_1 F_1 \exp[(\beta_1 - j\Delta k/2)z] + \kappa_2 F_2 \exp[(\beta_2 - j\Delta k/2)z], \tag{9.21b}$$

where we have introduced the following new quantities:

$$q = [\alpha(\omega_1) + \alpha(\omega_2)]/4 - j\Delta k/2, \tag{9.22a}$$

$$\kappa_{1,2} = \pm\sqrt{1 - q^2/g^2} + jq/g, \tag{9.22b}$$

$$\beta_{1,2} = [\alpha(\omega_1) - \alpha(\omega_2)]/4 \pm j\sqrt{g^2 - q^2}, \tag{9.22c}$$

$$F_1 = \frac{a_{1L}\kappa_2 - a_{20}^* \exp[(\beta_2 + j\Delta k/2)L]}{\kappa_2 \exp[(\beta_1 + j\Delta k/2)L] - \kappa_1 \exp[(\beta_2 + j\Delta k/2)L]}, \tag{9.22d}$$

$$F_2 = -\frac{a_{1L}\kappa_1 - a_{20}^* \exp[(\beta_2 + j\Delta k/2)L]}{\kappa_2 \exp[(\beta_1 + j\Delta k/2)L] - \kappa_1 \exp[(\beta_2 + j\Delta k/2)L]}. \tag{9.22e}$$

From the preceding solution, it is possible to estimate the parametric gain seen by the signal wave owing to energy transfer from the pump. Noting that no idler wave exists at the input end ($a_{20} = 0$), the parametric gain seen by the signal wave can be written as [45]:

$$G = \left|\frac{a_{10}}{a_{1L}}\right|^2 = \left|\frac{\exp[-q^*L + \alpha(\omega_2)L/2]}{\cos(RL) + q\sin(RL)/R}\right|^2, \tag{9.23}$$

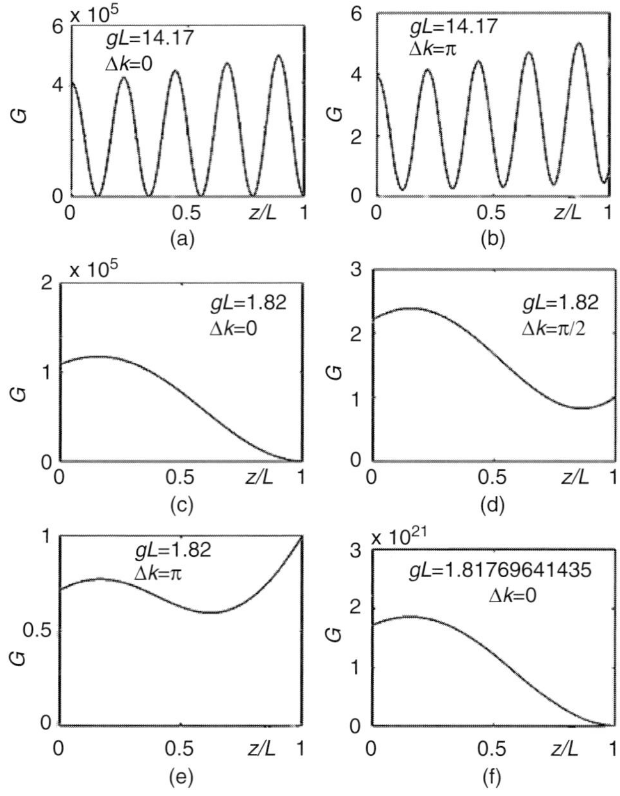

Figure 9.4 Gain G seen by the signal wave for several values of Δk and gL assuming $\alpha_1 L = 1$ and $\alpha_2 L = 0.5$. (After Ref. [46]; © OSA 2007)

where $R = \sqrt{g^2 - q^2}$. Owing to the presence of this distributed parametric gain, efficient loss compensation in the negative-index region is possible in metamaterials when $R > 0$. Figure 9.4 shows the strong dependence of the amplification factor G on phase matching for both large ($gL = 14.17$) and small ($gL = 1.82$) pump powers.

9.4 Resonant four-wave mixing using dopants

In Section 9.3, $\chi^{(2)}$ nonlinearity was exploited for loss compensation in a metamaterial through three-wave mixing. One drawback of the three-wave mixing process is that we have to rely on an external pump medium to alter the magnitude of the gain. A simple way to overcome this deficiency is to dope the metamaterial with suitable atoms and use a resonant four-wave mixing (FWM) process for signal amplification by transferring power from a pump beam [46]. As before, we assume

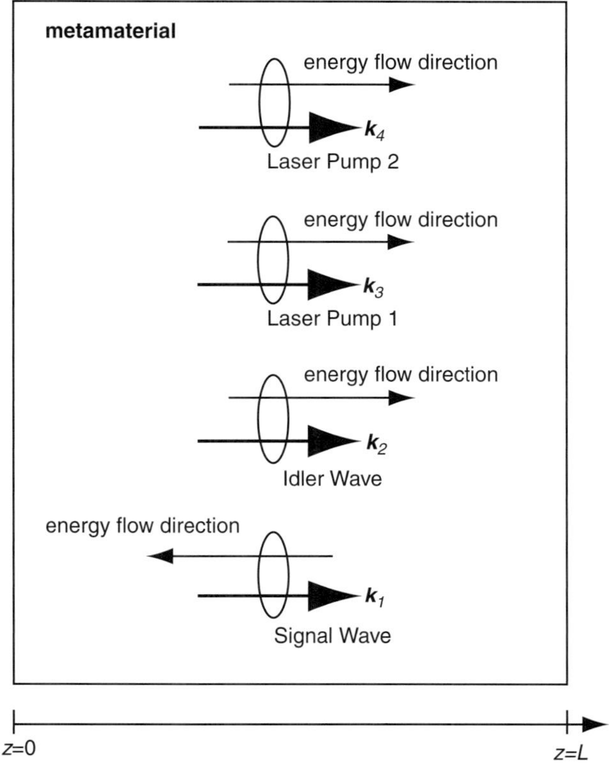

Figure 9.5 Schematic of the four energy levels involved in the resonant four-wave mixing process.

that the signal propagates in the negative-index region of a metamaterial, whereas all other waves, including the pump, see a positive refractive index.

9.4.1 Density-matrix equations for the dopants

Suppose the signal and idler waves have frequencies ω_1 and ω_2, respectively. The signal receives a gain due to FWM which is initiated by two pumps at frequencies ω_3 and ω_4. Because of the resonant nature of this process, one has to employ a 4×4 density matrix whose diagonal elements represent population densities and whose off-diagonal elements correspond to atomic transitions among the four levels. The corresponding density-matrix equations can be written in the following form [47]:

$$\frac{d\rho_{nn}}{dt} + \Gamma_{nn}\rho_{nn} = q_n - j\,[V,\,\rho]_{nn} + \gamma_{mn}\rho_{mm}, \tag{9.24a}$$

$$\frac{d\rho_{lm}}{dt} + \Gamma_{lm}\rho_{lm} = -j\,[V,\,\rho]_{lm} + \gamma_{mn}\rho_{mm}, \tag{9.24b}$$

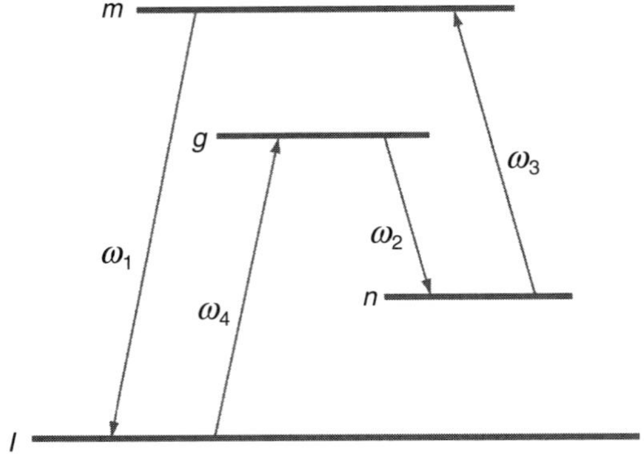

Figure 9.6 Illustration of the propagation vectors of signal wave, idler wave and two pump waves, together with the associated energy flow directions, in a metamaterial when only the signal wave experiences a negative refractive index.

$$V_{lm} = -\frac{\boldsymbol{E}_1 \cdot \boldsymbol{d}_{lm}}{\hbar} \exp\left[j(\omega_1 t - \omega_{ml} t - kz)\right], \tag{9.24c}$$

where γ_{mn} is the rate of relaxation from level m to n, the population relaxation rates Γ_n satisfy the relation $\Gamma_n = \sum_j \gamma_{nj}$, $q_n = \sum_j w_{nj} r_j$ is the rate of incoherent excitation to the energy state n, and Γ_{mn} is the homogeneous linewidth associated with the transition $m \rightarrow n$. In the absence of atomic collisions, $\Gamma_{mn} = (\Gamma_m + \Gamma_n)/2$. The equations for other elements of the density matrix can be written in the same way.

Figure 9.6 shows the directions of wave propagation and energy flow within the metamaterial for the four waves involved in the FWM process. We assume that the wave vectors $\boldsymbol{k}_j$ at frequencies ω_j (with $j = 1$ to 4) point in the same direction and satisfy the phase-matching condition required for FWM to occur. The interaction of the signal with the two pump fields generates an idler wave at a frequency $\omega_2 = \omega_3 + \omega_4 - \omega_1$ to in order satisfy energy conservation. Owing to the negative refractive index at the signal frequency, the signal energy flows in a direction opposite to that of the other three waves. Similarly to the three-wave mixing case discussed in Section 9.2, the signal wave must enter the medium at $z = L$ and leave it at $z = 0$.

9.4.2 Iterative solution for the induced polarization

The analysis of FWM requires a solution of Maxwell's equations in which the induced polarization and the associated third-order susceptibility can be calculated using the density-matrix equations given earlier. The induced polarization $\mathscr{P}(t)$

along a given direction direction r can be written as [48]

$$\mathscr{P}_r = \varepsilon_0 \sum_s \chi_{rs}^{(1)} E_s + \varepsilon_0 \sum_{st} \chi_{rst}^{(2)} E_s E_t + \varepsilon_0 \sum_{stu} \chi_{rstu}^{(3)} E_s E_t E_u + \cdots , \qquad (9.25)$$

where $\chi^{(1)}$, $\chi^{(2)}$, and $\chi^{(3)}$ are the first-, second-, and third-order susceptibilities, respectively, with subscripts representing the various directions in a Cartesian coordinate system.

In the quantum-mechanical description of nonlinear susceptibilities, different nonlinearities originate as perturbative solutions of different orders during an iterative solution of the density-matrix equations. We assume that the density-matrix element $\rho_{\alpha\beta}$ with $\alpha, \beta \in \{l, m, n, g\}$ (see Fig. 9.5 for energy levels) can be written as an iterative sequence $\rho_{\alpha\beta}^{(i)}$, where the index i refers to the iteration order. Using Eq. (9.24), the ith-order solution is used to calculate the $(i + 1)$-order terms as follows:

$$\frac{d\rho_{nn}^{(i+1)}}{dt} + \Gamma_{nn}\rho_{nn}^{(i)} = q_n - j\left[V, \rho^{(i)}\right]_{nn} + \gamma_{mn}\rho_{mm}^{(i)}, \qquad (9.26a)$$

$$\frac{d\rho_{lm}^{(i+1)}}{dt} + \Gamma_{lm}\rho_{lm}^{(i)} = -j\left[V, \rho^{(i)}\right]_{lm} + \gamma_{mn}\rho_{mm}^{(i)}, \qquad (9.26b)$$

$$V_{lm} = -\frac{E_1 \cdot d_{lm}}{\hbar} \exp\left(j\left[\omega_1 t - \omega_{ml} t - kz\right]\right). \qquad (9.26c)$$

These iterative solutions are then used to construct the original density matrix $\rho(t)$ using the relation

$$\rho(t) = \sum_i \rho^{(i)}(t). \qquad (9.27)$$

The macroscopic polarization $\mathscr{P}(t)$ induced by the presence of optical fields in a medium with dopant density N can be written as [44]

$$\mathscr{P}(t) = \text{Tr}(N d \rho) = N \sum_\alpha \sum_\beta d_{\alpha\beta}\rho_{\beta\alpha}. \qquad (9.28)$$

Introducing the ith-order polarization term as

$$\mathscr{P}^{(i)}(t) = \text{Tr}(N d \rho^{(i)}), \qquad (9.29)$$

it is possible to construct the polarization terms iteratively to any desired order. These expressions are then used to extract the corresponding susceptibility values, as discussed in Refs. [44] and [49].

Analysis of FWM is quite complicated in the most general case in which multiple fields interact with a multi-level atom, but can be simplified considerably if we assume that the two pump fields remain undepleted in spite of energy transfer from them to the signal and idler waves. This is a reasonably good approximation if the pump powers are initially much higher than the signal power, even when the signal sees a very high net gain. This is because the amount of energy the signal receives from the pump is quite low when a low-power signal enters the amplifier. In other words, a larger signal gain does not translate into a signal power which is comparable to the pump power level inside the amplifier. This approximation is known as the undepleted-pump approximation.

9.4.3 Coupled signal and idler equations

In the three-wave case, we used the magnetic field components of the optical waves. However, in the FWM case, it is more convenient to deal with the corresponding electric fields E_m associated with the four waves ($m = 1$ to 4). Using a procedure similar to that employed in Section 9.2, we obtain the following two equations for the signal and idler waves [47]:

$$\frac{dE_1}{dz} = -j\frac{\omega_1^2 \varepsilon(\omega_1)}{c^2 k_1}\chi^{(3)}(\omega_1)E_3 E_4 E_2^* \exp(j\Delta kz) + \frac{1}{2}\alpha(\omega_1)E_1, \quad (9.30a)$$

$$\frac{dE_2}{dz} = +j\frac{\omega_2^2 \varepsilon(\omega_2)}{c^2 k_2}\chi^{(3)}(\omega_2)E_3 E_4 E_1^* \exp(j\Delta kz) - \frac{1}{2}\alpha(\omega_2)E_2, \quad (9.30b)$$

where $\Delta k = k_3 + k_4 - k_1 - k_2$ is the phase mismatch, $\alpha(\omega)$ is the loss coefficient of the medium, $\mu(\omega)$ is the permeability of the metamaterial, and $\chi^{(3)}(\omega)$ is its third-order nonlinear susceptibility.

The following example taken from Ref. [46] shows the effectiveness of this scheme. It uses the following vales for the relaxation rates of the four levels involved in the FWM process: $\Gamma_n = 20 \times 10^6$ s^{-1}, $\Gamma_g = \Gamma_m = 120 \times 10^6$ s^{-1}, $\gamma_{gl} = 7 \times 10^6$ s^{-1}, $\gamma_{gn} = 4 \times 10^6$ s^{-1}, $\gamma_{ml} = 10 \times 10^6$ s^{-1}, $\Gamma_{lg} = 1 \times 10^{12}$ s^{-1}, $\Gamma_{lm} = 1.9 \times 10^{12}$ s^{-1}, $\Gamma_{ng} = 1.5 \times 10^{12}$ s^{-1}, $\Gamma_{nm} = 1.8 \times 10^{12}$ s^{-1}, $\Gamma_{gm} = 0.05 \times 10^{12}$ s^{-1}, and $\Gamma_{ln} = 0.01 \times 10^{12}$ s^{-1}. The signal wavelength is 480 nm, and wavelength of the idler corresponds to 756 nm. The pump powers are chosen such that the Rabi frequencies Ω_3 and Ω_4 are equal to one another, with the value 50 GHz. The pumps are detuned from exact resonance such that $\omega_3 - \omega_{mn} = 2.5\Gamma_{lg}$ and $\omega_4 - \omega_{gl} = 2.5\Gamma_{lg}$.

Figure 9.7 shows the absorption, reflection, and the nonlinear response of signal and idler as a function of a normalized detuning, defined as $y_4 = (\omega_1 - \omega_{ml})/\Gamma_{ml}$. The quantities γ_2 and γ_4 represent effective nonlinear parameters. Notice the sharp, narrow resonances at $y > 0$ that affect the transmission of the signal through the metamaterial. As an example, Figure 9.8(a) shows the narrow transmission

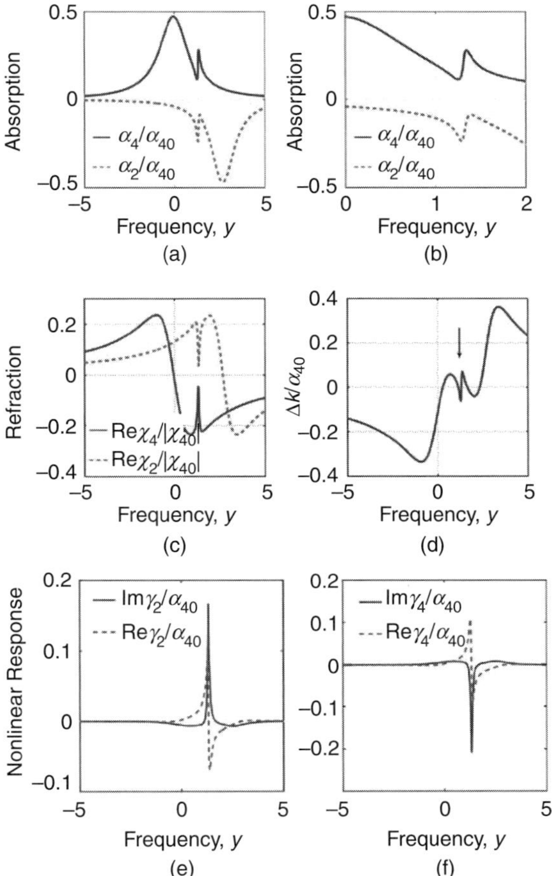

Figure 9.7 Nonlinear resonances induced by the pump fields in the case of resonant FWM. Solid and dashed lines correspond to signal and idler fields, respectively. (After Ref. [46]; © OSA 2007)

resonance occurring for $y = 2.5266$ or $L/L_{ra} = 36.52$, where L_{ra} is the absorption length, defined as $L_{ra} = 1/\alpha_{40}$. At the peak of the resonance taking place when $\alpha_{40}L = 37.02$, $T > 1$ because of the amplification provided by FWM. The corresponding intensity distribution of the signal with in the slab is shown in Figure 9.8(b). It clearly shows that the intensity of the signal within the slab significantly exceeds its output value at $z = 0$. We have used the definition $\eta_2(z) = |E_2(z)/E_1(L)|^2$ to denote the idler efficiency.

9.5 Backward self-induced transparency

In some cases, energy losses resulting from absorption can be overcome using features of the medium. For example, if a pulse of coherent light propagates through

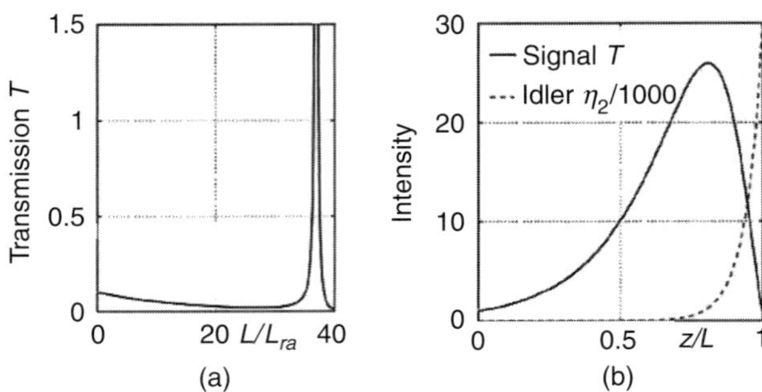

Figure 9.8 FWM-induced transmission resonances in a metamaterial. Solid and dashed lines correspond to the signal and idler fields, respectively. (After Ref. [46]; © OSA 2007)

a two-level medium, the energy absorbed from the first half of the pulse can be returned to the the pulse during its second half. This can occur only if the wavelength of the pulse is at or near a resonance peak of an effective two-level medium and the pulse is very short in duration compared to the material's lifetime (the time scale over which its properties can change). This striking phenomena, known as self-induced transparency (SIT), was first predicted in 1967 [50] and has been studied extensively since then [51]. SIT can be applied to metamaterials to overcome moderate absorption and strong dispersion when the signal sees a negative refractive index. Since the signal propagates backward in such a medium, this effect is called backward SIT [52].

9.5.1 Mathematical approach

To describe backward SIT mathematically, we assume that the metamaterial is homogenously doped with two-level centers whose concentration is low enough that the doping has negligible effect on the metamaterials properties (such as absorption and dispersion). We also assume that the frequency dependence of the electric permittivity $\varepsilon(\omega)$ and magnetic permeability $\mu(\omega)$ of the metamaterial can be described using a lossy Drude–Lorentz model, so that [52]

$$\varepsilon(\omega) = \varepsilon_0 \left[1 - \frac{\omega_{pe}^2}{\omega(\omega + j\Gamma_e)} \right], \tag{9.31a}$$

$$\mu(\omega) = \mu_0 \left[1 - \frac{\omega_{pm}^2 - \omega_{0m}^2}{\omega(\omega + j\Gamma_m) - \omega_{0m}^2} \right], \tag{9.31b}$$

where ω_{pe} is the electric plasma frequency, ω_{pm} is the magnetic plasma resonance frequency, Γ_e is the electric loss rate, Γ_m is the magnetic loss rate, and ω_{0m} is the magnetic resonance frequency. Figure 9.9 shows the real and imaginary parts of

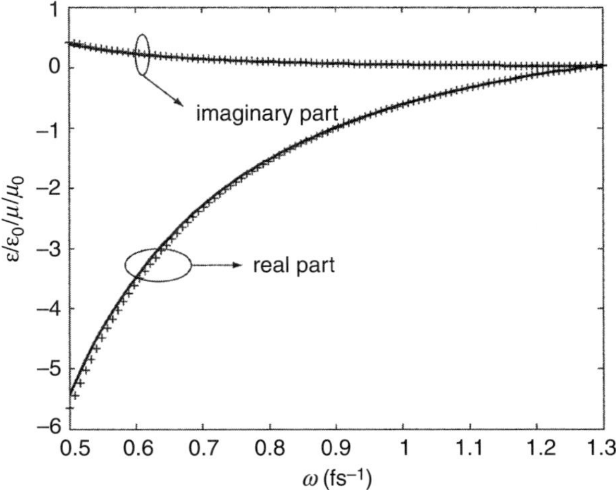

Figure 9.9 Frequency dependence of the real and imaginary parts of the permittivity ε (solid lines) and permeability μ (crosses) for a metamaterial whose dielectric response follows the Drude–Lorentz model.

the permittivity $\varepsilon(\omega)$ and the permeability $\mu(\omega)$ using $\omega_{pm} = 1.27$ fs^{-1}, $\omega_{0m} = 0.1$ fs^{-1}, and $\Gamma_e = 0.03$ fs^{-1}, $\Gamma_m = 0.03$ fs^{-1}. Negative values for the real parts of both these quantities confirm that such a metamaterial has a negative refractive index over a wide frequency range.

Consider a plane-wave pulse propagating along the z axis within such a doped metamaterial and described using the four electromagnetic field variables E_x, D_x, H_y, and B_y. The pairs (D_x, H_y) and (E_x, B_y) are related to one another through Maxwell's equations as follows:

$$\frac{\partial D_x}{\partial t} = -\frac{\partial H_y}{\partial z}, \tag{9.32a}$$

$$\frac{\partial B_y}{\partial t} = -\frac{\partial E_x}{\partial z}. \tag{9.32b}$$

At the same time, the pairs (E_x, D_x) and (H_y, B_y) are related to one another through the constitutive relations,

$$\frac{\partial D_x}{\partial t} = \varepsilon_0 \frac{\partial E_x}{\partial t} + \omega_{pe}^2 \frac{\partial K_x}{\partial t}, \tag{9.33a}$$

$$\frac{\partial B_y}{\partial t} = \mu_0 \frac{\partial H_y}{\partial t} + \left(\omega_{pm}^2 - \omega_{0m}^2\right) \frac{\partial M_y}{\partial t}, \tag{9.33b}$$

where K_x and M_y are the electric and magnetic material polarizations induced in the medium.

K_x and M_y can be obtained using the following two equations derived from the permittivity and permeability values given in (9.31):

$$\frac{\partial^2 K}{\partial t^2} + \Gamma_e \frac{\partial K}{\partial t} = \Omega_e, \tag{9.34a}$$

$$\frac{\partial^2 M}{\partial t^2} + \Gamma_m \frac{\partial M}{\partial t} + \omega_{0m}^2 M = \Omega_m, \tag{9.34b}$$

where $K = dK_x/\hbar$ is the electrification and $M = \sqrt{\varepsilon_0/\mu_0}\,dM_y/\hbar$ is the magnetization of the medium,[1] $\Omega_e = dE_x/\hbar$ is the Rabi frequency of the electric field E_x, $\Omega_m = \sqrt{\varepsilon_0/\mu_0}\,dH_y/\hbar$ is the Rabi frequency of the magnetic field H_y, and d is the resonant dipole moment.

9.5.2 Bloch equations and Rabi frequencies

Combining Eqs. (9.32) and (9.33), one can obtain the following two equations for the two Rabi frequencies:

$$\frac{\partial \Omega_e}{\partial t} + 2\beta \frac{\partial \mathrm{Re}(\rho_{12})}{\partial t} + \omega_{pe}^2 \frac{\partial K}{\partial t} = -\frac{\partial \Omega_m}{\partial z}, \tag{9.35a}$$

$$\frac{\partial \Omega_m}{\partial t} + (\omega_{pm}^2 - \omega_{0m}^2)\frac{\partial M}{\partial t} = -\frac{\partial \Omega_e}{\partial z}, \tag{9.35b}$$

where ρ_{12} is the off-diagonal element of the density matrix governing the dynamics of our two-level system. It is obtained by solving the usual Bloch equations [51]:

$$\frac{\partial W}{\partial t} = -\frac{(W+1)}{T_1} - 2j\Omega_e\left(\rho_{12}^* - \rho_{12}\right), \tag{9.36a}$$

$$\frac{\partial \rho_{12}}{\partial t} = j\omega_a\rho_{12} - \frac{\rho_{12}}{T_2} + j\Omega_e W, \tag{9.36b}$$

where $W = \rho_{22} - \rho_{11}$ represents the degree of population inversion, ω_a is the atomic resonance frequency, and T_1 and T_2 are the population and polarization relaxation times, respectively.

The nonlinear polarization P_x of the two-level system can now be calculated using the relation $P_x = 2Nd\,\mathrm{Re}(\rho_{12})$. From Eq. (9.36b), we can obtain $\mathrm{Re}(\rho_{12})$ to calculate P_x:

$$\frac{\partial^2 \mathrm{Re}(\rho_{12})}{\partial t^2} + 2\frac{\omega_a}{T_2}\frac{\partial \mathrm{Re}(\rho_{12})}{\partial t} + \omega_a^2 = -\omega_a\Omega_e W. \tag{9.37}$$

---

[1] $c = 1$ was used in these definitions to reduce the complexity of the equations.

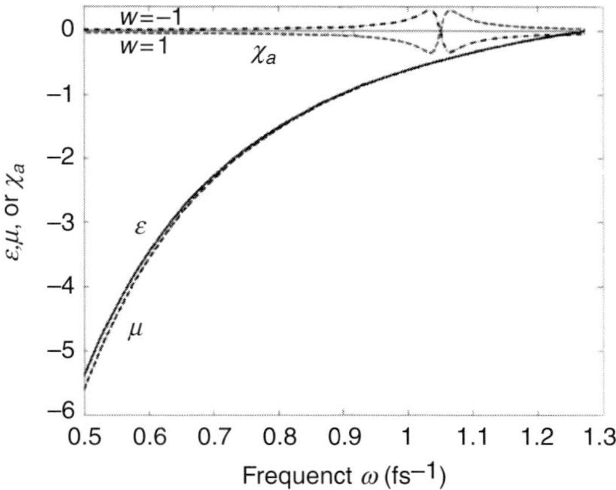

Figure 9.10 Real parts of the permittivity ε (solid line) and permeability μ (dashed line) as a function of frequency for a doped metamaterial. The electric susceptibility $\chi_a(\omega)$ is shown by a dashed-dotted, line for $W = \pm1$. (After Ref. [52]; © APS 2009)

This equation can be solved in the frequency domain to get the susceptibility $\chi_a(\omega)$ of the two-level system in the form

$$\chi_a(\omega) = \frac{2\varepsilon_0\omega_a\beta W}{\omega^2 - \omega_a^2 + 2j\omega/T_2}, \tag{9.38}$$

where $\beta = Nd^2/\varepsilon_0\hbar$ is the coupling constant representing interaction of the metamaterial with the dopant atoms, and N is the dopant density.

As an example, Figure 9.10 shows the real part of the electric susceptibility $\chi_a(\omega)$ when $\omega_a = 1.05$ fs^{-1}, $T_2 = 5$ ps, $d = 1 \times 10^{-29}$ cm, and $N = 1 \times 10^{20}$cm^{-3}. The real parts of the permittivity and the permeability are also shown for comparison in this figure. The top two curves shown the variation of χ_a for $W = \pm1$.

It is possible to find an analytical solution to the preceding equations by making the transformation $\tau = t - z/v_g$, where v_g is the group velocity of light propagating in the doped metamaterial, given by

$$v_g \equiv \frac{\partial\omega}{\partial k} = -\left[\frac{1}{c}\frac{\partial}{\partial\omega}\sqrt{[\varepsilon(\omega) + \chi_a(\omega)]\mu(\omega)}\right]^{-1}. \tag{9.39}$$

In this equation, the negative branch of the square root was chosen because the medium has a negative refractive index. Owing to the very low density of doping, we assume that the condition $|\chi_a(\omega)| \ll |\varepsilon(\omega)|$ is always satisfied.

In terms of the new variable τ, Eqs. (9.35) and (9.36) take the following form:

$$\frac{\partial\Omega_e}{\partial\tau} + 2\beta\frac{\partial\mathrm{Re}(\rho_{12})}{\partial\tau} + \omega_{pe}^2\frac{\partial K}{\partial\tau} = \frac{1}{v_g}\frac{\partial\Omega_m}{\partial\tau}, \tag{9.40a}$$

$$\frac{\partial \Omega_m}{\partial \tau} + (\omega_{pm}^2 - \omega_{0m}^2)\frac{\partial M}{\partial \tau} = \frac{1}{v_g}\frac{\partial \Omega_e}{\partial \tau}, \tag{9.40b}$$

$$\frac{\partial^2 M}{\partial \tau^2} + \Gamma_m \frac{\partial M}{\partial \tau} + \omega_{0m}^2 M = \Omega_m, \tag{9.40c}$$

$$\frac{\partial^2 K}{\partial \tau^2} + \Gamma_e \frac{\partial K}{\partial \tau} = \Omega_e, \tag{9.40d}$$

$$\frac{\partial W}{\partial \tau} = -\frac{(W+1)}{T_1} - 2j\Omega_e\left(\rho_{12}^* - \rho_{12,}\right) \tag{9.40e}$$

$$\frac{\partial \rho_{12}}{\partial \tau} = j\omega_0 \rho_{12} + j\Omega_e W, \tag{9.40f}$$

where $\omega_0 = \omega_a - 1/T_2$.

9.5.3 Solution in the form of a solitary wave

Using our understanding of SIT in a uniform medium, we seek a solution of Eqs. (9.40) in the following form [52]:

$$\Omega_e = A_0\left[1 - \Theta(\tau)\right]\,\text{sech}(\alpha\tau), \tag{9.41a}$$

$$\Omega_m = \Omega_e v_g, \tag{9.41b}$$

$$K = K_0\,\text{sech}(\alpha\tau), \tag{9.41c}$$

$$M = M_0\,\text{sech}(\alpha\tau), \tag{9.41d}$$

$$\text{Re}(\rho_{12}) = C_0\,\text{sech}(\alpha\tau), \tag{9.41e}$$

$$W = -1 - \frac{A_0^2\omega_{pe}^2}{\beta\omega_0\alpha^2}\,\text{sech}^2(\alpha\tau), \tag{9.41f}$$

where A_0 is the amplitude of the solitary wave, α is related inversely to its width, and Θ is a real perturbation parameter that quantifies deviations from the standard SIT solution.

These parameters can be calculated by substituting the assumed solution into Eqs. (9.40), resulting in the following relations:

$$2\beta C_0 + \omega_{pe}^2 K_0 = 0, \tag{9.42a}$$

$$\omega_0 C_0 + \alpha^2\frac{C_0}{A_0} - 1 = 0, \tag{9.42b}$$

$$\left(\omega_{pm}^2 - \omega_{0m}^2\right)M_0 + A_0\left(v_g - \frac{1}{v_g}\right) = 0, \tag{9.42c}$$

$$\pm\Gamma_e K_0\alpha + A_0(\Theta - 1) + K_0\alpha^2 = 0, \tag{9.42d}$$

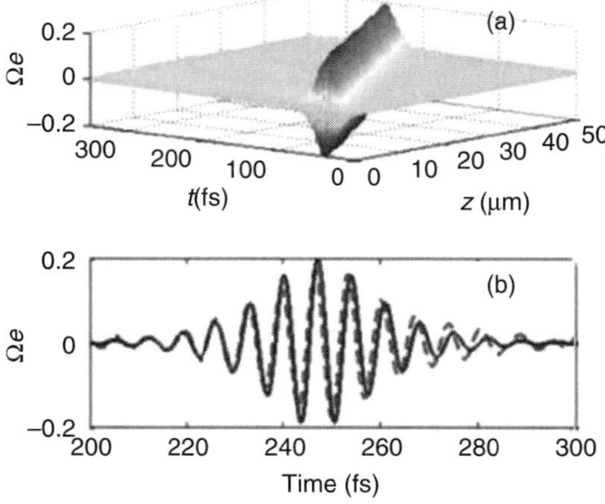

Figure 9.11 (a) A sech pulse propagating as a backward SIT soliton in a metamaterial with a negative refractive index ($n = -1 + j0.067$). (b) Comparison of the electric field after 50 μm (dashed line) with that of the input pulse (solid line). (After Ref. [52]; © APS 2009)

$$\pm\Gamma_m M_0 \alpha + A_0 v_g (\Theta - 1) + M_0 \alpha^2 + \omega_{0m}^2 M_0 = 0. \tag{9.42e}$$

The preceding equations can be solved under the assumptions that ensure $\Theta \ll 1$:

$$\Gamma_e \approx \frac{A_0 \Theta}{K_0 \alpha} \ll \alpha, \qquad \Gamma_m \approx \frac{A_0 v_g \Theta}{M_0 \alpha} \ll \alpha. \tag{9.43}$$

They provide the following simple analytical solution:

$$C_0 = -\frac{\omega_{pe}^2}{2\beta} K_0, \qquad K_0 = \frac{A_0}{\alpha^2}, \tag{9.44a}$$

$$\alpha = \sqrt{\frac{\omega_0 (\omega_0 C_0 - A_0)}{C_0}}, \tag{9.44b}$$

$$M_0 = \frac{(1 - v_g^2)}{v_g (\omega_{pm}^2 - \omega_{0m}^2)} A_0 = \frac{v_g}{\alpha^2 + \omega_{0m}^2} A_0. \tag{9.44c}$$

We know that, for SIT to occur, the area, defined as $\theta_e = \int_{-\infty}^{\infty} \Omega_e(\tau)\, d\tau$, must be an integer multiple of 2π [50]. Keeping the perturbation terms up to first order in Θ, we obtain the relation,

$$\theta = A_0 \int_{-\infty}^{\infty} \operatorname{sech}(\alpha\tau)\, d\tau = \frac{A_0}{\alpha}\pi. \tag{9.45}$$

Thus, the condition $|A_0| = 2\alpha$ must be satisfied for a SIT soliton to exist. To judge the accuracy of these results, one must perform numerical simulations. Figure

9.11(a) shows the evolution of a femtoseond pulse in a metamaterial medium with a negative refractive index ($n = -1 + j0.067$) using the input field in the form

$$\Omega_e(z = 0, t) = \Omega_0 \cos(\omega_0 t)\, \mathrm{sech}(t/T_0), \tag{9.46}$$

with $\Omega_0 = 0.2$ fs^{-1}, $\omega_0 = 0.9$ fs^{-1} ($\lambda_0 = 2.1\ \mu$m), $T_0 = 10$ fs and $\Theta = 10^{-4}$. A Rabi frequency of $\Omega_0 = 0.2$ fs^{-1} corresponds to an electric field of 1.1×10^{10} V m^{-1} or an intensity of 32 TW cm^{-2}. The pulse was assumed to have an area of 2π, enabling SIT propagation. As expected, the pulse does not change much, even after traveling a distance of 50 μm. Figure 9.11(b) compares the electric field after 50 μm with that of the input pulse. Except for minor changes in the trailing parts of the pulse, the electric field remains nearly unchanged in the metamaterial.

References

[1] S. Zhang, W. J. Fan, B. K. Minhas, A. Frauenglass, K. J. Malloy, and S. R. J. Brueck, "Mid-infrared resonant magnetic nanostructures exhibiting a negative permeability," *Phys. Rev. Lett.*, vol. 94, p. 037402 (4 pages), 2005.

[2] W. Cai, U. K. Chettiar, H.-K. Yuan, *et al.*, "Meta-magnetics with rainbow colors," *Opt. Express*, vol. 15, pp. 3333–3341, 2007.

[3] R. A. Shelby, D. R. Smith, and S. Schultz, "Experimental verification of a negative index of refraction," *Science*, vol. 292, pp. 77–79, 2001.

[4] A. V. Kildishev, W. S. Cai, U. K. Chettiar, *et al.*, "Negative refractive index in optics of metal-dielectric composites," *J. Opt. Soc. Am. B*, vol. 23, pp. 423–433, 2006.

[5] A. J. Hoffman, L. V. Alekseyev, S. S. Howard, *et al.*, "Negative refraction in semiconductor metamaterials," *Nat. Mater.*, vol. 6, pp. 946–950, 2007.

[6] M. A. Noginov, Y. A. Barnakov, G. Zhu, T. Tumkur, H. Li, and E. E. Narimanov, "Bulk photonic metamaterial with hyperbolic dispersion," *Appl. Phys. Lett.*, vol. 94, p. 151105 (3 pages), 2009.

[7] J. B. Pendry, "Negative refraction makes a perfect lens," *Phys. Rev. Lett.*, vol. 85, pp. 3966–3969, 2000.

[8] Z. Jacob, L. V. Alekseyev, and E. E. Narimanov, "Optical hyperlens: Far-field imaging beyond the diffraction limit," *Opt. Express*, vol. 14, pp. 8247–8256, 2006.

[9] A. Salandrino and N. Engheta, "Far-field subdiffraction optical microscopy using meta-material crystals: Theory and simulations," *Phys. Rev. B*, vol. 74, p. 075103 (5 pages), 2006.

[10] J. B. Pendry, D. Schurig, and D. R. Smith, "Controlling electromagnetic fields," *Science*, vol. 312, pp. 1780–1782, 2006.

[11] W. Cai, U. K. Chettiar, A. V. Kildishev, and V. M. Shalaey, "Optical cloaking with metamaterials," *Nat. Photonics*, vol. 1, pp. 224–227, 2007.

[12] K. R. Catchpole and A. Polman, "Plasmonic solar cells," *Opt. Express*, vol. 16, pp. 21 793–21 800, 2008.

[13] E. Plum, V. A. Fedotov, P. Kuo, D. P. Tsai, and N. I. Zheludev, "Towards the lasing spaser: Controlling meta-material optical response with semiconductor quantum dots," *Opt. Express*, vol. 17, pp. 8548–8551, 2009.

[14] M. A. Noginov, G. Zhu, A. M. Belgrave, *et al.*, "Demonstration of a spaser-based nanolaser," *Nature*, vol. 460, pp. 1110–1112, 2009.

[15] V. M. Shalaev, "Optical negative index metamaterials," *Nature Photonics*, vol. 1, pp. 41–47, 2007.

[16] C. Kittel, *Introduction to Solid State Physics*, 7th ed. Wiley, 1995.

[17] C. M. Bingham, H. Tao, X. Liu, R. D. Averitt, X. Zhang, and W. J. Padilla, "Planar wallpaper group metamaterials for novel terahertz applications," *Opt. Express*, vol. 16, pp. 18 565–18 575, 2008.

[18] D. Schattschneider, "The plane symmetry groups: Their recognition and notation," *Am. Math. Mon.*, vol. 85, pp. 439–450, 1978.

[19] N. Fang, H. Lee, C. Sun, and X. Zhang, "Sub–diffraction-limited optical imaging with a silver superlens," *Science*, vol. 308, pp. 534–537, 2005.

[20] K. L. Tsakmakidis, A. D. Boardman, and O. Hess, "Trapped rainbow storage of light in metamaterials," *Nature*, vol. 450, pp. 397–401, 2007.

[21] A. K. Popov and S. A. Myslivets, "Transformable broad-band transparency and amplification in negative-index films," *Appl. Phys. Lett.*, vol. 93, p. 191117 (3 pages), 2008.

[22] N. M. Litchinitser and V. M. Shalaev, "Metamaterials: Loss as a route to transparency," *Nat. Photonics*, vol. 3, p. 75, 2009.

[23] A. A. Govyadinov and V. A. Podolskiy, "Active metamaterials: Sign of refractive index and gain-assisted dispersion management," *Appl. Phys. Lett.*, vol. 91, pp. 191103/1–191103/3, 2007.

[24] N. Liu, L. Langguth, T. Weiss, *et al.*, "Plasmonic analogue of electromagnetically induced transparency at the Drude damping limit," *Nat. Mater.*, vol. 8, pp. 758–762, 2009.

[25] S. Zhang, D. A. Genov, Y. Wang, M. Liu, and X. Zhang, "Plasmon-induced transparency in metamaterials," *Phys. Rev. Lett.*, vol. 101, p. 047401 (4 pages), 2008.

[26] P. Tassin, L. Zhang, T. Koschny, E. N. Economou, and C. M. Soukoulis, "Low-loss metamaterials based on classical electromagnetically induced transparency," *Phys. Rev. Lett.*, vol. 102, p. 053901 (4 pages), 2009.

[27] M. I. Stockman, "Criterion for negative refraction with low losses from a fundamental principle of causality," *Phys. Rev. Lett.*, vol. 98, p. 177404 (4 pages), 2007.

[28] A. Reza, M. M. Dignam, and S. Hughes, "Can light be stopped in realistic metamaterials?" *Nature*, vol. 455, pp. E10–E11, 2008.

[29] W. Cai and V. Shalaev, *Optical Metamaterials: Fundamentals and Applications.* Springer, 2009.

[30] M. A. Noginov, G. Zhu, M. Bahoura, *et al.*, "Enhancement of surface plasmons in an Ag aggregate by optical gain in a dielectric medium," *Opt. Lett.*, vol. 31, pp. 3022–3024, 2006.

[31] N. I. Zheludev, S. L. Prosvirnin, N. Papasimakis, and V. A. Fedotov, "Lasing spaser," *Nat. Photonics*, vol. 2, pp. 351–354, 2008.

[32] R. F. Oulton, V. J. Sorger, T. Zentgraf, *et al.*, "Plasmon lasers at deep subwavelength scale," *Nature*, vol. 461, pp. 629–632, 2009.

[33] A. Fang, T. Koschny, M. Wegener, and C. M. Soukoulis, "Self-consistent calculation of metamaterials with gain," *Phys. Rev. B*, vol. 79, p. 241104(R) (4 pages), 2009.

[34] M. I. Stockman, "The spaser as a nanoscale quantum generator and ultrafast amplifier," *J. Opt.*, vol. 12, p. 024004 (13 pages), 2010.

[35] J. A. Gordon and R. W. Ziolkowski, "The design and simulated performance of a coated nano-particle laser," *Opt. Express*, vol. 15, pp. 2622–2653, 2007.

[36] M. Wegener, J. L. Garcia-Pomar, C. M. Soukoulis, N. Meinzer, M. Ruther, and S. Linden, "Toy model for plasmonic metamaterial resonances coupled to two-level system gain," *Opt. Express*, vol. 16, pp. 19 785–19 798, 2008.

[37] Z.-J. Yang, N.-C. Kim, J.-B. Li, "Surface plasmons amplifications in single Ag nanoring," *Opt. Express*, vol. 18, pp. 4006–4011, 2010.

[38] M. Ambati, D. A. Genov, R. F. Oulton, and X. Zhang, "Active plasmonics: Surface plasmon interaction with optical emitters," *IEEE J. Sel. Top. Quantum Electron.*, vol. 14, pp. 1395–1403, 2008.

[39] Y. Fu, L. Thylen, and H. Agren, "A lossless negative dielectric constant from quantum dot exciton polaritons," *Nano Lett.*, vol. 8, pp. 1551–1555, 2008.

[40] Y. Fu, H. Agren, L. Hoglund, *et al.* "Optical reflection from excitonic quantum-dot multilayer structures," *Appl. Phys. Lett.*, vol. 93, p. 183117 (3 pages), 2008.

[41] D. J. Bergman and M. I. Stockman, "Surface plasmon amplification by stimulated emission of radiation: Quantum generation of coherent surface plasmons in nanosystems," *Phys. Rev. Lett.*, vol. 90, p. 027402 (4 pages), 2003.

[42] M. I. Stockman, S. V. Faleev, and D. J. Bergman, "Localization versus delocalization of surface plasmons in nanosystems: Can one state have both characteristics?" *Phys. Rev. Lett.*, vol. 87, p. 167401 (4 pages), 2001.

[43] A. K. Popov and V. M. Shalaev, "Compensating losses in negative-index metamaterials by optical parametric amplification," *Opt. Lett.*, vol. 31, pp. 2169–2171, 2006.

[44] R. W. Boyd, *Nonlinear Optics*, 3rd ed. Academic Press, 2008.

[45] A. K. Popov and V. M. Shalaev, "Negative-index metamaterials: Second-harmonic generation, Manley–Rowe relations and parametric amplification," *Appl. Phys. B*, vol. 84, pp. 131–137, 2006.

[46] A. K. Popov, S. A. Myslivets, T. F. George, and V. M. Shalaev, "Four-wave mixing, quantum control, and compensating losses in doped negative-index photonic metamaterials," *Opt. Lett.*, vol. 32, pp. 3044–3046, 2007.

[47] A. K. Popov, S. A. Myslivets, and T. F. George, "Nonlinear interference effects and all-optical switching in optically dense inhomogeneously broadened media," *Phys. Rev. A*, vol. 71, p. 043811 (13 pages), 2005.

[48] Y. R. Shen, *The Principles of Nonlinear Optics.* Wiley, 1991.

[49] E. Roshencher and P. Bois, "Model system for optical nonlinearities: Asymmetric quantum wells," *Phys. Rev. B*, vol. 44, pp. 11 315–11 327, 1991.

[50] S. L. McCall and E. L. Hahn, "Self-induced transparency by pulsed coherent light," *Phys. Rev. Lett.*, vol. 18, pp. 908–911, 1967.

[51] L. Allen and J. H. Eberly, *Optical Resonance and Two-Level Atoms.* Wiley InterScience, 1975.

[52] K. Zeng, J. Zhou, G. Kurizki, and T. Opartrny, "Backward self-induced transparency in metamaterials," *Phys. Rev. A*, vol. 80, p. 061806(R) (4 pages), 2009.

Index